WNA
-75

MASS SPECTROMETRY

IN BIOCHEMISTRY AND MEDICINE

MONOGRAPHS OF THE MARIO NEGRI INSTITUTE FOR PHARMACOLOGICAL RESEARCH, MILAN

Amphetamines and Related Compounds
Edited by E. Costa and S. Garattini

The Benzodiazepines
Edited by S. Garattini, E. Mussini, and L. O. Randall

Chemotherapy of Cancer Dissemination and Metastasis
Edited by S. Garattini and G. Franchi

Drug Interactions
Edited by P. L. Morselli, S. Garattini, and S. N. Cohen

Insolubilized Enzymes
Edited by M. Salmona, C. Saronio, and S. Garattini

Isolated Liver Perfusion and Its Applications
Edited by I. Bartošek, A. Guaitani, and L. L. Miller

Mass Spectrometry in Biochemistry and Medicine
Edited by A. Frigerio and N. Castagnoli, Jr.

MONOGRAPHS OF THE MARIO NEGRI INSTITUTE
FOR PHARMACOLOGICAL RESEARCH, MILAN

MASS SPECTROMETRY
IN BIOCHEMISTRY
AND MEDICINE

Editors:

A. Frigerio, Ph.D.

Laboratory of Mass Spectrometry
Mario Negri Institute for
* Pharmacological Research*
Milan

N. Castagnoli, Jr., Ph.D.

Department of Pharmaceutical Chemistry
School of Pharmacy
University of California
San Francisco, California

Raven Press, Publishers ▪ New York

Made in the United States of America

International Standard Book Number 0-911216-53-7
Library of Congress Catalog Card Number 73-91164

Preface

Progress in medicine has always been characterized by new ideas or by new methodology.

In recent years there has been a constant trend to utilize, in the medical field, the technology developed by basic sciences, with particular reference to physical chemistry. The symposium upon which this volume is based was organized at the Mario Negri Institute in Milan to focus attention on the use of gas-liquid chromatography, combined with mass spectrometry.

This technique has gained wide acceptance because it permits separation of the molecular species present, identification of the compound to be measured, and very high sensitivity. Specificity and sensitivity are two highly desirable characteristics in medicine, where endogenous or exogenous substances are frequently present at submicrogram levels.

This resulting volume provides a variety of examples to illustrate the potential applications of the technique. Of particular interest are the studies concerning the identification of metabolites formed from the ingestion of nutrients, pollutants, or drugs. Other investigations concern the identification of endogenous metabolites in living organisms, either animal or vegetable. Finally, there are presented medical applications of this technique for diagnostic purposes, which will, in time, make mass spectrometry a necessary tool in every major hospital.

It is hoped that this volume may stimulate researchers in utilizing mass spectrometry as a tool for attacking some of the many unsolved problems in biology and medicine.

S. Garattini

Contents

Mass Spectrometry in Biochemistry and Medicine,
edited by A. Frigerio and N. Castagnoli.
Raven Press, New York © 1974

A Mass Spectrometric Study of the Metabolism of Piribedil

B. J. Millard, D. B. Campbell,* P. Jenner,* and A. R. Taylor*

*School of Pharmacy, University of London, London, England, and *Servier Laboratories Ltd., Harrow, England*

I. INTRODUCTION

Piribedil (ET 495), 1-(3,4-methylenedioxybenzyl)-4-(2-pyrimidinyl)piper-azine (I), one of a series of 38 analogues (Regnier, Canevari, Laubie, and Le Douarec, 1968), was found to have peripheral vasodilator properties on preliminary pharmacological screening. Its efficacy has been confirmed subsequently in clinical trials (Cristol, 1969; Kollitsch, 1969).

(I)

It has also been shown that at intraperitoneal doses greater than 12 mg/kg a marked stereotyped behavior is provoked in rats, suggesting a central dopaminergic stimulation (Corrodi, Fuxe, and Ungerstedt, 1971; Poignant, Laubie, Tscouris-Kupfer, and Schmitt, 1972; Costall and Naylor, 1973). It has been suggested that the drug may be useful in the treatment of Parkinson's disease.

Studies into the distribution and metabolism of ET 495 labeled with ^{14}C at the benzylic group (Thomasett, Benakis, and Glasson, 1971) showed that the drug is readily absorbed after oral administration, rapidly entering the kidney, liver, heart, bladder, and intestine. Only a small percentage of the drug is located in the central nervous system and peripheral tissues. Although six metabolites were observed in urine by these workers, their identity remained unresolved.

The present work describes the structural investigation of some of these metabolites by high-resolution mass spectrometry.

II. EXPERIMENTAL

A. Mass Spectrometry

Mass spectra were obtained on an AEI MS902 instrument operating at a source temperature of 220°C and beam energy of 70 eV using a direct inlet system. High-resolution accurate mass measurements were carried out at a resolving power of 20,000 (10% valley) using the peak-matching technique. The compositions of all ions discussed in this chapter were confirmed by such determinations.

For quantitative work, standard mixtures containing 100 μg/ml of proma-zine and varying amounts (0.5 to 20 μg/ml) of synthetic 1-(3,4-methylene-dioxybenzyl)-4-(5-hydroxypyrimidin-2-yl)piperazine (II) were prepared. Five-μl aliquots of these solutions were evaporated on to the direct inlet probe. With the instrument set at 10,000 resolving power the peak-matching unit was set to m/e 284.1347 (M^+ of promazine) and m/e 314.1379 (M^+ of II). The signal was passed to an ultraviolet chart recorder while the sample was evaporated in the source over a period of 3 to 4 min. The intensities of each ion were summed and a calibration curve constructed (Fig. 1).

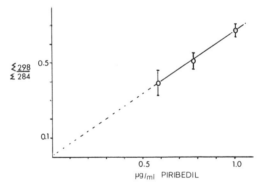

FIG. 1. Calibration curve from the sum of the intensities of the two ions.

B. Isolation of Metabolites

1. In Vitro Metabolites

ET 495 methane sulfonate labeled with ^{14}C in the benzylic position was incubated in 10,000 × g hepatic supernatant preparation. Complete micro-somal incubates from male rabbits, hamsters, and guinea pigs were examined by two-way thin-layer chromatography (TLC) (butanol/methanol/water; 90/10/100 and pentan-1-ol/ammonia; 85/15) after extraction with chloro-form against a pH gradient. Examination of the TLC plates by autoradiog-raphy gave four major components for all species.

2. *In Vivo Metabolites*

(a) After oral administration of ^{14}C-labeled ET 495 to male Wistar rats (40 mg/kg; 30 μC per animal), the urine was collected over 72 hr. After enzymatic (suc Helix Pomatia) and acidic hydrolysis, the urine was extracted at pH 12.8 and 1 with chloroform. Two-way TLC as above yielded three basic metabolites and two acidic metabolites.

(b) The urine from healthy male volunteers receiving 40 mg of ET 495 orally was treated in the same way to yield two basic metabolites.

C. Synthesis of Metabolites

The synthesis of metabolites and potential metabolites of ET 495 will be published elsewhere.

III. RESULTS AND DISCUSSION

A. In Vitro Metabolism

A mass spectrometric study was carried out in order to obtain preliminary information on metabolic pathways and to isolate quantities of metabolites suitable for the usual physicochemical techniques of identification. Of the four compounds isolated, the first one, extractable at pH 12, had an R_f value identical with that of ET 495. The mass spectrum of this compound was also identical to the mass spectrum of ET 495 (Fig. 2). The major ions in the spectrum are believed to arise as follows:

(I) M^+ m/e 298

m/e 164   m/e 108   m/e 135

From this simple fragmentation scheme, it can be seen that metabolic changes in the molecule can be located by comparison of the mass spectra of the metabolites with the spectrum of (I).

The second compound (metabolite 1) was a highly fluorescent material

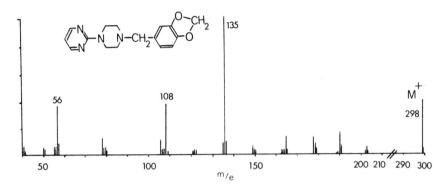

FIG. 2. Mass spectrum of ET 495.

extractable at pH 9. Its mass spectrum is shown in Fig. 3. Comparison of this spectrum with that of (I) leads to the following conclusions: (1) The molecular ion at m/e 314 corresponds to the presence of an extra oxygen atom. This was confirmed by accurate mass measurement. (2) The presence of m/e 135 indicates that the methylenedioxybenzyl group is intact. (3) The m/e 108 has been shifted to m/e 124; therefore, the extra oxygen is located in this part of the molecule. (4) An N-oxide structure is unlikely since neither oxygen nor water is lost from the molecular ion and fragment ions rarely retain this oxygen, contrary to the present case.

This leads to the conclusion that the metabolite is hydroxylated at either the 4 or 5 position of the pyrimidine ring. The 4-hydroxy compound when synthesized was not fluorescent and gave the mass spectrum shown in Fig. 4. Although this is similar to that of metabolite 1, there are differences, especially in intensities. Therefore it is concluded that metabolite 1 is the 5-hydroxy compound (II). This was later confirmed by synthesis.

The third component (metabolite 2) was extractable at pH 9, and gave the mass spectrum shown in Fig. 5. Comparison of this spectrum with that of (I) leads to the following conclusions: (1) The molecular ion at m/e 314

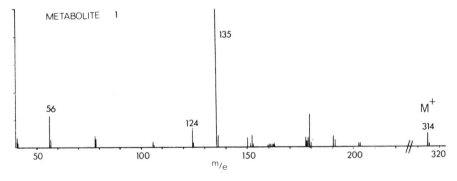

FIG. 3. Mass spectrum of metabolite 1.

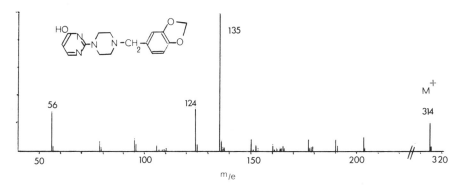

FIG. 4. Mass spectrum of the 4-hydroxy compound.

corresponds to the presence of an extra oxygen atom in the molecule. This was confirmed by accurate mass measurement. (2) The presence of m/e 135 shows the methylenedioxybenzyl group to be intact. (3) The loss of both oxygen and water from the molecular ion suggests an N-oxide structure.

The two most likely compounds 1-(3,4-methylenedioxybenzyl)-4(2-pyrimidinyl)piperazine-2-N-oxide and the 4-N-oxide were synthesized. The mass spectra of these are shown in Fig. 6. Although the 4-N-oxide shows loss of oxygen and water from the molecular ion, the 2-N-oxide does not. The spectrum of metabolite 2 is virtually identical with that of the 4-N-oxide. It is therefore concluded that metabolite 2 is the 4-N-oxide (III).

The fourth component (metabolite 3), a more water-soluble material, was extractable from the incubates only with methanol. Its mass spectrum is given in Fig. 7. The following conclusions can be drawn: (1) The molecular ion at m/e 286 contains one carbon atom less than ET 495. This was confirmed by accurate mass measurement. (2) The ion at m/e 135 is missing, suggesting modification of the methylenedioxybenzyl part of the molecule. (3) The pyrimidinylpiperazine part of the molecule is intact since m/e 164 is present.

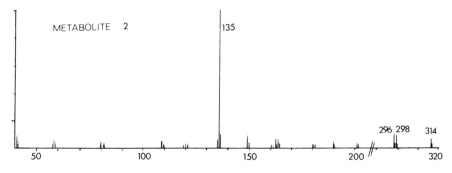

FIG. 5. Mass spectrum of metabolite 2.

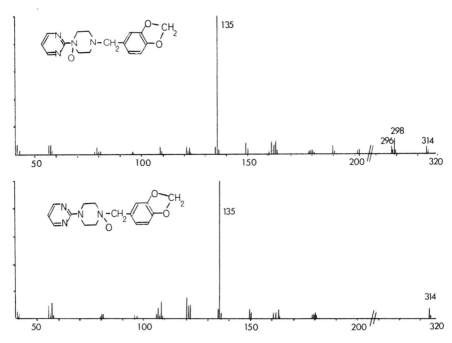

FIG. 6. Mass spectra of 1-(3,4-methylenedioxybenzyl)-4(2-pyrimidinyl)piperazine-2-N-oxide (upper) and the 4-N-oxide (lower).

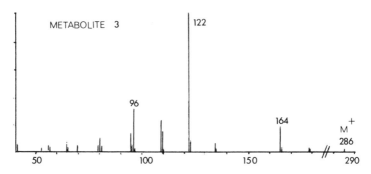

FIG. 7. Mass spectrum of metabolite 3.

The most likely explanation for the loss of a carbon atom from the methylenedioxybenzyl group is that it has been converted to a dihydroxybenzyl group. The structure IV for metabolite 3 was confirmed by synthesis.

The *in vitro* metabolism of ET 495 can be summarized as follows:

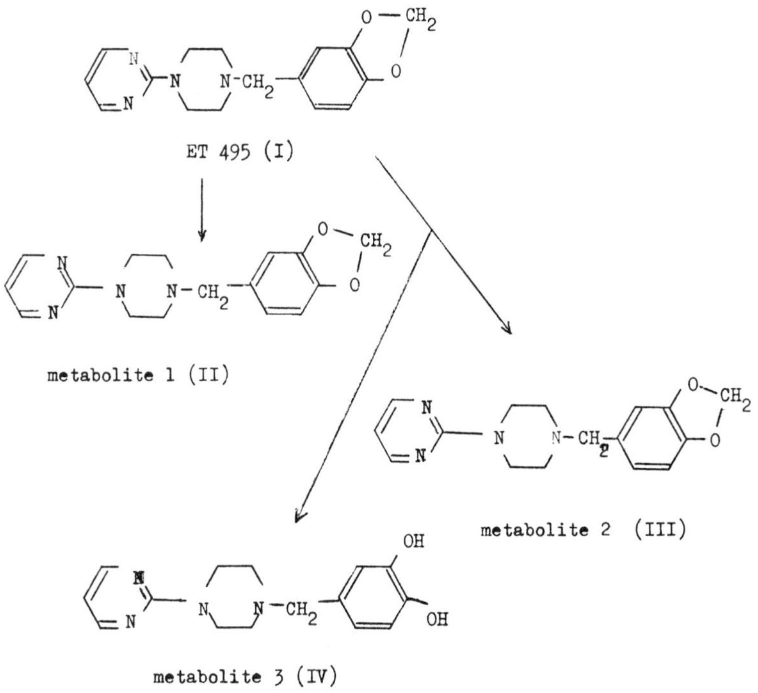

ET 495 (I)

metabolite 1 (II)

metabolite 2 (III)

metabolite 3 (IV)

B. In Vivo Metabolism in Rat

After oral administration of ^{14}C-labeled ET 495 methane sulfonate, 45% of the radioactivity was recovered in the urine and 30% in the feces over a 72-hr period. Direct chloroform extraction against a pH gradient before and after enzymatic hydrolysis, acidic hydrolysis, and reduction with titanium chloride removed only 50% of the total radioactivity in both urine and feces. These results indicate the presence of as yet unidentified highly polar metabolites. In urine 84% of the extractable radioactivity was present in conjugated form. Examination of extracts of enzymatically hydrolyzed rat urine showed that all the *in vitro* metabolites, 1, 2, and 3, were present, together with a small amount of unchanged drug. Metabolites 1 and 3 were excreted mainly as the glucuronide or sulfate conjugates.

C. Metabolism in Man

Using the TLC procedure developed for the *in vitro* studies, the presence of two major metabolites was confirmed in the urine of patients taking up to 120 mg of ET 495 daily. No unchanged drug was detected. On enzymatic hydrolysis the metabolites had the same TLC characteristics as metabolites 1 and 3 and gave identical mass spectra. The metabolites were not detected in unhydrolyzed urine, and so were excreted exclusively as the glucuronide and sulfate conjugates. No N-oxide was detectable in the urine.

D. Investigation of Human Plasma by Quantitative Mass Spectrometry

An attempt was made to quantify the amount of the hydroxy metabolite (II) present in plasma by high-resolution mass spectrometry. The technique was to add 100 μg of promazine as internal standard to 1 ml of blank plasma and to 1 ml of plasma from patients taking the drug. Five-μg aliquots of these solutions were evaporated on to the probe and the relevant molecular ions were monitored as described in the experimental section. The blank plasma gave readings for the ion $C_{16}H_{18}N_4O_3^{+\cdot}$ which from the calibration curve were equivalent to 0.6 ± 0.3 μg/ml of (II), i.e., 3 ng on the probe, whereas plasma containing the metabolite gave readings equivalent to 8.1 ± 1.5 μg/ml of (II), i.e., 40 ng on the probe.

Although the metabolite is present in conjugated form in plasma, under the mass spectrometer source conditions used (220°C) this evidently breaks down to generate the molecular ion of (II), making the analysis feasible.

REFERENCES

Corrodi, H., Fuxe, K., and Ungerstedt, V. (1971): Evidence for a new type of dopamine receptor stimulating agent. *Journal of Pharmacology and Pharmacology,* 23:989–991.

Cristol, R. (1969): Essai clinique d'une nouvelle médication dans les artérites des membres inférieurs. *Le Progrès Médical,* 97:189–198.

Kollitsch, M. (1969): Une thérapeutique originale à effet prolongé des artérites obliterantes des membres inférieurs. *Therapeutique,* 45:766–768.

Costall, B., and Naylor, R. J. (1973): Neuropharmacological studies on the site and mode of action of ET 495. *Advances in Neurology, Vol. 3: Progress in the Treatment of Parkinsonism,* edited by D. B. Calne, pp. 281–293. Raven Press, New York.

Poignant, J. M., Laubie, M., Tscouris-Kupfer, D., and Schmitt, H. (1972): Action de divers réactif pharmacologiques sur les mouvements stereotypés produits chez le rat par l'ingestion de 1-(2'-pyrimidyl)-4-piperonylpiperazine (Piribedil). *Comptes Rendus de l'Académie des Sciences* (Paris), 275:715–717.

Regnier, G. L., Canevari, R. J., Laubie, M. J., and Le Douarec, J. C. (1968): Synthesis and vasodilator activity of new piperazine derivatives. *Journal of Medicinal Chemistry,* 11:1151–1155.

Thomasett, M., Benakis, A., and Glasson, B. (1971): Distribution, elimination and metabolism of (pyrimidyl-2″)-1-(methylene ¹⁴C dioxy-3',4'-benzyl)-4-piperazine (ET 495). *Réunion Commune de la British Pharmacological Society et de l'Association des Pharmacologiestes,* Paris, C. 17.

Mass Spectrometry in Biochemistry and Medicine,
edited by A. Frigerio and N. Castagnoli.
Raven Press, New York © 1974

Applications of Gas Chromatography–Mass Spectrometry in Quantitative Drug and Metabolite Measurement

G. H. Draffan, R. A. Clare, Faith M. Williams, E. Emons,
and J. L. Jackson

*Department of Clinical Pharmacology, Royal Postgraduate Medical School, Ducane Road,
London, W. 12, England, and John Wyeth and Brother Ltd., Taplow, Maidenhead, England*

I. INTRODUCTION

Selective ion monitoring or "mass fragmentography" in combined gas chromatography–mass spectrometry (GC–MS) (Hammar, Holmstedt, and Ryhage, 1968) is well established as a highly sensitive means of identification of minor components in complex biological mixtures. Our interests are in quantitative aspects of drug metabolism and in the application of single- and multiple-ion monitoring GC–MS as a precise measuring technique where more conventional methods lack the necessary sensitivity or specificity. Quantitative assay from biological material normally requires that response to sample be measured relative to a reference compound added at the earliest stage in the separation sequence. Ideally the reference may be the same compound but with a stable isotopic label and thus may potentially act also as a carrier in extraction and in minimizing adsorptive loss of the sample in gas chromatography (Samuelsson, Hamberg, and Sweeley, 1970; Gaffney, Hammar, Holmstedt, and McMahon, 1971). Alternatively a structural analogue may serve as the internal standard. In this case where chromatographic problems are encountered, these must be overcome by conventional protective derivatization.

In this chapter, we describe two distinct applications in which techniques representative of the above general standardization methods have been employed. The first concerns quantitative aspects of barbiturate metabolism. Barbiturates (I) and their metabolites, after N-methylation to improve chromatographic behavior, are particularly amenable to the separate structural analogue approach. Single-ion monitoring techniques were developed initially for studies of amobarbital $[Ia, R = CH_2CH_2C(CH_3)_2]$ elimination in newborn infants (Draffan, Clare, and Williams, 1973; Krauer, Draffan, Williams, Clare, Dollery, and Hawkins, 1973) after administration of the drug to the mother during labor. The plasma levels of the major metabolite, hydroxyamobarbital $[Ib, R = CH_2CH_2C(OH)(CH_3)_2]$, were also determined in the infants. We have extended these methods to the investigation of enzyme activity in human liver by measuring the kinetics of formation of hydroxyamobarbital from amobarbital *in vitro*. Using 5 to 20 mg of needle biopsy tissue, conversion by the liver of a 9,000 × g fraction at different substrate concentrations may be determined. Conversions in the range of 0.01 to 0.1% are measurable. This is projected as a possible means of defining oxidative capacity in liver disease.

As an illustration of the stable isotope dilution approach to quantitation, a feasibility study is described involving the hypotensive drug indoramin (II,R = H), 3-[2-(benzamido-piperid-1-yl)ethyl] indole and its deuterated analogue (II,R = ^{2}H) acting both as carrier and internal standard. To complete our pharmacokinetic studies in man we required a specific plasma assay. This drug, although chemically stable and readily recoverable from biological fluid, cannot be chromatographed carrier-free without variable adsorptive loss.

I   (a) Amobarbital

         R = $CH_2CH_2CH(CH_3)_2$

    (b) 3'-hydroxyamobarbital

         R = $CH_2CH_2C(OH)(CH_3)_2$

    (c) 3'-hydroxypentobarbital

         R = $CH(CH_3)CH_2CH(OH)CH_3$

II

II. METHODS

A. Instrumentation

A Varian 1400 gas chromatograph coupled via a silicone membrane separator to an AEI MS12 is employed in GC–MS. The mass spectrometer is operated at a resolving power of 800, source temperature 260°C, basic accelerating voltage 8 kV, and at an ionizing voltage of 12, 20, or 23 eV. The trap current used at 12 eV is 100 μA and at 20 or 23 eV, 250 μA.

The mass spectrometer has been substantially modified to facilitate multiple-ion monitoring. Sample and hold circuitry and independent channel gain and filtering have been incorporated following general principles adopted by Kelly (1972). The signal from the electron multiplier is fed into a varactor bridge amplifier (Analogue Devices type 310J), which was built together with a power supply (Analogue Devices model 902H) as a complete replacement (feedback resister 1,000 MΩ, bandwidth 30 Hz) for the existing valve head amplifier. The signal is then taken to a processing unit consisting of a low-pass filter from which signal is distributed to a number of preamplifiers (type 741) providing a voltage gain of one. Backing-off facilities have been added. The output of each preamplifier is fed to a sample and hold circuit controlled by a reed relay in turn controlled by a logic circuit. The logic circuit consists of a clock which is locked to the mains, a reset and gate control circuit, and a counter. The outputs from the counter are gated and provide channel selection. An improvement in signal/noise is obtained by signal averaging achieved by a resistor in series with the reed relay. The sampling time per peak, set by the logic circuit (five times the averaging time constant), is 50 msec. The accelerating voltage switching rate is 2 cps and the mass range 10%. The output from each sample and hold circuit is fed via another low-pass filter (1, 0.5, or 0.2 sec time constant) to a multichannel pen recorder. Only two channels are used in the work described in this communication.

B. Barbiturates

N,N'-Dimethyl derivatives of hydroxyamobarbital (Ib) and the reference hydroxypentobarbital (Ic) were chromatographed at 170°C on a 6 ft × $\frac{1}{8}$ in 3% OV-1 (Gas Chrom Q, 100 to 120 mesh) column at a helium flow rate of 40 ml/min. Retention times were 2.4 and 3.0 min respectively. Mass 184 was monitored during elution.

The preparation method used for liver tissue was as follows. Liver biopsy samples (5 to 20 mg tissue) were homogenized in 100 to 150 μl ice-cold buffer (Tris KCl). The 9,000 × g supernatant was prepared by centrifugation at 12,000 rpm for 20 min in a "Superspeed 50" centrifuge (Measuring and Scientific Equipment Ltd.) at 4°C. The protein content of the 9,000 × g supernatant was measured by a modification of the method of Lowry (1951). All activities were related to 9,000 × g protein content.

The hydroxylation of amobarbital was carried out by incubation at 37°C for 15 min in a 50-μl volume medium containing the required cofactors (Davies, Gigon, and Gillette, 1969), approximately 100 μg 9,000 × g protein and the substrate, amobarbital, in concentrations ranging from 0 to 4 mM. Under these conditions, the hydroxylation reaction was linear with time and protein concentration. Reaction was terminated by the addition of 10 μl 4 M NaH$_2$PO$_4$ and placing on ice. Starting with a 20-mg biopsy, six to eight incubations were possible.

Hydroxypentobarbital (10 ng), the internal reference, was added to each 50-μl incubation mixture which was then extracted with 1.5% isoamyl alcohol in heptane (3 × 1 ml), substantially removing unchanged amobarbital. The hydroxylated barbiturates were recovered into ether (1 ml), methylated with diazomethane, and finally reconstituted for GC–MS in methanol (10 μl). Quantitation is effected by measurement of peak height response ratio of sample to internal standard. Calibration was effected in the range 0.25 to 4 ng of hydroxyamobarbital added to 9,000 × g homogenates prepared from fresh rat tissue. Control (substrate free) incubations were carried out on all human biopsy samples assayed.

C. Indoramin

Deuterium-labeled indoramin (II,R = ^{2}H) was prepared from the amine (III). To a suspension of the dihydrochloride of III (0.005 mole) in chloroform a solution of K_2CO_3 (0.01 mole) in water was added followed by dropwise addition of $^2H_5$benzoyl chloride (0.005 mole nominally 99 atom% ^{2}H) (Merck Sharpe & Dohme, Ltd., Canada). After 1.5 hr at room temperature, the product II,R = ^{2}H precipitated. Its homogeneity was confirmed by thin-layer chromatography.

GC–MS of II was effected via a 3 ft × $\frac{1}{8}$ in 3% OV-1 column (Gas Chrom Q, 100 to 120 mesh) at 260°C (helium flow rate, 30 ml/min); the retention time was 4.5 min.

The procedure adopted for assay of II (R = H) from plasma was as follows. Plasma (3 ml) with added deuterated carrier (30 μg), made basic with NaOH (1 ml, 1.0 N), was extracted with heptane 1.5% isoamyl alcohol (20 ml). The organic layer was back extracted into HCl (0.01 N, 3 ml); the pH of the aqueous phase adjusted to 12 and II re-extracted into ether (2 × 4 ml). This extract was concentrated by evaporation (nitrogen) and reconstituted in methanol (10 μl). Aliquots (2 to 3 μl) were assayed by GC–MS. Recovery at this carrier level and plasma volume was checked with ^{14}C-labeled drug and found to be 33 to 40%.

III. RESULTS AND DISCUSSION

A. Barbiturates

The mass spectra of the N,N'-dimethyl derivatives of barbiturates with an alkyl side chain R (I) have fragment ions in common at m/e 169 and 184 (Gilbert, Millard, and Powell, 1970). These ions carry a significant fraction of the total ion current. At 23 eV for dimethyl-3'-hydroxyamobarbital (Ib) %Σ_{40} at m/e 169 is 28.5 and at m/e 184 is 31.0 and for 3'-hydroxypento-

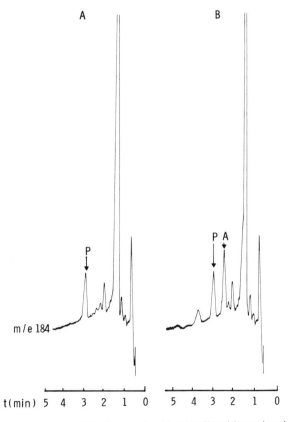

barbital (Ic) is 22.0 at 169 and 22.7 at 184. Monitoring m/e 184, 0.02 ng of dimethyl Ib was detectable, response was linear to 10 ng, and with a well-conditioned column and inlet system, there was no significant adsorptive loss.

A linear calibration was obtained for hydroxyamobarbital recovered from incubations (in the range 0.25 ng to 4.0 ng) against peak height ratio of dimethylhydroxyamobarbital to dimethylhydroxypentobarbital at m/e 184.

FIG. 1. Chromatograms at m/e 184 of extracts of human liver biopsy incubations (100 μg 9,000 × g protein 50 μl incubation). A. Control incubation. Internal reference, dimethyl-hydroxypentobarbital (P). B. Dimethylhydroxyamobarbital (A) produced by incubation with 1 mM amobarbital.

The reproducibility of measurement at 1 ng per extract was ± 7.6% (S.D. six determinations). One ng hydroxyamobarbital per 50 μl incubation is equivalent to 0.01% conversion at 1 mM amobarbital (11.3 μg/50 μl). Figure 1 shows the m/e 184 chromatograms of a control incubation with human liver 9,000 × g and an incubation with 1 mM amobarbital. In this incubation 3.4 ng hydroxyamobarbital was produced equivalent to 0.03% conversion.

The conversion to hydroxyamobarbital with increasing substrate concentration for two biopsies from human liver compared to one from the male rat is shown in Fig. 2. The activity of the human liver enzymes was one-fourth to one-third that of the male rat. At 1.0 mM the percentage conversion by the human livers was 0.03 and 0.02% and by the rat, 0.10%. The relative activities are similar to those observed by other workers in human liver samples removed during surgery (Kuntzman, March, Brand, Jacobson, Levin, and Conney, 1966; Davies and Thorgeirsson, 1972).

The activity of an enzyme preparation can be defined by the V_{max} (maximal velocity) and K_m (substrate concentration producing half-maximal velocity). These are calculated from the rate of formation of product with increasing substrate concentration (Michaelis-Menton Kinetics). The K_m and V_{max} values for the human liver biopsies in Fig. 2 are I:K_m 1.1 mM and V_{max} 7.6 ng hydroxyamobarbital formed/100 μg 9,000 × g protein/15 min, and II:K_m 6.5 mM and V_{max} 14.4 ng formed/100 μg protein/15 min.

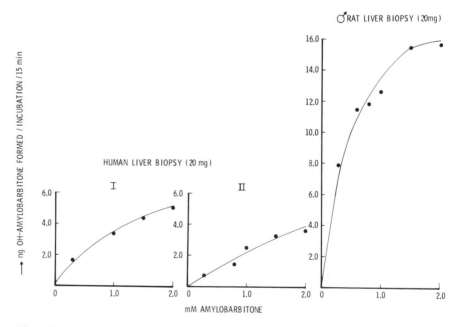

FIG. 2. Conversion of amobarbital to hydroxyamobarbital by human and rat liver biopsies.

Measurements of K_m and V_{max} values for amobarbital hydroxylation by normal and diseased human liver are presently being made.

When we first became interested in *in vitro* studies of amobarbital hydroxylation, ^{14}C-labeled substrate was used and conversion determined by addition of unlabeled carrier hydroxyamobarbital, preparative thin-layer chromatography, and scintillation counting of labeled product. The intention was to use GC–MS simply to validate this procedure. However, radioactive impurities in the substrate contributed to a high uncertainty at the 0.1% conversion level even after exhaustive purification by solvent extraction. An assessment of the two approaches, ^{14}C and GC–MS, will be published in detail elsewhere.

B. Indoramin (II,R = H)

Indoramin is relatively involatile and has poor gas chromatographic properties with peak tailing at low concentrations. Formation of the indole perfluoropropionyl derivative improves the peak profile to some extent and increases volatility. Residual chromatographic problems, difficulties in achieving complete reaction, and the lack of a suitable internal reference have not encouraged us to attempt an assay based on electron capture detection. The limit of detection of II in GC–MS using a silicone membrane separator was of the order of 100 ng, somewhat higher than with a flame ionization detector and indicating a further, probably simple adsorptive, loss in the interface. With a Watson-Biemann glass frit separator, the limit was substantially higher. The perfluoropropionyl derivative did not improve inlet transfer efficiency. An approach based on the coinjection of a large excess of deuterium-labeled material (II,R = [2]H) proved effective in carrying the underivatized protium form.

The mass spectra of indoramin and its deuterated analogue are shown in Fig. 3. The principal fragment ions in the spectrum of the protium form, m/e 105, 174, and 217, are attributable to the benzamidopiperidine portion of the molecule and are shifted by 5 mass units on deuteration. The residual contribution in the spectrum of the deuterated analogue at mass 217 is less than 0.1% of the base peak, m/e 222. In quantitative measurement of the protium form, m/e 217 was used. In this preliminary study, the reference ion in the deuterated material was either 223 or 224 (16.6 and 3.3% relative abundance respectively). There appeared to be no theoretical objection to this practical expedient which was adopted to avoid the possible approach of the reference signal to the saturation level of the head amplifier (10 v). Linearity of absolute response to carrier was confirmed in the 1 to 4 μg sample range. Residual contributions at m/e 223 and 224 in the spectrum of the deuterium form due to the protium form did not significantly affect measurements made at the protium-deuterium dilutions in the range of interest (1 in 2,000 to 1 in 100). Measuring peak height response ratio at m/e 217 to

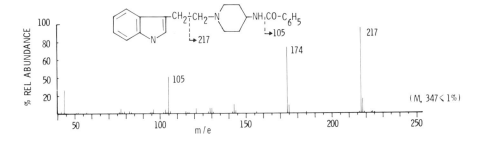

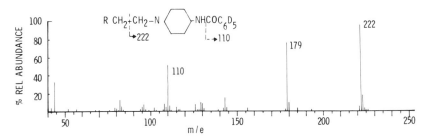

FIG. 3. Mass spectra of (H₅) indoramin (upper) and (D₅) indoramin (lower) at 20 eV.

m/e 224, linear calibration was obtained for the protium form from 0.5 to 10 ng on coinjection with 2 μg of deuterated carrier.

Indoramin is a chemically stable lipid soluble compound. Concentrated extracts from plasma are readily made by simple organic solvent extraction and back extraction methods which, in the presence of carrier, may be made highly reproducible. Interference from endogenous material in chromatograms from plasma was significant on the 217 channel (protium form) at 20 eV. This was effectively eliminated at 12 eV. Calibration for the protium form in 3-ml aliquots of plasma was effected from 5 to 100 ng/ml in the presence of 10 μg/ml of deuterated carrier. Fractions of the final concentrates assayed by GC–MS were equivalent to 1.5 to 3 μg carrier injection ensuring maximum carrier efficiency and good peak profile. A highly correlated linear regression ($R = 0.999$) was obtained, the values at each plasma concentration being within the standard deviation (5 to 14%) of measurement of corresponding standard dilutions of protium in deuterium form. The practical value of this method is being evaluated in studies of the drug's pharmacokinetics in man.

Indoramin is an interesting, if rather extreme, example of a compound which cannot be effectively chromatographed at low concentrations carrier-free. Novel inlet system effects are encountered when using low absolute amounts of carrier. These, and other observations which are relevant to the general problem of carrier compensation for poor gas chromatographic behavior, will be the subject of a further communication.

ACKNOWLEDGMENTS

We thank Professor C. T. Dollery and Dr. D. S. Davies for their interest and encouragement; Dr. M. Davis, the Liver Unit, Kings College Hospital, for clinical assistance; and L. Davies and H. Sahyoun for skilled technical assistance.

REFERENCES

Davies, D. S., Gigon, P. L., and Gillette, J. R. (1969): Species and sex differences in electron transport systems in liver microsomes and the relationship to ethylmorphine N-demethylation. *Life Sciences,* 8:85–91.

Davies, D. S., and Thorgeirsson, S. S. (1972): Mechanism of hepatic drug oxidation and its relationship to individual differences in rates of oxidation in man. *Annals of the New York Academy of Sciences,* 179:411–419.

Draffan, G. H., Clare, R. A., and Williams, F. M. (1973): Determination of barbiturates and their metabolites in small plasma samples by gas chromatography–mass spectrometry. *Journal of Chromatography,* 75:45–53.

Gaffney, T. E., Hammar, C.-G., Holmstedt, B., and McMahon, R. E. (1971): Ion specific detection of internal standards labelled with stable isotopes. *Analytical Chemistry,* 43:307–310.

Gilbert, J. N. T., Millard, B. J., and Powell, J. W. (1970): Combined gas-liquid chromatography–mass spectrometry in the study of barbiturate metabolism. *Journal of Pharmacy and Pharmacology,* 22:897–901.

Hammar, C.-G., Holmstedt, B., and Ryhage, R. (1968): Mass fragmentography: Identification of chlorpromazine and its metabolites in human blood by a new method. *Analytical Biochemistry,* 25:337–339.

Kelly, R. W. (1972): A simple device for analogue recording of the abundance of selected ions from a combined gas chromatograph–mass spectrometer. *Journal of Chromatography,* 71:337–339.

Krauer, B., Draffan, G. H., Williams, F. M., Clare, R. A., Dollery, C. T., and Hawkins, D. F. (1973): Elimination kinetics of amylobarbital in mothers and their newborn infants. *Clinical Pharmacology and Therapeutics,* 14:442–447.

Kuntzman, R., March, L. C., Brand, L., Jacobson, M., Levin, W., and Conney, A. H. (1966): Metabolism of drugs and carcinogens by human liver enzymes. *Journal of Pharmacology and Experimental Therapeutics,* 152:151–156.

Lowry, D. H., Rosenbrough, N. J., Farr, A. L., and Randall, R. J. (1951): Protein measurement with the folin phenol reagent. *Journal of Biological Chemistry,* 193:263–270.

Samuelsson, B., Hamberg, M., and Sweeley, C. C. (1970): Quantitative gas chromatography of prostaglandin E, at nanogram level: Use of a deuterated carrier and multiple-ion analyser. *Analytical Biochemistry,* 38:301–304.

Mass Spectrometry in Biochemistry and Medicine,
edited by A. Frigerio and N. Castagnoli.
Raven Press, New York © 1974

The Metabolism of Cyclophosphamide

T. A. Connors, P. J. Cox, P. B. Farmer, A. B. Foster, and
M. Jarman

Chester Beatty Research Institute, Institute of Cancer Research: Royal Cancer Hospital,
Fulham Road, London SW3 6JB, England

I. INTRODUCTION

Cyclophosphamide {2-[*bis*(2-chloroethyl)amino]tetrahydro-2H-1,3,2-ox-azaphosphorine 2-oxide, I} is a widely used antitumor agent. It is relatively nontoxic to cancer cells *in vitro*, and the conversion, *in vivo*, into a cytotoxic compound occurs principally in the liver (Arnold and Bourseaux, 1958). In this activation process (Fig. 1) 4-hydroxycyclophosphamide (IIa) and its acyclic tautomer aldophosphamide (IIb) are believed to be key intermediates (Hill, Laster, and Struck, 1972); and indirect evidence for their presence in microsomal incubates was the conversion, mediated by aldehyde oxidase, of two cytotoxic metabolites isolated therefrom into carboxyphosphamide [2-carboxyethyl-N,N-*bis*(2-chloroethyl)phosphorodiamidate, III]. This compound and 4-ketocyclophosphamide {2-[*bis*(2-chloroethyl)amino]-tetrahydro-2H-1,3,2-oxazaphosphorine-4-one 2-oxide, IV}, which could be formed respectively from aldophosphamide (IIb) and 4-hydroxyphosphamide (IIa), are urinary metabolites of cyclophosphamide (Hill, Kirk, and Struck, 1970) which are relatively nontoxic toward tumor cells both *in vivo* and *in vitro*. In contrast, a purely chemical process, the β-elimination of acrolein from aldophosphamide (IIb), would give the cytotoxic (Goldin and Wood, 1969) phosphoramide mustard [N,N-*bis*(2-chloroethyl)phosphorodiamidic acid, V]. Although the latter has not been detected either *in vivo* or in microsomal incubates of cyclophosphamide, indirect evidence for its formation under these last conditions was the isolation of acrolein as its 2,4-dinitrophenylhydrazone (Alarcon and Meienhofer, 1971). Very recently, both 4-hydroxycyclophosphamide (IIa) (Takamizawa, Matsumoto, Iwata, Katagiri, Tochino, and Yamaguchi, 1973) and aldophosphamide (IIb) (Struck and Hill, 1972) have been synthesized and shown to be sufficiently cytotoxic toward tumor cell lines *in vitro* for them, or substances spontaneously formed therefrom, to be the active metabolites of cyclophosphamide.

The objective of the present studies was to seek further evidence in support of the unifying scheme (Fig. 1) interrelating previous experimental

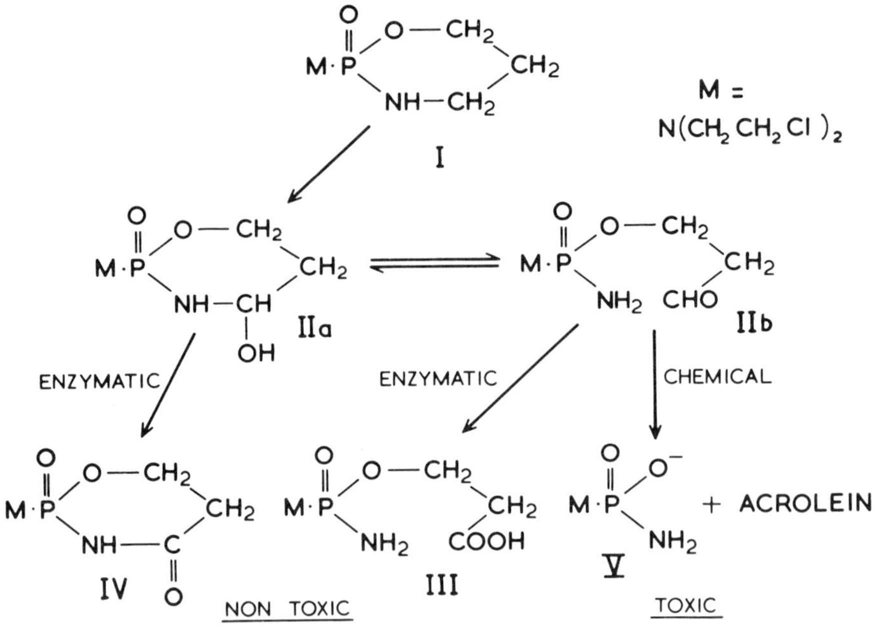

FIG. 1. Scheme for the metabolic activation of cyclophosphamide.

findings. A parallel investigation of the closely related drug isophosphamide [3-(2-chloroethyl)-2-(2-chloroethylamino)tetrahydro-1,3,2-oxazaphosphorine 2-oxide, VI] (Brock, 1967) was also undertaken.

II. MATERIALS AND METHODS

The method for the preparation of rat liver microsomes and for the incubation with the drug (either [^{32}P]-cyclophosphamide, specific activity 2.3 mC/mmole, or [^{32}P]-isophosphamide, specific activity 13.4 mC/mmole) was that previously described (Connors, McLoughlin, and Grover, 1970) with the additional proviso that here, special care was taken in washing the microsomes thoroughly, since the soluble fraction of the cell is believed to promote the conversion of aldophosphamide (IIb) into carboxyphosphamide (III). After incubation at 37°C for 45 min, protein was precipitated with ethanol (this precipitating agent is essential) and the concentrated supernatant was extracted with chloroform. Chromatography of extracts was conducted on glass plates (20 × 5 cm) coated with Merck Kieselgel G. After development, radioactive components were located with a Berthold radiochromatogram scanner. The radioactive components were eluted in 0.1-ml ethanol by the method of Rix, Webster, and Wright (1969). 2,4-Dinitrophenylhydrazones were located visually (orange or yellow spots).

Mass spectra were determined with an A.E.I. MS-12 spectrometer, using the direct insertion technique, an ionizing voltage of 70 eV, and a source temperature not exceeding 100°C above ambient.

III. RESULTS AND DISCUSSION

A chloroform solution of the products extracted, following the microsomal incubation of cyclophosphamide, was subjected to thin-layer chromatography (TLC) in two solvent systems. TLC in chloroform-ethanol (19:1, System A) revealed four radioactive components, R_f values 0.00, 0.11, 0.23, and 0.34. The slowest mobile component (R_f 0.11) gave a mass spectrum identical with that reported for 2-(2-chloroethylamino)tetrahydro-2H-1,3,2-oxazaphosphorine 2-oxide (VII), a compound which has been isolated from the urine of sheep following the administration of cyclophosphamide (Bakke, Feil, Fjelstul, and Thacker, 1972). The component which was immobile in System A was eluted with methanol and conventionally methylated with diazomethane. TLC of the product in chloroform-methanol (9:1, System B) revealed a single radioactive component, of R_f 0.41, which gave a mass spectrum (Fig. 2) appropriate for the methyl ester of the previously mentioned phosphoramide mustard (V, Fig. 1), this being the first direct evidence for the metabolic production of this substance from cyclophosphamide.[1] Thus a small, 2 Cl-containing molecular ion (m/e 234/236) underwent the usual (Bakke et al., 1972) loss of 49 mass units (CH_2Cl) to give m/e 185/187 as the principal fragment. Another conventionally observed diagnostic fragment was ($ClCH_2CH_2NH{=}CH_2$)$^+$ at m/e 92/94. The addi-

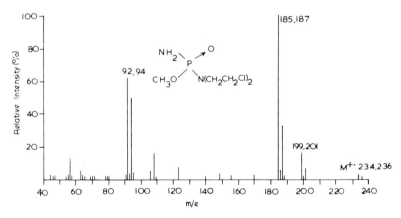

FIG. 2. Mass spectrum of methylated cyclophosphamide metabolite: believed to be the methyl ester of phosphoramide mustard.

[1] Since this chapter was prepared, an identical finding has been published [Colvin, M., Padgett, C. A., and Fenselau, C. (1973): A biologically active metabolite of cyclophosphamide. *Cancer Research*, 33:915–918].

tional contribution to m/e 94 of the ion $(CH_3O\!-\!\underset{\underset{O}{\downarrow}}{P}\!-\!NH_2)^+$ accounts for the apparently anomalous intensity of this peak.

The component of R_f 0.23 in System A was composite, and separated into two components on rechromatography in ethanol-ether (1:9, System C). The slower component was cyclophosphamide. The faster component gave a mass spectrum (Fig. 3b) which was qualitatively similar to that (Fig. 3a) given by the most mobile component in the System A (R_f 0.34) and the two

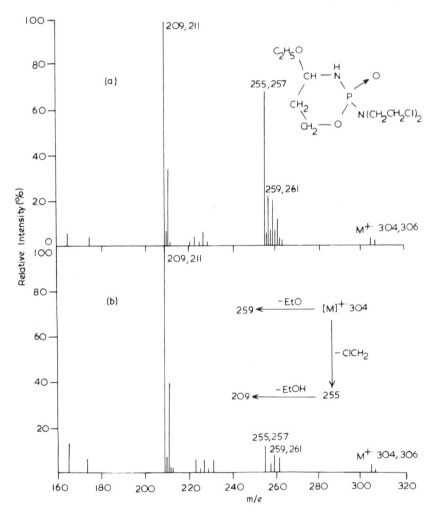

FIG. 3. Mass spectra of cyclophosphamide metabolites produced with washed microsomes: (a) Component of R_f 0.34 in chloroform-ethanol, 19:1. (b) Component of R_f 0.23 in chloroform-ethanol, 19:1 (after further resolution from cyclophosphamide in ethanol-ether, 1:9).

substances therefore appear to be structurally related. The structural assignments given in Fig. 3 were consistent with the observed losses from the molecular ions (m/e 304/306) of fragments attributable to EtO (45 mass units) and EtOH (46 mass units), either preceded by, or followed by the previously mentioned elimination of CH_2Cl. Isophosphamide (VI) afforded metabolite derivatives with TLC mobilities and mass spectra (Fig. 4) similar to the last-mentioned cyclophosphamide products. Since these ethoxy derivatives were not observed when precipitating agents other than ethanol were used to remove the protein following microsomal incubation, the ethyl substituents were presumably introduced at this stage, or during the subsequent extraction into chloroform. The implicit relationship between these products and the appropriate 4-hydroxy derivatives was further validated by

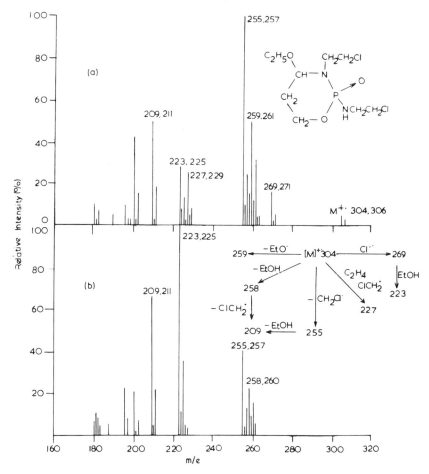

FIG. 4. Mass spectra of isophosphamide metabolites produced with washed microsomes: (a) Component of R_f 0.33 in chloroform-ethanol, 19:1. (b) Component of R_f 0.25 in chloroform-ethanol, 19:1.

the formation of acrolein 2,4-dinitrophenylhydrazone therefrom. Thus all four derivatives afforded this compound when treatment with 2,4-dinitrophenylhydrazine in 2 M hydrochloric acid was followed by adjustment to alkaline pH (9 to 10). In Fig. 5 the mass spectrum of the product thus obtained from the more mobile (TLC) isophosphamide derivative (which gave the mass spectrum shown in Fig. 3a) is compared with that of an authentic sample. The fragmentation pathways are based on those postulated (Seibl, 1970) for crotonaldehyde 2,4-dinitrophenylhydrazone.

4-Substitution of the ethoxy-group is not proved by the foregoing observations, since the corresponding 6-substituted derivatives could likewise yield an acrolein derivative via β-elimination from the appropriate aldehyde (Alarcon and Meienhofer, 1971). Evidence for 4-substitution was, however, afforded by deuterium-labeling experiments. Thus cyclophosphamide-4-D_2 was prepared (Fig. 6a) and metabolized in the same manner as was cyclophosphamide. The number of deuterium atoms in the subsequently isolated ethoxy derivatives (Fig. 6b), and in the derived acrolein 2,4-dinitrophenylhydrazones (Fig. 6c), should differ according to the position of the ethoxy substituent. In the mass spectra of the two ethoxy derivatives and of the

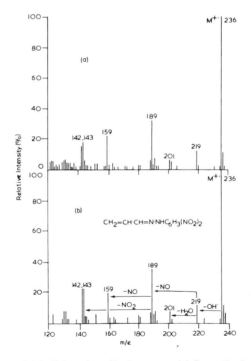

FIG. 5. Mass spectra of 2,4-dinitrophenylhydrazones: (a) From the isophosphamide metabolite of R_f 0.33 in chloroform-ethanol, 19:1. (b) Authentic acrolein 2,4-dinitrophenylhydrazone. Fragmentation scheme based on that for crotonaldehyde 2,4-dinitrophenylhydrazone.

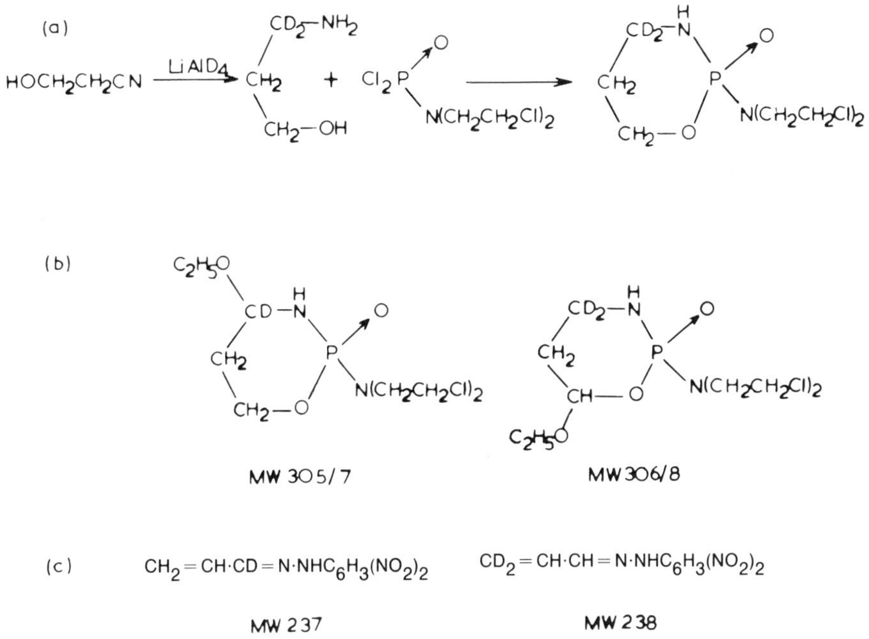

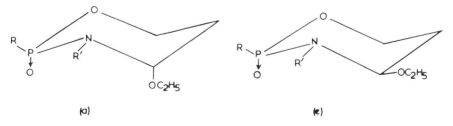

FIG. 6. (a) Synthesis of cyclophosphamide-4-D$_2$. (b) Ethoxy derivatives from cyclophosphamide-4-D$_2$. (c) Acrolein 2,4-dinitrophenylhydrazones which could be isolated following 4-hydroxylation (MW 237) and 6-hydroxylation (MW 238) of cyclophosphamide-4-D$_2$.

derived acrolein derivatives, only one deuterium atom was indicated, an observation consistent with 4-substitution in both ethoxy derivatives. The stereochemistry of the oxazaphosphorine ring in cyclophosphamide (Clardy, Mosbo, and Verkade, 1972) allows for either an axial or an equatorial orientation for a 4-substituent (Fig. 7).

The toxicity of each of the ethoxy derivatives of cyclophosphamide toward Walker tumor cells *in vitro* was comparable with values reported for 4-hydroxycyclophosphamide when tested against other cell lines (Takamizawa et al., 1973). In water, each ethoxy derivative was apparently converted within 48 hr into a single product (R_f 0.10 in System A) giving a

FIG. 7. Possible conformations for 4-ethoxycyclophosphamide (R=N(CH$_2$CH$_2$Cl)$_2$, R'=H) and isophosphamide (R=NHCH$_2$CH$_2$Cl, R'=CH$_2$CH$_2$Cl): (a) axial, (e) equatorial.

positive reaction with 2,4-dinitrophenylhydrazine, but which was insufficiently thermally stable for characterization by mass spectrometry. Conceivably, this reaction represents the hydrolysis of the ethoxy derivatives to 4-hydroxycyclophosphamide, an explanation which would account for the toxic effects of the ethoxy derivatives *in vitro*. If so, the ethoxy derivatives may be a convenient latent form in which to administer the active metabolite of cyclophosphamide, since 4-hydroxycyclophosphamide itself was reported by the Japanese workers to be unstable.

FORMULAE

VI   VII

ACKNOWLEDGMENTS

This investigation was supported by grants to this Institute from the Medical Research Council and the Cancer Research Campaign. The A.E.I. MS-12 mass spectrometer was purchased on a special grant from the Medical Research Council. The following fellowships are gratefully acknowledged: a Ward Blenkinsop Fellowship (to P.J.C.), a William Shepherd Fellowship (to P.B.F.), and a Ludwig Fellowship (to M.J.).

REFERENCES

Alarcon, R. A., and Meienhofer, J. (1971): Formation of the cytotoxic aldehyde acrolein during *in vitro* degradation of cyclophosphamide. *Nature New Biology,* 233:250–252.

Arnold, H., and Bourseaux, F. (1958): Neuartige Krebs-Chemotherapeutika aus der Gruppe der zyklischen N-Lost-Phosphamidester. *Naturwissenschaften,* 45:64–66.

Bakke, J. E., Feil, V. J., Fjelstul, C. E., and Thacker, E. J. (1972): Metabolism of cyclophosphamide by sheep. *Journal of Agriculture and Food Chemistry,* 20:384–388.

Brock, N. (1967): Pharmakologische Untersuchungen mit neuen N-Chloroäthylphosphorauester-Diamiden. In: *Proceedings of the International Congress of Chemotherapy. 5th Congress, Vienna,* edited by K. U. Spitzy and H. Haschek. Wiener Medizinischer Akademie.

Clardy, J. C., Mosbo, J. A., and Verkade, J. G. (1972): Crystal and molecular structure of cyclophosphamide hydrate. *Chemical Communications,* 20:1163–1164.

Connors, T. A., McLoughlin, A. M., and Grover, P. L. (1970): Microsomal activation of cyclophosphamide *in vitro. Biochemical Pharmacology,* 19:1533–1535.

Goldin, A., and Wood, H. B. (1969): Preclinical investigation of alkylating agents in cancer chemotherapy. *Annals of the New York Academy of Sciences,* 163:954–1005.

Hill, D. L., Kirk, M. C., and Struck, R. F. (1970): Isolation and identification of 4-ketocyclophosphamide, a possible active form of the antitumor agent cyclophosphamide. *Journal of the American Chemical Society,* 92:3207–3208.

Hill, D. L., Laster, W. R., and Struck, R. F. (1972): Enzymic metabolism of cyclophosphamide and nicotine and production of a toxic cyclophosphamide metabolite. *Cancer Research,* 32:658–665.

Rix, M. J., Webster, B. R., and Wright, I. C. (1969): A technique for obtaining the mass spectra of substances isolated by thin layer chromatography. *Chemistry and Industry (London)*, 452.

Seibl, J. (1970): Zur Massenspektrometrie von Nitrophenylverbindungen – II. Nitrophenyl-hydrazone Aliphatischer Aldehyde und Ketone. *Organic Mass Spectrometry*, 3:417–432.

Struck, R. F., and Hill, D. L. (1972): Investigation of the synthesis of aldophosphamide, a toxic metabolite of cyclophosphamide. *Proceedings of the American Association of Cancer Research*, 13: Abs. 199.

Takamizawa, A., Matsumoto, S., Iwata, T., Katagiri, K., Tochino, Y., and Yamaguchi, K. (1973): Studies on cyclophosphamide metabolites and their related compounds. II. Preparation of an active species of cyclophosphamide and some related compounds. *Journal of the American Chemical Society*, 95:985–986.

Mass Spectrometry in Biochemistry and Medicine,
edited by A. Frigerio and N. Castagnoli.
Raven Press, New York © 1974

Contribution of Gas Chromatographic–Mass Spectrometric Techniques in the Identification of Urinary Conjugated Metabolites of 4-Allyloxy-3-Chlorophenylacetic Acid (Alclofenac)*

R. Roncucci, M.-J. Simon, G. Lambelin, and K. Debast

Continental Pharma Research Laboratories, 30 Steenweg op Haacht, 1830-Machelen (Bt), Belgium

I. INTRODUCTION

The metabolic fate of alclofenac (A) has been investigated in several animal species and in man (Roncucci, Simon, Gillet, Lambelin, and Kaisin, 1972). These studies have shown that three major metabolites can be found in the hydrolyzed urine of all the species considered (man, monkey, pig, rabbit, rat, and mouse). These metabolites, which have been identified by mass spectrometry (Table 1), are the unchanged drug, the dealkylated com-

TABLE 1   METABOLITES OF ALCLOFENAC

Structure		Description
$CH_2=CH-CH_2-O$—(ring, Cl)—CH_2-COOH	**A**	4-allyloxy-3-chlorophenylacetic acid $C_{11}H_{11}O_3Cl_1$ MW = 226.039
$CH_2=CH-CH_2-O$—(ring, Cl)—CH_2-CO (glucuronide)	**A-G**	1'-β-D-glucuronic acid-4-allyloxy-3-chlorophenylacetate $C_{17}H_{19}O_9Cl_1$ MW = 402.071
$\overset{HO}{}\overset{OH}{}$ $CH_2-CH-CH_2-O$—(ring, Cl)—CH_2-COOH	**DHA**	4-(γ.β-dihydroxy)propyloxy-3-chlorophenylacetic acid $C_{11}H_{13}O_5Cl_1$ MW = 260.040
HO—(ring, Cl)—CH_2-COOH	**4-HCPA**	4-hydroxy-3-chlorophenylacetic acid $C_8H_7O_3Cl_1$ MW = 186.008
$CH_2=CH-CH_2-O$—(ring, Cl)—CH_2-CO $NH-CH_2-COOH$	**HAA**	(N-carboxymethyl)-4-allyloxy-3-chlorophenylacetamide $C_{13}H_{14}N_1O_4Cl_1$ MW = 283.061

---

* Mervan®, Continental Pharma s.a., Brussels, Belgium.

pound (4-hydroxyphenylacetic acid, 4-HCPA), and the dihydroxy derivative of A [racemic 4-(γ,β-dihydroxy)propyloxy-3-chlorophenylacetic acid,
DHA]. However, dramatic interspecies differences have been observed
when the amounts of these end products in the urine were compared: man
and monkey excrete very large amounts of A whereas in the other species
DHA and 4-HCPA are the major metabolites. On the other hand, it appeared that complete quantitation of A in man and monkey by GLC (Roncucci, Simon, and Lambelin, 1971) required the hydrolysis of the urine
specimen. This fact led us to suppose that A was partially present in the
urine of these species in a conjugated form.

This chapter describes the characterization of two urinary conjugated
forms of A, i.e., the ester glucuronide (A–G) found in man and monkey and
the glycine conjugate (HAA) detected in pig.

II. MATERIAL AND METHODS

A. Drug and Reagents

The structures of the metabolites of A discussed in this chapter are shown
in Table 1. Except in man, the ^{14}C-carbonyl-labeled A was used throughout
the study (Gillet, Gautier, Roncucci, Simon, and Lambelin, 1973). Unless
otherwise stated analytical grade solvents and reagents were purchased
from Merck, Darmstadt.

B. Biological Sampling

In man, urine was obtained from one healthy volunteer (M, 27 years,
65 kg) who took 500 mg (i.e., about 7.7 mg/kg) of A as a commercial tablet.
The 0- to 12-hr urine was analyzed.

Three rhesus monkeys (*Macaca mulata;* no. 15, M, 3.18 kg; no. 16, F,
3.44 kg; and no. 18, F, 3.42 kg) were studied. The drug (200 mg/kg; 25 μC/
animal) was administered in a mucilage (1% tragacanth gum) using a naso-
gastric tube. The animals were housed in appropriate metabolic cages and
the 0- to 24-hr urine was collected.

In the pig (Large White breed), the 0- to 24-hr urine was analyzed; one
animal (no. 3, F, 84.5 kg) received a 500 mg/kg (50 μC) oral dose administered in water suspension by stomach tube. Three other pigs were used for
analytical purposes (see Table 7).

In all the experiments, urine was kept frozen ($-20°C$) until analyzed.

C. Extraction Procedures

In all the species considered urine was extracted with methyl isobutyl
ketone (MIBK; U.C.B.) in acidic conditions (pH 2; three times, 1:1 v/v).
This solvent yielded 95 to 100% extraction on the basis of ^{14}C determinations. The metabolites, which were adsorbed on the silica gel that was

scraped off the plates used for purification purposes, were generally dissolved in water prior to extraction with MIBK at acidic pH. Unless otherwise stated, extracts were dried (Na_2SO_4), evaporated under vacuum (Rotavapor, Büchi), and residues dissolved in 1-4-dioxane.

D. Purification Techniques

Preparative layer chromatography (PLC) and thin-layer chromatography (TLC) were carried out using silica gel GF 254 (Merck, Darmstadt). Unless otherwise stated, chloroform, cyclohexane, and acetic acid (6:4:1, v/v) were employed as the chromatographic mixture.

In the case of the urine of the pig, preparative ion-exchange chromatography on DEAE-sephadex A25 (Pharmacia) was used for purification. Metabolites were eluted with a linear NaCl gradient (up to 1 M) in Tris-HCl buffer (0.01 M) at pH 7.

E. Detection and Quantitation Techniques

[14]C-Material was quantitated directly or after combustion by liquid scintillation counting (LSC) using a Tri-Carb spectrometer (Packard Inst. model 3375) with a 1,4-dioxane-based scintillation mixture (Roncucci, Lambelin, Simon, and Soudyn, 1968). TLC plates were scanned with a radiochromatogram scanner (Packard Inst. model 7201). Radioactive spots on PLC plates were located by autoradiography (ARG) using X-ray films (Structurix D10, Gevaert).

Quantitation of free metabolites was achieved by GLC (Packard Inst. model 7200), as described previously (Roncucci et al., 1971), using appropriate derivatizing techniques. A flow splitter placed between the column and the FID detector was used to determine the specific radioactivities of the metabolites after collection of the eluted fractions.

F. Hydrolysis

In order to determine the drug moiety in the conjugated molecules analyzed, HCl- or NaOH-hydrolyses were achieved in the aqueous phase before extraction. Five percent (v/v) 12 N HCl or NaOH was added and the mixture heated at 100°C for 20 min. Enzymic hydrolyses were carried out for the specific case of A–G using β-glucuronidase (Sigma Chemical Co., Type V-A, bacterial origin, 6,000 Fishman units/ml). Incubations in acetate buffer (0.2 M, pH 6) were performed at 37°C for at least 4 days.

G. Chemical Tests

According to Dische (1947), in addition to the Benedict's solution (see Vogel, 1956), hexuronic acids can be chemically characterized by the color reaction they develop with carbazole.

H. Derivative Formation for GLC and Mass Spectral Analysis

A, 4-HCPA, and DHA were silylated according to the procedure pre-viously described (Roncucci et al., 1971) or by derivatization with N,O-*bis*-(trimethylsilyl)trifluoroacetamide (BSTFA; Pierce Chemical Co.).

The ester glucuronide of A (A–G) was derivatized in two ways:

1. Tris(trifluoroacetyl)methyl ester: A–G was dissolved in methanol and diazomethane added. The excess of N_2CH_2 was rapidly removed under vacuum or by a stream of N_2. To the residue, trifluoroacetic anhydride (Fluka) and 1,4-dioxane (1:1, v/v) were added and the mixture was allowed to stand at room temperature for 1 hr. The residue obtained after evapora-tion under vacuum is dissolved in 1,4-dioxane.

2. Tris(trimethylsilyl)methyl ester: the compound was dissolved in methanol and methylated as described above. The residue was dissolved in 1,4-dioxane and evaporated to dryness (this operation was repeated twice in order to eliminate methanol completely). The residue was then dissolved in a mixture of pyridine-hexamethyldisilazane (HMDS; Pierce Chemical Co.) and trifluoroacetic anhydride (100:100:1, v/v) and allowed to react for 2 hr. The excess of reagents can be removed by evaporation under vacuum.

The deuterium-labeled trimethylsilyl derivative of A–G was obtained in the same way but using hexamethyl-d_{18}-disilazane (purchased from Supelco Inc.) as the derivatizing agent. D-Glucuronic acid (Sigma Chemical Co.) was used as the reference compound. It has been derivatized as described for A–G. The glycine conjugate of A (HAA) was dissolved in methanol and methylated as described above. N_2CH_2 was allowed to react for several hours prior to evaporation.

III. MASS SPECTROMETRY

A single focusing, low-resolution Varian-MAT CH7 mass spectrometer was used. The inlet of the unit was fitted either with a temperature-controlled probe for solid samples or with a line-of-slight probe which allowed for coupling with a Varian Aerograph series 2700 GC-system via a two-stage Bieman-Watson type separator. Unless otherwise stated, source settings for solid sample analyses were electron energy 70 eV (emission current 300 μA) and ion-accelerating voltage 3kV. The temperature of the source was fixed at about 100°C.

In the combined GC–MS mode of operation, a glass column (2 m by 2 mm) packed with 3% OV-1 on Gas-Chrom Q 100–120 mesh (Serva) was used. The carrier gas was He ($\pm$ 12 ml/min). The adjustable slit was fixed to obtain a resolution of about 1,500. The temperatures were inlet 260°C, column oven 180 up to 250°C (according to the case), separator 250°C, and source 260°C.

Mass-fragmentography (MF) studies were achieved thanks to the peak-matcher unit of the mass spectrometer. Two masses were focused and their intensities recorded alternatively during the GC run. The lower mass was focused by regulating the magnet current with the accelerating voltage fixed at 3 kV whereas the higher mass was focused by reduction of the accelerating voltage and by keeping the magnet current constant. The frequency of the voltage alternations was about 2/sec.

In peak-matching studies high-boiling point perfluorokerosine (PFK, Merck) was used as the reference for precise mass determinations. PFK was introduced in the source via a separate heated inlet. The resolution was about 5,000.

IV. RESULTS

A. Identification of the Ester Glucuronide of A in the Urine of Rhesus Monkey

Figure 1a shows that TLC allows a quite good separation of the metabolites of ^{14}C-A extracted from the unhydrolyzed urine. The fraction with $R_f = 0$ disappears when TLC is carried out on extracts of hydrolyzed urine; with these conditions larger amounts of free A are obtained. It is therefore concluded that A is present in urine partially in a conjugated form. With a more appropriate solvent system DHA can be separated from the conjugate. The separation of this unresolved mixture scraped from the plate is shown in Fig. 1b.

An aliquot of the purified conjugate obtained by extraction of the gel of the plates with water is submitted to HCl-hydrolysis and chromatographed: free A is obtained (Fig. 1e). A second aliquot is submitted to incubation with β-glucuronidase and, after TLC, it appears that only free A is present (Figs. 1c and 1d). A third and a fourth aliquot are used to test the compound in the Benedict's and Dische's reactions. Both are clearly positive. At this stage it can be thought that the conjugate is an ester glucuronide of A.

Larger amounts of A–G are purified by PLC and the conjugate is extracted from the gel of the plates with water. The aqueous phase is then extracted with MIBK and the extract concentrated. Adding a mixture of ether and pentane at low temperature causes a brown oily compound to precipitate. The specific radioactivity of this compound is found to be consistent with the structure proposed. The underivatized compound is then submitted to mass spectral analysis using the temperature-controlled solid sample probe with the ion source at a temperature of about 100°C. The spectrum recorded is shown in Fig. 2.

In spite of the fact that no molecular ions are observed the two sets of chlorinated ions at m/e 384/386 and 366/368 are consistent with the proposed structure of the conjugate (expected MW 402/406) if one considers the possibility of the loss of one and two molecules of water respectively

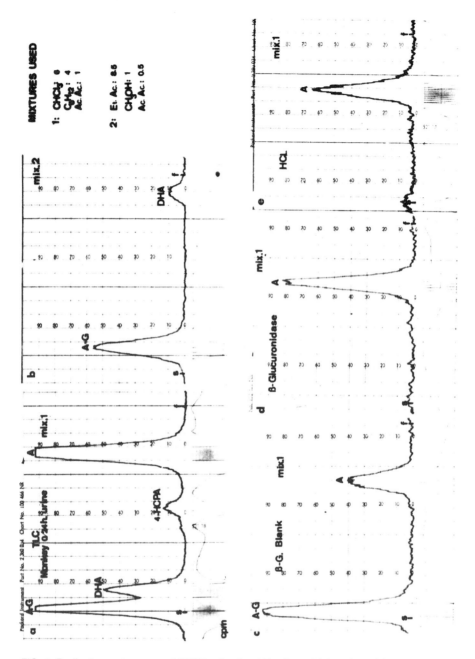

FIG. 1. Radiochromatograms of MIBK extracts of the O- to 24-hr urine of rhesus monkey after oral intake of ^{14}C–A (200 mg/kg). Upper chromatograms (from the left to the right): (a) total MIBK extract chromatographed with solvent mixture 1. (b) The ^{14}C-spot with $R_f = 0$ in Fig. 1a is extracted from the gel and rechromatographed with solvent mixture 2. Lower chromatograms (from the left to the right): (c) Shows the action of β-glucuronidase (37°C, pH 6, 4 days) on A–G. (d) The TLC (Fig. 1c) of the MIBK extract of the blank (without enzyme) shows that only small amounts of free A can be obtained by the effect of the incubation (probably formed by acid hydrolysis in the medium at pH 6). On the contrary, in the presence of β-glucuronidase (Fig. 1d) A–G is completely split into free A. (e) Shows that HCl-hydrolysis completely splits A–G into free A [in the "blank" (without HCl) extract, only A–G is present (not shown)].

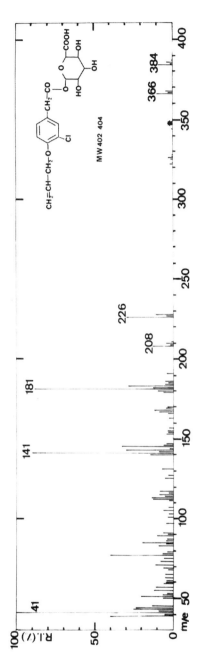

FIG. 2. Mass spectrum of underivatized A–G isolated from the urine of rhesus monkey. Solid sample probe. Resolution: 1,500. Source and probe temperature: 100°C. * = observed metastable ions.

TABLE 2. *Determination by peak-matching of the accurate masses of fragments m/e 384 and 366 of underivatized A–G[a]*

Fragment	Proposed formula	Calculated mass	PFK reference mass	Theoretical Ms/M$_L$	Observed Ms/M$_L$	Δ (ppm)
384 (M-H$_2$O)	C$_{17}$H$_{17}$O$_8$Cl	384.060699 (Ms)	380.976071 (M$_L$)	1.008097	1.008099	2
366 (M-2H$_2$O)	C$_{17}$H$_{15}$O$_7$Cl	366.050134 (M$_L$)	380.976071 (Ms)	1.040776	1.040709	67

[a] See Fig. 2; electron energy: 70 eV; resolution: 5,000.

from the unseen molecular ions. Indeed the determination of the accurate masses of fragments 366 and 384 by the peak-matching method confirms their formula (Table 2). On the other hand, the mass spectrum of underivatized D-glucuronic acid obtained in the same conditions also does not show the presence of a molecular ion but peaks at -18 and -36 mass units (Fig. 3).

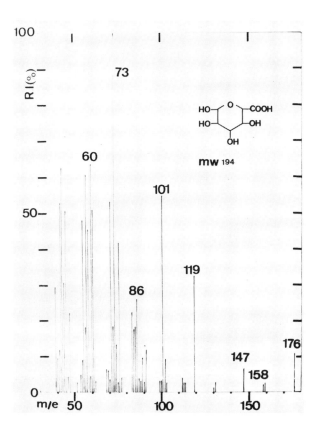

FIG. 3. Mass spectrum of underivatized D-glucuronic acid. Solid sample probe. Resolution: 1,500. Source and probe temperature: 100°C.

Finally all the fragments typical for the aglycon moiety are also present in the spectrum of A–G (see Roncucci et al., 1972). The structure of the compound is further investigated by MS using derivatization techniques. The Tris(trimethylsilyl)methyl ester derivative of A–G (expected MW 632/634) as well as its d_{27} analog (expected MW 659/661) are prepared and their respective mass spectrum recorded using the solid sample probe (Figs. 4 and 5). For comparison purposes, the tetra(trimethylsilyl)methyl ester of

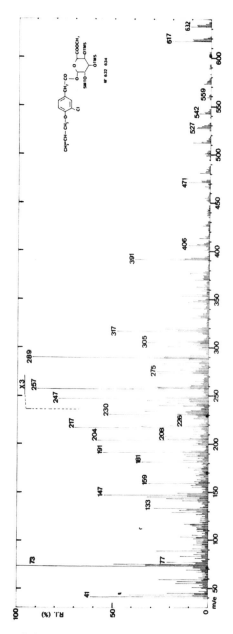

FIG. 4. Mass spectrum of the ester glucuronide of alclofenac (A–G) isolated from the urine of the rhesus monkey. Tris(trimethylsilyl)methyl ester derivative. Conditions: solid sample probe; resolution: 1,500; source and probe temperature: 100°C; ● = observed metastable ions; TMS: trimethylsilyl.

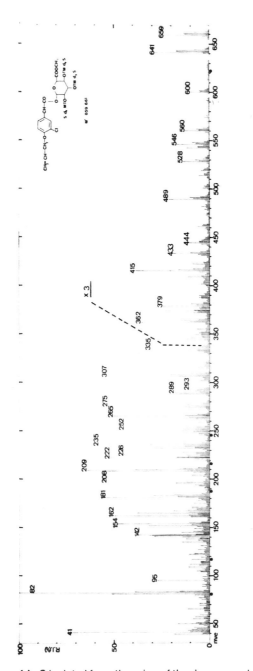

FIG. 5. Mass spectrum of A—G isolated from the urine of the rhesus monkey. Tris(trimethyl-d₉-silyl)methyl ester derivative. Same conditions as in Fig. 4.

D-glucuronic acid (expected MW 496) and its d_{36} analog (expected MW 496) are also analyzed (Figs. 6 and 7).

The fragmentation patterns observed for the derivatized A–G (Fig. 4) are in good agreement with those reported for synthetic ether glucuronides (Billets, Lietman, and Fenselau, 1973). The "molecular ion set" is formed by chlorinated peaks at M(632/634), M–15(617/619), M–59(573/575), M–73 (559/561), M–90(542/544), and M–105(527/529). The proposed formula of these fragments is shown in Table 3. The structure of the compound is further confirmed by the presence of ions at m/e 406 (407) which corresponds to the loss of the aglycon (226/228) from the molecular ions (cleavage of the exocyclic acetal bound).

The ion at m/e 406 (407) fragments similarly to the ions giving rise to peaks at m/e 392 (406–15), 347 (406–59) etc. (see Table 3). Many other fragments are formed from the ion at m/e 406 (407) and they are generally common to most silylated carbohydrates and especially to silylated D-glucuronic acid (see Fig. 6 and Radford and De Jongh, 1972, for review). These fragmentation patterns are confirmed by the mass shifts observed in the spectrum of the perdeuterated TMS-derivative (Figs. 5 and 7; Table 3). On the basis of the presence of chlorinated ions at m/e 208/210 it seems that electron impact produces a second major cleavage between the aglycon and the sugar moieties as indicated in the following scheme:

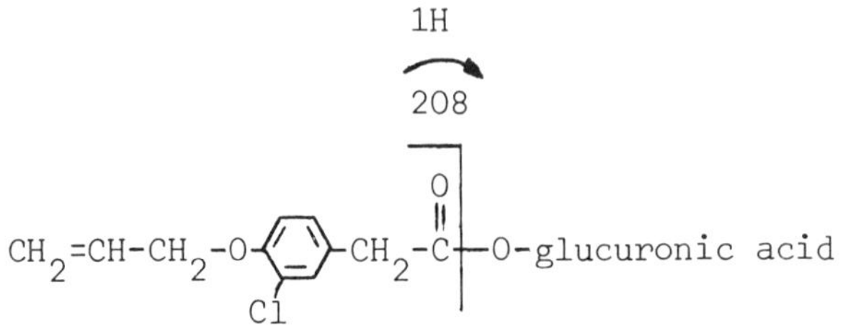

$$\text{CH}_2=\text{CH}-\text{CH}_2-\text{O}-\underset{\text{Cl}}{\underbrace{\bigcirc}}-\text{CH}_2-\overset{\text{O}}{\overset{\|}{\text{C}}}-\text{O-glucuronic acid}$$

The fragments at m/e 208/210 can also be seen in the mass spectrum of the underivatized A–G (see Fig. 2). This kind of fragmentation has also been observed in the case of testosterone ether glucuronide-TMS-methyl ester derivative (Billets et al., 1973).

The Tris(trifluoroacetyl)methyl ester derivative of A–G (expected MW 704/706) was also studied. Its mass spectrum is shown in Fig. 8. The mass spectrum of D-glucuronic acid derivatized in the same way is shown in Fig. 9. The A–G derivative shows a very intense chlorinated molecular ion (55%) at m/e 704/706.

The fragmentation pattern is summarized in Table 4. The molecular ion loses the "C_6-methyl ester group" and gives the chlorinated ions at m/e 645/647. Losses of TFA and OTFA from M lead to the fragments at m/e 608/610 and 591/593 respectively. As reported for the TMS-methyl ester derivative of A–G, two kinds of cleavage seem to occur between the drug and the sugar moieties.

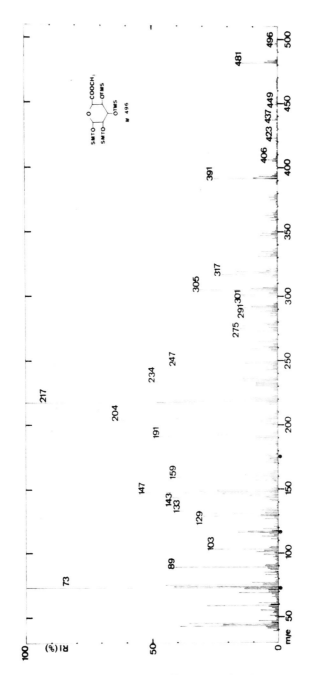

FIG. 6. Mass spectrum of D-glucuronic acid. Tetra(trimethylsilyl)methyl ester derivative. Same conditions as in Fig. 4.

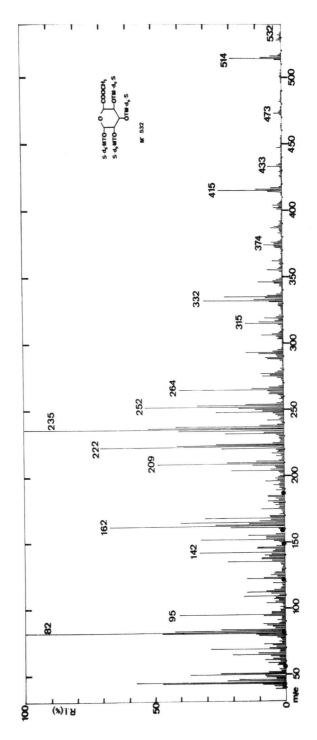

FIG. 7. Mass spectrum of D-glucuronic acid. Tetra(trimethyl-d₉-silyl)methyl ester derivative. Same conditions as in Fig. 4.

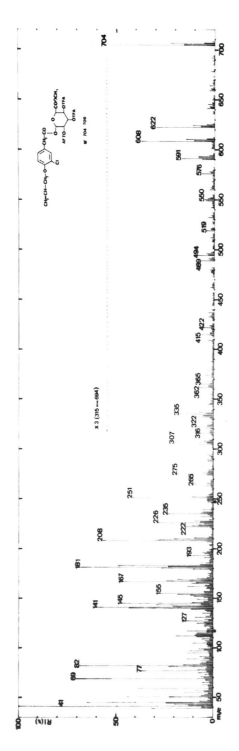

FIG. 8. Mass spectrum of the ester glucuronide of alclofenac isolated from the urine of rhesus monkey. Tris(trifluoroacetyl)methyl ester derivative. Conditions: solid sample probe; resolution: 1,500; source and probe temperature: 100°C; ● = observed metastable ions; TFA = trifluoroacetyl.

TABLE 3. Main fragments of A–G as Tris(trimethylsilyl)methylester derivative[a]

m/e	Proposed formula	Deuterated derivative m/e (mass shift)
	Hydrogenated derivative	
632/634 (M)	$CH_2=CH—CH_2—O—CH_2—CO—O—$ [ring structure: TMSO, OTMS, OTMS, COOCH₃; aromatic ring with Cl]	659/661 (+27)
617/619 (M–15)	M—(CH₃)	641/643 (+24)
573/575 (M–59)	M—(COOCH₃)	600/602 (+27)
559/561 (M–73)	M—[Si(CH₃)₃]	
542/544 (M–90)	M—[HO—Si—(CH₃)₃]	560/562 (+18)
527/529 (M–105)	M—[HO—Si—(CH₃)₃ + (CH₃)]	542/544 (+15)
407* (M–225)=S	[ring structure: H, SMTO, O, OTMS, OTMS, COOCH₃]	434 (+27)
406* (M–226)=S'	—	433 (+27)
391* (S'–15)	S'—(CH₃)	415 (+24)
347 (S'–59)	S'—(COOCH₃)	374 (+27)
317* (S–90) or (S'–89)	S or S'—[(H)O—Si—(CH₃)₃]	335 (+18)
301* S'(–105)	S'—[HO—Si(CH₃)₃ + (CH₃)]	316 (+15)
305*	TMSOCH=C—CH—OTMS, OTMS	332 (+27)

$COOCH_3$

275* $TMSO^+\!\!=\!CH\!-\!CH\!=\!C\!-\!OTMS$ 293 (+18)

217* $TMSO^+\!\!=\!CH\!-\!CH\!=\!CH\!-\!OTMS$ 235 (+18)

204* $[TMSO\!-\!CH\!=\!CH\!-\!OTMS]^{+\cdot}$ 222 (+18)

191* $TMSO\!-\!CH^+\!\!-\!OTMS$ 209 (+18)

147* $(CH_3)_2\!-\!Si^+\!\!-\!O\!-\!Si(CH_3)_3$ 162 (+15)

133* $CH_3\!-\!Si^+ \overset{O}{\underset{O}{\diamondsuit}} Si\!-\!(CH_3)_2$ 142 (+9)

73* $(CH_3)_3\!-\!Si^+$ 82 (+9)

Fragments at m/e 226/228, 181/183, 145, 141/143, 105, 77, 51, and 41 are related to the aglycon moiety (Roncucci et al., 1972).

Fragments at m/e 289, 257, 247, 234, 230, 184, and 159 and those indicated by * in the table are also present in the mass spectrum of D-glucuronic acid tetra(trimethylsilyl) methyl ester derivative (see Figs. 6 and 7 for the d₃₆-analog). For comparison with the spectra of synthetic ether glucuronides see Billets et al. (1973).

ᵃ See Figs. 4 and 5.

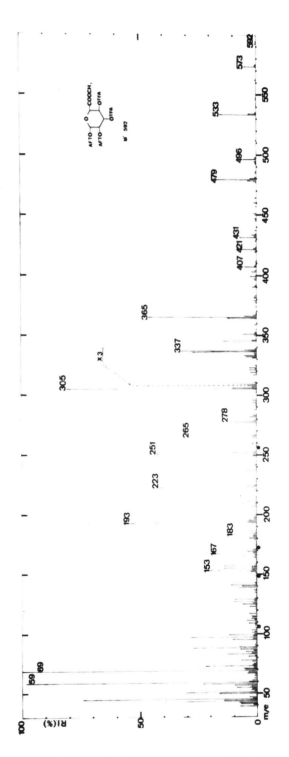

FIG. 9. Mass spectrum of D-glucuronic acid. Tetra(trifluoroacetyl)methyl ester derivative. Same conditions as in Fig. 8.

TABLE 4. *Characteristic fragments of the tris(trifluoroacetyl)methyl ester derivative of A–G[a]*

m/e[b]	Fragment	Corresponding fragment in ether glucuronides Tris(acetyl)methyl ester derivatives[c]
704/706	M	
645/647 (M–59)	M—COOCH$_3$	
608/610 (M–96)	M—TFA	
591/593 (M–113)	M—OTFA	
495 (M–209)	M—(CH$_2$=CH—CH$_2$—O—⟨ring,Cl⟩—CH$_2$—CO)	
479 (M–225)=S	M—aglycon moiety	317
365 (S–114)	S—HOTFA	257
251 (S–228)	S—2(HOTFA)	197
155 (S–324)	S—[2(HOTFA) + TFA]	155
127 (S–352)	S—[2(HOTFA) + TFA + CO]	127

[a] See Fig. 8.
[b] The ions corresponding to the fragmentation of the aglycon are listed in Table 3.
[c] From Paulson et al. (1973); fragment at m/e 215 is not observed in the mass spectrum of A–G.

Loss of 209 mass units (see fragment at m/e 208/210 obtained by rearrangement) gives the ion at m/e 495 (494) and loss of 225 mass units (see fragment at m/e 226/228) leads to the low-intensity ion at m/e 479 which appears to fragment further as described by Paulson, Zaylskie, and Dockter (1973) for synthetic ether glucuronides (acetyl-methyl ester derivatives). It must be noted, however, that the ion at m/e 215 is not observed in the mass spectrum of A–G and that the ion at m/e 622/624 remains unexplained.

Finally, it appears that in the case of the TFA-methyl ester derivatives the similarities between the mass spectra of A–G (Fig. 8) and D-glucuronic acid (Fig. 9) are less evident.

B. Identification of A–G in the Urine of Man

The conjugate of A (not ^{14}C-labeled) was purified from human urine using the methods previously described for the urine of monkeys. β-Glucuronidase was quite effective in splitting the conjugate (free A was obtained as the result of the enzymic hydrolysis) and both the Benedict's and the Dische's reactions were also positive. MS analyses carried out on the underivatized conjugate and on the TMS-methyl ester derivative as well as on the TFA-methyl ester derivative of the compound lead to the conclusion that the conjugated molecule is the ester glucuronide of A just as found in the urine of rhesus monkeys. Tables 5 and 6 seem, however, to indicate that the ratio of the amounts A–G and free A found in the 0- to 24-hr urine is higher for humans (approximate average values 3.4 for men versus 0.7 for monkeys).

TABLE 5. Urinary excretion of free A and A–G in man after intake of the drug by different routes – typical examples

Subject (sex)	Dose (mg)	Route of administration	0- to 24-hr urinary excretion[a]	% of total urinary metabolites			A–G / A
				A	A–G	A + A–G	
MG (F)	600	rectal	94.22	20.35	64.45[b]	84.80	3.2
SM (M)	600	i.m.	78.88	11.48	55.07[b]	66.55	4.8
KD (M)	500	p.o.	37.99[c]	20.30	46.40[d]	66.70	2.3

[a] GLC determinations (figures calculated from A equivalents).
[b] A–G is calculated on the basis of the difference between the amounts of A found in the unhydrolyzed and hydrolyzed urine.
[c] 0- to 12-hr urine.
[d] Direct measurement.

TABLE 6. Urinary excretion of free A and A–G in the rhesus monkey after oral intake of 100 mg/kg of ^{14}C–A

Monkey no. (sex)	0- to 24-hr urinary excretion (% of the dose)[a]	% of total urinary metabolites			A–G / A
		A	A–G	A + A–G	
15 (M)	58	42.3	33.3	75.6	0.79
16 (F)	60.4	48.2	28.3	76.5	0.58
18 (F)	65.6	41	34.6	75.6	0.84

[a] ^{14}C determinations.

C. Identification of the Glycine Conjugate of A (HAA) in the Urine of the Pig

The purification of this minor metabolite (MW 283/285) from the urine of pig has already been discussed (Roncucci et al., 1972). The pure metabolite, obtained by crystallization at low temperature from water extracts of gel (PLC)saturated with $(NH_4)_2SO_4$ and slow addition of HCl, has been submitted to direct mass spectral analysis (see Fig. 7 in the above-mentioned paper). The molecular ions (RI = 18%) were observed at m/e 283/285; the loss of the glycine moiety leads to the fragment at m/e 208/210. From this point on the main fragments are identical to those observed for A (see Table 3). The compound has now been synthetized (the acylchloride of A is allowed to react with glycine in alkaline medium) and its mass spectrum is quite similar to that of the metabolite (not shown).

Accurate GLC analyses show that in the 0- to 48-hr urine of the pig the ratio between the amounts of HAA and free A is approximately 1.1 (0.92 to 1.63) (Table 7). The amounts of HAA detected in 0- to 48-hr urine of the pig are a little bit lower than those announced previously [maximum 12% (versus 17%) of the total urinary metabolites].

TABLE 7. Urinary excretion of free A and HAA in pigs after oral intake of ^{14}C–A

Pig no. (sex)	Dose (mg/kg)	0- to 48-hr urinary excretion (% of the dose)[a]	% of total urinary excretion[b]			HAA / A
			A	HAA[c]	A + HAA	A
3 (F)	500	65.70	7.2	11.7	18.9	1.63
4 (F)	500	41.37	7.9	7.3	15.2	0.92
5 (F)	10	77.69	6.6	6.1	12.7	0.92
6 (F)	10	78.44	3.9	4.5	8.4	1.15

[a] [14]C determinations.

[b] GLC determinations (figures calculated from A equivalents).

[c] HAA is calculated on the basis of the difference between the amounts of A found in the unhydrolyzed and hydrolyzed urine.

D. Quantitation of Alclofenac Metabolites by Mass Fragmentography (MF)

Quantitation of the metabolites of A by GLC has already been described (Roncucci et al., 1971). A (TMS-derivative, MW 298/300, Fig. 10), 4-HCPA [bis(TMS)-derivative, MW 330/332, Fig. 11], DHA [Tris(TMS)-derivative, MW 476/478, Fig. 12], HAA (methyl ester derivative, MW 297/299 Fig. 13), and A–G [Tris(trifluoroacetyl)methyl ester derivative, MW 704/706, Fig. 8] can also be easily quantitated at the nanogram level by MF (recording of two masses).

For the examples shown in Fig. 14, the chlorinated molecular ions have been chosen except for 4-HCPA where $M(^{35}Cl)$ and $M(^{35}Cl)$-15 fragments are focused. If higher sensitivities are required, other ion sets can be selected, e.g., for A (trimethylsilyl derivative) by focusing the peaks at m/e 213 and 215 (Fig. 11) amounts as low as 50 pg have been detected.

Finally, the chromatograms of the A–G derivatives obtained at relatively high temperatures (≥250°C) show additional small peaks close to that of the actual compound. This fact suggests thermal degradation of the derivatives described.

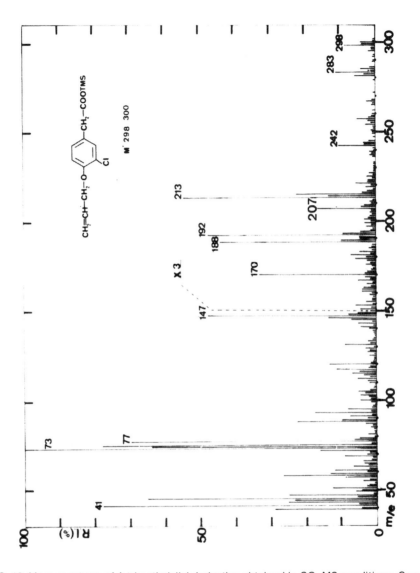

FIG. 10. Mass spectrum of A trimethylsilyl derivative obtained in GC–MS conditions. Conditions: OV-1 3%; column oven: 210°C; He flow: 12 ml/min; ion source: 270°C; electron energy: 70 eV.

V. DISCUSSION

The conjugated form of A present in the urine of rhesus monkeys and humans given the native compound by various route was purified by chromatography (TLC and PLC) and obtained in analytical amounts by

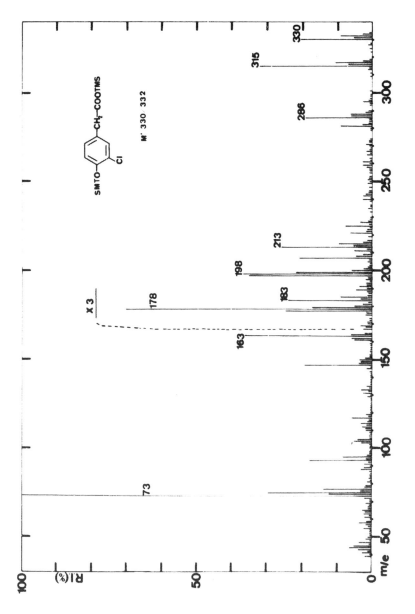

FIG. 11. Mass spectrum of 4-HCPA *bis*(trimethylsilyl) derivative obtained in GC–MS conditions. Same conditions as in Fig. 10.

precipitation from concentrated MIBK extracts. The compound present in the urine of the rhesus monkey was ^{14}C-labeled.

Its structure was elucidated by the following methods: 1. The compound was split by HCl-hydrolysis which led to the formation of free A. 2. Incu-

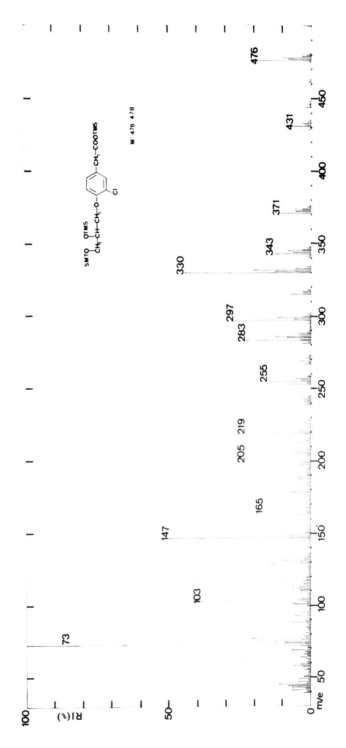

FIG. 12. Mass spectrum of DHA Tris(trimethylsilyl) derivative obtained in GC–MS conditions. Same conditions as in Fig. 10 except column oven: 240°C.

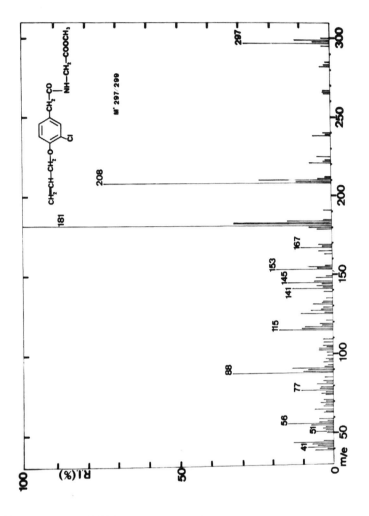

FIG. 13. Mass spectrum of HAA methyl ester derivative obtained in GC–MS conditions. Same conditions as in Fig. 10 except column oven: 250°C.

bation with β-glucuronidase also freed the aglycon moiety. 3. The compound reacted positively with the Benedict's and Dische's solutions. 4. Mass spectral analyses allowed clear-cut identification of the conjugate as the ester glucuronide of A.

Interpretation of fragmentation patterns was greatly facilitated by the presence of a halogen atom in the drug moiety and also by the fact that the carbohydrate attachment can be characterized by appropriate derivatization techniques (methylation/silylation and methylation/trifluoroacetylation). In the specific case of the silylated derivative, the structure of the major fragments containing residues of the sugar moiety has been confirmed

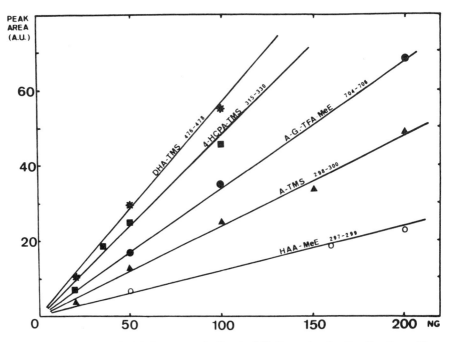

FIG. 14. Quantitation of alclofenac metabolites by MF. Example of calibration lines. Mass fragmentograms are obtained recording, during the GC run, at intervals of 0.5 sec, the intensity of two specific ions (indicated in the graph) by alternation of the accelerating voltage (peak-matching unit of the Varian-MAT CH7 used). The total ion current is recorded simultaneously. The area under the curve drawn by the successive signals of the more intense mass is measured after injection of different amounts of the compounds studied. The results for the following sets of ions are illustrated: m/e 476/478 (R = 1.77), 315/330 (R = 1.66), 704/706 (R = 2.43), 298/300 (R = 2.2), 297/299 (R = 2.4) for the indicated derivatives of DHA, 4-HCPA, A-G, A, and HAA respectively.

using the perdeuterated silylating agent. The derivatization with trifluoro-acetic anhydride was very helpful for the quantitation of A-G by MF since the derivative shows very intense molecular ions at m/e 704/706 (RI of the ion at m/e 704:55%).

The fragmentation patterns of the Tris(trimethylsilyl)methyl ester derivatives agree with the "schemes" reported by Billets et al. (1973) and by Paulson et al. (1973) for the characterization of derivatized synthetic ether glucuronides. However, slight differences have been encountered comparing these patterns with those obtained for the ester glucuronide of A. First, in the case of A-G derivatives more intense molecular ions are observed; this fact could indicate that the fragmentation is less intense for ester glucuronides. Secondly, using the direct inlet probe, the mass spectra of A-G derivatives apparently show two types of fragmentation at the level of the ester bound between the drug and the sugar moieties; for instance, in the case of the silylated derivative, intense chlorinated ions are observed at $M(^{35}Cl)$-423 (m/e 208/210) and at $M(^{35}Cl)$-406 (m/e 226/228). Since traces of free A (probably formed by chemical instability of A–G) have been found by GLC analyses in the purified A–G, it can be thought that the ions at $M(^{35}Cl)$-406 are due to the presence of free alclofenac (MW 226/228). Nevertheless the fragments at M-406 have also been found in mass spectra of silylated A–G obtained in GC–MS conditions. This supports the hypothesis of the occurrence in the molecule of the ester glucuronide of A in the two above-mentioned cleavages.

From the results obtained in this study it appears that glucuronates present in biological fluids can be easily characterized by mass spectrometry provided they are appropriately purified and derivatized.

From the biological point of view, the mode of conjugation of substituted phenylacetic acids varies from one animal species to another and even within a given species, and the conjugating agent appears to be different (glycine or glutamine) according to the nature of the radicals attached to the ring (James, Smith, and Williams, 1972). On the basis of the results reported in this chapter it is concluded that 4-allyloxy-3-chlorophenylacetic acid conjugates with glucuronic acid in man and monkey but not in the pig where the glycine conjugate was found. The amounts of A–G in comparison with those of free A seem to be higher in humans than in monkeys. These are however only approximate data since the stability of A–G in the urine of the two species considered is not precisely known.

ACKNOWLEDGMENTS

The authors are indebted to Dr. M. Kaisin for his valuable comments on the fragmentation of A–G derivatives and to Dr. C. Gillet for the synthesis of HAA. Thanks are also addressed to Mr. R. Vandriessche for his skillful technical assistance.

The work reported in this chapter was supported by a grant from the Institut pour l'Encouragement de la Recherche Scientifique dans l'Industrie et l'Agriculture (IRSIA) of the Belgian government (IRSIA convention 1991).

REFERENCES

Billets, S., Lietman, P. S., and Fenselau, C. (1973): Mass spectral analysis of glucuronides. *Journal of Medicinal Chemistry,* 16:30–33.

Dische, Z. (1947): A new specific color reaction of hexuronic acids. *Journal of Biological Chemistry,* 167:189–198.

Gillet, C., Gautier, M., Roncucci, R., Simon, M.-J., and Lambelin G. (1973): Synthesis of 4-allyloxy-3-chlorophenylacetic-1-^{14}C-acid. *Journal of Labelled Compounds,* 9:167–169.

James, M. O., Smith, R. L., and Williams, R. T. (1972): The conjugation of 4-chloro- and 4-nitro-phenylacetic acids in man, monkey and rat. *Xenobiotica,* 2:499–506.

Paulson, G. D., Zaylskie, R. G., and Dockter, M. M. (1973): Characterization of aryl glucuronic acid conjugates by derivatization and mass spectral analysis. *Analytical Chemistry,* 45:21–27.

Radford, T., and De Jongh, D. C. (1972): Carbohydrates. In: *Biochemical Application of Mass Spectrometry*, edited by G. R. Waller. Willey-Interscience, New York.

Roncucci, R., Lambelin, G., Simon, M.-J., and Soudyn, W. (1968): Simultaneous liquid scintillation determination of tritium and sulfur-35 in biological low-level samples using the oxygen flask method. *Analytical Biochemistry*, 26:118–136.

Roncucci, R., Simon, M.-J., Gillet, C., Lambelin, G., and Kaisin, M. (1972): Characterization of alclofenac metabolites in various animal species by gas chromatography and mass spectrometry. In: *Proceedings of the International Symposium on Gas Chromatography Mass Spectrometry—Isle of Elba, Italy*, edited by A. Frigerio, pp. 431–449. Tamburini Editore, Milano.

Roncucci, R., Simon, M.-J., and Lambelin, G. (1971): Gas chromatographic determination of 4-allyloxy-3-chlorophenylacetic acid (alclofenac) and its metabolites. *Journal of Chromatography*, 62:135–137.

Vogel, A. I. (1956): *A Text-Book of Practical Organic Chemistry*. Longmans, Green and Co, Ltd., London.

Mass Spectrometry in Biochemistry and Medicine,
edited by A. Frigerio and N. Castagnoli.
Raven Press, New York © 1974

Mass Spectrometric Analyses of Stable Isotopes in Drug Metabolism Studies

Neal Castagnoli, Jr., Ermias Dagne, and Larry D. Gruenke

Department of Pharmaceutical Chemistry, School of Pharmacy, University of California, San Francisco, California 94143

Studies in a number of laboratories strongly suggest that the cytochrome P_{450} mediated oxidations of drugs involve an electron-deficient oxygen species (Ullrich and Staudinger, 1971). For example, arene oxide formation (Jerina, Daly, Landis, Witkop, and Udenfriend, 1971) and N-oxidations (Weisburger and Weisburger, 1971) can be readily understood in terms of nucleophilic attack by substrate on some form of electron-deficient oxygen. However, other frequently encountered metabolic oxidations, such as oxidative N- and O-dealkylations (Gram, 1971) and carbon oxidations α to a carbonyl group (Keberle, Riess, and Hoffman, 1963), are mechanistically less easily rationalized. With the aim of gaining a better understanding of metabolic conversions involving introduction of oxygen α to a lactam carbonyl group, we have undertaken studies of the α-oxidations of cotinine (1a), the principal mammalian metabolite of the tobacco alkaloid nicotine (2), and the 1,4-benzodiazepine, diazepam.[1]

Our studies (Dagne and Castagnoli, 1972*a,b*), as well as those of others (McKennis, Turnball, Bowman, and Tamaki, 1963), have led to the structural elucidation of several urinary metabolites of cotinine including desmethylcotinine (3a), cotinine-N-oxide (4a), the stereospecific C-hydroxylation product, 3-(R)-hydroxy-S-(5)-3'-pyridyl-1-methyl-2-pyrrolidinone (*trans*-3-hydroxycotinine, 5a), and 5-hydroxycotinine (6a). The stability, mass spectral characteristics (see below), and ease of isolation of compound 5a, together with the stereospecificity of the conversion and the ready availability of cotinine-3,3-d_2 (1b) prompted us to investigate the influence on product formation of replacing, with deuterium, the two protons α to the carbonyl group.

---

[1] Studies on the metabolism of diazepam will not be included here since the data are published (Sadeé, Garland, and Castagnoli, 1970; Castagnoli and Sadeé, 1972).

	R$_1$	R$_2$	R$_3$	R$_4$	
1a:	CH$_3$	H	H	H	
1b:	CH$_3$	H	D	D	
3a:	H	H	H	H	
3b:	H	H	D	D	
4a:	CH$_3$	H	H	H	N → O
4b:	CH$_3$	H	D	D	N → O
5a:	CH$_3$	H	OH	H	
5b:	CH$_3$	H	OH	D	
6a:	CH$_3$	OH	H	H	
6b:	CH$_3$	OH	D	D	

Cotinine-3,3-d$_2$ (**1b**) could be easily obtained by deuterium exchange of cotinine in D$_2$O in the presence of K$_2$CO$_3$. An electron impact mass spectrum of cotinine shows in addition to the parent ion (m/e 176, 32%) major fragments at m/e 119 (11%) for the nitrilium ion **7** and at m/e 98 (100%) for the pyrrolinonium ion **8a.** As expected, the mass spectrum of the dideuterio compound **1b** shows shifts in the parent ion (m/e 178, 32%) and pyrrolinonium ion **8b** (m/e 100, 100%) by two mass units. The percent deuterium incorporation in **1b** cannot be calculated from the parent ion since there is a relatively large M-1 ion (8% of the base) due to loss of a hydrogen radical. However, the region of the base peak is exceptionally

7

8a: R=H
8b: R=D

free of interfering ions, and, since C–D bonds are not broken in forming this ion, isotope effects due to loss of deuterium do not occur. Thus, the ion intensities at m/e 100, 99, and 98 in the mass spectrum of cotinine can be used to calculate the relative amounts of the d$_2$, d$_1$, and d$_0$ species present. A calculation done in this manner shows that the **1b** obtained was 99% d$_2$-enriched.

The dideuterio compound **1b** was administered intravenously to a male rhesus monkey, the 24-hr urine was collected, and the organic soluble base fraction was isolated. Analytical TLC showed five fluorescent spots corresponding in R$_f$ values to cotinine, 5-hydroxycotinine, desmethylcotinine, *trans*-3-hydroxycotinine, and cotinine-N-oxide. The mass spectrum of the recovered cotinine-3,3-d$_2$ isolated from preparative TLC was identical with the spectrum of the administered compound, which establishes that the starting lactam had not suffered loss of deuterium while in the animal or during work-up. Further evidence that exchange of the deuterium

atoms of **1b** had not occurred came from mass spectral analysis of the
dideuterio-N-oxide **(4b)** and the dideuterio-desmethyl **(3b)** metabolites
which were also isolated from the preparative TLC plate. The parent ions
and fragments containing the pyrrolidinone moiety were all shifted two
mass units higher. Furthermore, the relative intensities of the major ions
for the protio and deuterio compounds were essentially identical and the M-1
and M-2 ions for the parent and base peaks in the spectra of the protio and
deuterio compounds were also identical.

The mass spectrum of the deuterio compound **5b** from the preparative
TLC plate which corresponds to the metabolite *trans*-3-hydroxycotinine
(5a) is reproduced in Fig. 1 (lower) together with the corresponding spectrum
of **5a** (Fig., upper). The relatively small ion intensity at m/e 192, corresponding to the M-1 ion for **5b** and the protio metabolite **5a**, shows that there is
less than 15% of **5a** present.

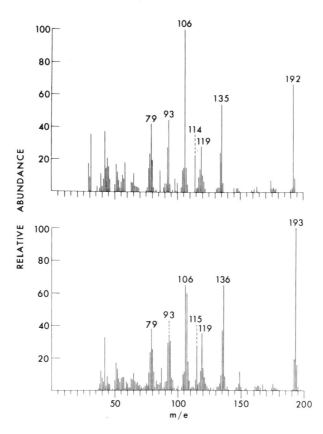

FIG. 1. (Upper): Electron impact mass spectrum of *trans*-3-hydroxycotinine **(5a)** isolated
from the urine of a rhesus monkey receiving cotinine-d_0 **(1a).** (Lower): Electron impact
mass spectrum of deuterio-*trans*-3-hydroxycotinine **(5b)** isolated from the urine of a
rhesus monkey receiving cotinine-3,3-d_2 **(1b).**

The mass spectrum of **5** shows an unexpected feature in that the ions corresponding to m/e 135/134, 106, and 93/92 in the protio compound **5a** are partially but not wholly shifted to one mass unit higher in the deuterio compound **5b**. Furthermore, when the hydroxyl group of **5a** was labeled with deuterium, it was found that the above ions were almost completely shifted to one mass higher. Some retention of deuterium was also seen in both cases in the m/e 79 ion. This suggests that extensive ion rearrangements are occurring in the mass spectrometer. Figure 2 presents a rationalization of the above data and is only meant to be suggestive of ion identities and frag-

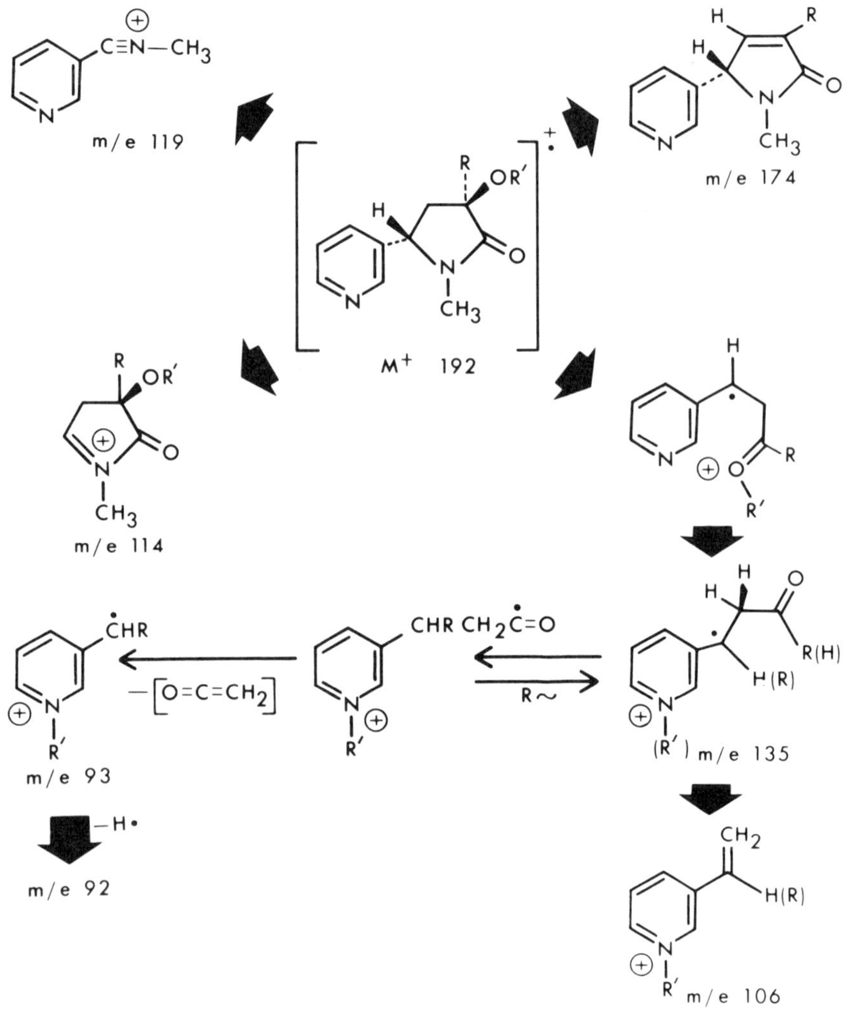

FIG. 2. Proposed pathway for the electron impact mass fragmentation pattern of protio- and deuterio-*trans*-3-hydroxycotinine (**5a** and **5b**, respectively).

mentation pathways. Metastable scans show the existence of pathways from m/e 135 to 106 to 79 in the protio compound **5a.**

In any case we felt justified in using the intensities of the m/e 193–192 ions of *trans*-3-hydroxycotinine to calculate the ratio of the d_1 to d_0 species present. If a mixture of protio and dideuterio cotinine is administered to the monkey, the ratio of the d_1 to d_0 *trans*-3-hydroxycotinine recovered can, therefore, be used to calculate the deuterium isotope effect for the metabolism. This, of course, assumes that the ratio of d_2 to d_0 cotinine administered does not change during the course of the experiment. It also assumes that further metabolic or spontaneous conversions of **5a** and **5b** proceed at similar rates. Although we have not investigated the metabolism of **5a/5b,** it seemed reasonable to assume that the experimentally determined k_H/k_D would be a minimum value since an isotope effect involved in the further breakdown of **5** would tend to increase the concentration of the deuterated product and hence decrease the experimentally measured deuterium isotope effect.

We prepared an approximately equal mixture of **1a** and **1b** which by mass spectral analysis was found to be 55% d_0 and 45% d_2. This mixture was administered intravenously to a male rhesus monkey and the 24-hr urine was collected and worked up. Unchanged cotinine (**1**), cotinine-N-oxide (**4**), desmethylcotinine (**3**), and *trans*-3-hydroxycotinine (**5**) were isolated by preparative TLC and were identified by mass spectrometry. The data pertaining to the calculation of the isotope effect are summarized in Table 1.

TABLE 1. *Ion intensities for deuterium isotope effect calculation (**1a/1b** = 1.0/0.82)*

Administered cotinine		Recovered cotinine	Recovered *trans*-3-hydroxycotinine	
m/e 100	80%	100%	m/e 193	21%
m/e 99	10%	14%	m/e 192	86%
m/e 98	100%	80%		

It can be seen that, whereas the administered substrate had more cotinine-d_0 than cotinine-d_2, the recovered substrate was richer in the dideuterio species showing that the rate of metabolism of **1a** and **1b** is not the same; clearly, cotinine-d_0 is metabolized faster than cotinine-d_2. The mass spectra of the metabolites cotinine-N-oxide and desmethylcotinine showed d_0/d_2 isotope ratios near one, consistent with that of cotinine itself. However, the mass spectrum of *trans*-3-hydroxycotinine showed m/e 193/192 ratio of 1/4 which, when corrected for the contribution to the m/e 193 ion intensity by the ^{13}C-isotope natural abundance of the m/e 192 ion, indicates an unexpectedly high k_H/k_D of about 7. If one estimates a 15% conversion of **5b** to **5a,** this value decreases to about 6.

The most significant error in the above value of k_H/k_D comes from the determination of the relative amount of **5b** since the normal statistical and measurement errors in the intensity of the m/e 193 ion are magnified by the subtraction of the large ^{13}C correction. In order to avoid this problem and as a check on our results, we repeated the above experiment using a much lower ratio of d_0 to d_2 cotinine. A mixture which by mass spectral analysis contained 1.0 d_0 to 4.6 d_2 cotinine was administered to the monkey as above. Preparative TLC of the 24-hr urine extract gave bands which by retention time and mass spectra were identified as cotinine, 5-hydroxy-cotinine, desmethylcotinine, *trans*-3-hydroxycotinine, and cotinine-N-oxide. The data for cotinine and *trans*-3-hydroxycotinine are summarized in Table 2. Again the faster metabolism of the d_0 as compared with the d_2

TABLE 2. *Ion intensities for deuterium isotope effect calculation (1a/1b = 1.0/4.6)*

| m/e | Cotinine | | m/e | *trans*-3-hydroxycotinine |
	Administered	Recovered		
100	100%	100%	193	44.3%
99	7.3%	5.8%	192	58.9%
98	22.1%	14.6%		

cotinine is obvious; the ratio of d_0 to d_2 cotinine decreased from 1/4.6 in the administered mixture to 1/7.0 in the recovered material. The large isotope effect in the metabolism to *trans*-3-hydroxycotinine is also confirmed. After correction for the ^{13}C contribution, the d_0 to d_1 ratio of the metabolites is found to be 1.6/1 (5a/5b), which, based on the starting ratio of d_0 to d_2 cotinine, gives a deterium isotope effect of 7.4. This number is not corrected for the significant change in the ratio of d_0 to d_2 cotinine, nor for the possible further metabolism of 5. However, each of these corrections can only increase the calculated value of the isotope effect. This value could decrease to 6 if one assumes a 15% conversion of 5b to 5a. Thus maximizing corrections to minimize the deuterium isotope effect gives a value of 6 although the number could be significantly larger.

This is the first report in which a primary isotope effect is clearly involved in a drug metabolic reaction. Although these limited data do not permit any definitive mechanistic interpretation, it is tempting to point out the obvious analogy between this hydroxylation reaction and nonanxymatic oxidations α to a carbonyl group in which enolization is the rate-limiting step. In such reactions the deuterium isotope effect has been shown to be in the same range as we have observed in the formation of *trans*-3-hydroxycotinine (Wiberg, 1955). Perhaps an analogous reaction pathway is followed in the oxidation of cotinine. The fact that cotinine-d_2 does not undergo deuterium exchange while exposed to the animal means that the enolization would have to be irreversible or reversible with the original deuterium atom serving as the sole proton source in the back reaction. These speculations lead to the following proposed pathway.

This mechanism is consistent with the electron-deficient nature of **4** the oxidizing species $P_{450}\cdot O$ and also with the large isotope effect observed in the present study.

REFERENCES

Castagnoli, N., Jr., and Sadée, W. (1972): Mechanism of the reaction of a 1,4-benzodiazepine N-oxide with acetic anhydride. *Journal of Medicinal Chemistry,* 15:1076–1078.

Dagne, E., and Castagnoli, N., Jr. (1972*a*): Structure of hydroxycotinine, a nicotine metabolite. *Journal of Medicinal Chemistry,* 15:356–360.

Dagne, E., and Castagnoli, N., Jr. (1972*b*): "Cotinine N-oxide, a new metabolite of nicotine. *Journal of Medicinal Chemistry,* 15:840–841.

Gram, T. E. (1971): Enzymatic N-, O-, and S-dealkylation of foreign compounds by hepatic microsomes. In: *Concepts in Biochemical Pharmacology,* edited by B. B. Brodie and J. R. Gillette, pp. 334–348. Springer-Verlag, Berlin.

Jerina, D., Daly, J. W., Landis, W., Witkop, B., and Udenfriend, S. (1971): In: *The Role of Biogenic Amines and Physiological Membranes in Modern Drug Therapy,* edited by J. Biel, pp. 413–476. Marcel Dekker, New York.

Keberle, H., Reiss, W., and Hoffman, K. (1963): Ueber den Stereospezifischen Metabolismus der Optischen Antipoden von α-Phenyl-α-äthyl-Glutarimid (Doriden (R)). *Archives Internationales de Pharmacodynamie et de Thérapie,* 142:117–124.

McKennis, H., Jr., Turnball, L. B., Bowman, E. R., and Tamaki, E. (1963): The synthesis of hydroxycotinine and studies on its structure. *Journal of Organic Chemistry,* 28:383–387.

Sadée, W., Garland, W., and Castagnoli, N., Jr. (1971): Microsomal 3-hydroxylation of 1,4-benzodiazepines. *Journal of Medicinal Chemistry,* 14:643–645.

Ulbrich, V., and Staudinger, H. (1971): Model systems in studies of the chemistry and the enzymatic activation of oxygen. In: *Concepts in Biochemical Pharmacology,* edited by B. B. Brodie and J. R. Gillette, pp. 251–263. Springer-Verlag, Berlin.

Weisburger, J. H., and Weisburger, E. K. (1971): N-Oxidation enzymes. In: *Concepts in Biochemical Pharmacology,* edited by B. B. Brodie and J. R. Gillette, pp. 312–348. Springer-Verlag, Berlin.

Wiberg, K. B. (1955): The deuterium isotope effect. *Chemical Reviews,* 55:713–743.

Mass Spectrometry in Biochemistry and Medicine,
edited by A. Frigerio and N. Castagnoli.
Raven Press, New York © 1974

Mass Spectrometric Studies of Tricyclic Compounds Used in Psychiatry

A. Frigerio, K. M. Baker, and P. L. Morselli

Istituto di Ricerche Farmacologiche "Mario Negri," Via Eritrea 62, 20157 Milano, Italy

I. INTRODUCTION

A. Gas Chromatography–Mass Spectrometry

Gas chromatography is a technique developed during the last decade which has made the problems of the pharmacologist and toxicologist much easier. Many reviews and articles have been published on the separation of a wide range of organic compounds from biological sources (Goldbaum, Johnston, and Blumberg, 1963; Gudzinowicz, 1967; Kroman and Bender, 1968; Hammar, Holmstedt, Lindgren, and Tham, 1969; Sutherland, Williamson, and Theivagt, 1971; Taylor, 1971; Cram and Juvet, 1972).

A further more recent advance is the coupling of mass spectrometry with gas chromatography which offers many advantages, not least of which is a tentative identification of the drug corresponding to a gas chromatographic peak. The technique possesses the advantage that the analysis of very small quantities can be carried out completely, down to identification of the components. With a suitable molecular separator between the gas chromatograph and the mass spectrometer, the effluent from the gas chromatograph can be passed directly into the mass spectrometer, eliminating the necessity of an intermediate isolation procedure. Several reviews of the use of this technique have appeared and a detailed description of the instrumentation appears in a paper by Hammar et al. (1969).

The reader is referred to several articles which deal with the use of the technique in drug studies (Hammar et al., 1969; Brooks, Thawley, Rocher, Middleditch, Anthony, and Stillwell, 1970; Merritt, 1970; Brooks, 1971; Horning and Horning, 1971; Burlingame and Johanson, 1972; Junk 1972).

II. STUDIES ON TRICYCLIC DRUGS

The technique of gas chromatography–mass spectrometry is very useful for studying the metabolism and levels of tricyclic drugs because these drugs are usually very amenable to gas chromatographic separation either before or after derivatization, depending on the substitution and polarity of the molecule. Several works have appeared on the complete metabolic pathways of some tricyclic drugs, using mass spectrometry to identify the metabolites, e.g., chlorpromazine (Hammar, Holmstedt, and Ryhage, 1968), protriptyline (Sisenwine, Tio, Shrader, and Ruelius, 1970), and nortriptyline (Hammar, Alexanderson, Holmstedt, and Sjöqvist, 1971), and some estimating drug levels using mass spectrometry, e.g., nortriptyline and desmethylnortriptyline (Borgä, Palmér, Linnarsson, and Holmstedt, 1971).

This chapter presents some results related to the tricyclic psychotropic drugs. One case shows how the metabolic pathway of carbamazepine can be studied using gas chromatography–mass spectrometry, and the inherent pitfalls that can be encountered in using the technique. The other case is that of imipramine where the metabolic pathway is known, but gas chromatography–mass spectrometry is used to assess the imipramine levels at very low concentrations in biological samples from people being treated with this drug.

A. Carbamazepine

This section deals with use of gas chromatography–mass spectrometry to identify the metabolites of carbamazepine after administration to both humans and rats. The behavior of the metabolite on subjection to gas chromatography will also be described, and this will show some of the ambiguities and unexpected results which can be obtained using the combined technique.

Carbamazepine (I) which is chemically and structurally similar to the imipramine class of drugs, does not have any of the antidepressant activity normally associated with the tricyclic antidepressant drugs. In humans the compound is used as a treatment for grand mal seizures and psychomotor epilepsy (Bonduelle, Bouygues, Sallon, and Chemaly, 1964; Davis, 1964; Jongmans, 1964; Livingston, 1966; Fichsel and Heyer, 1970) and it is considered the drug of choice for the treatment of trigeminal neuralgia (Blom, 1962; Dalessio and Abbott, 1966). Several reports have described the pharmacokinetics of the drug in animals and humans (Meinardi, 1971; Morselli, Gerna, Frigerio, Zanda, and De Nadai, 1971; Morselli, Gerna, and Garattini, 1971; Frigerio, Fanelli, Biandrate, Passerini, Morselli, and Garattini, 1972; Morselli, Biandrate, Frigerio, and Garattini, 1972).

1. Metabolism

In spite of the wide use of the drug in treating the above-mentioned disorders, until very recently little was known about the metabolism of the drug. A report by Weist and Zicha (1967) has shown that when the urine of patients on carbamazepine was subjected to thin-layer chromatography, seven spots not present in the control urine were found. Several recent reports from this laboratory have shown that 10,11-dihydro-10,11-epoxy-5H-dibenz[b,f]azepine-5-carboxamide (carbamazepine-10,11-epoxide) (II) can be isolated as one of the metabolites of carbamazepine (Frigerio et al., 1972; Morselli et al., 1972; Frigerio, Biandrate, Fanelli, Baker, and Morselli, 1972) whereas more recent unpublished work has shown that iminostilbene (III) and 10,11-dihydro-10,11-dihydroxy-5H-dibenz[b,f] azepine-5-carboxamide (IV) are also metabolites, the latter having also been found very recently by other workers in the field (Goenechea and Hecke-Seibicke, 1972). A knowledge of the metabolic pathway for many drugs has proven useful in the interpretation of their pharmacological action, and can provide in many instances a rational basis for therapy. It is therefore of great interest to know in more detail the biotransformation pathway of this widely used drug. The description which follows shows the overall procedure used in identifying the metabolites of carbamazepine, and Fig. 1 shows the presently known metabolic pathway.

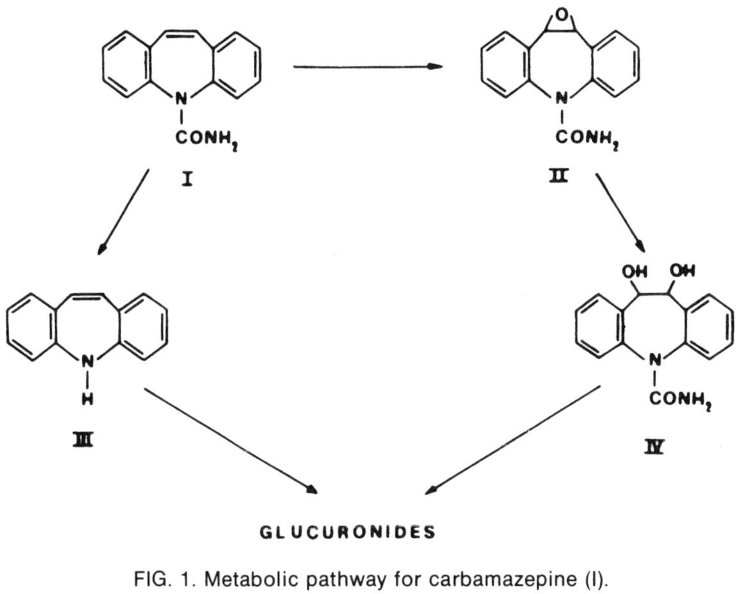

FIG. 1. Metabolic pathway for carbamazepine (I).

2. Human and Animal Studies

On animals the studies were carried out using ^{14}C-labeled carbamazepine (I). The carbamazepine was administered to rats which were housed in metabolic cages. On analysis of the waste products, 0.5% of the administered activity appeared in the respiratory carbon dioxide, whereas 27% was found in the 24-hr urine of the animals. The results of extraction of the urine with various organic solvents is shown in Table 1. It should be

STUDIES OF TRICYCLIC COMPOUNDS

TABLE 1. *Thin-layer chromatography of processed urine of rats and humans after carbamazepine administration*

Rats			Humans	
Extract A	Extract B	Extract C	Extract A'	Extract B'
	0.70^a	0.70^a		0.80^c
			0.43^b	
0.34^c	0.34^c		0.34^c	0.34^c
	0.12^d	0.12^d		0.12^d
			0.08^e	0.08^e
				0.04^e

R_f values of radioactive or U.V. quenching zones developed in benzene-ethanol-diethylamine (8:1:1).

A and A' are ethylene dichloride extracts; B and B' are ethyl acetate extracts; C is ethyl acetate extract after incubation with β-glucuronidase.

[a] U.V. quenching zone corresponding in R_f to iminostilbene (III).

[b] U.V. quenching zone corresponding in R_f to carbamazepine (I).

[c] U.V. quenching or radioactive zone corresponding in R_f to carbamazepine-10,11-epoxide (II).

[d] U.V. quenching or radioactive zone corresponding in R_f to 10,11-dihydro-10,11-dihydroxy-5H-dibenzo[b,f]azepine-5-carboxamide (IV).

[e] U.V. quenching zones also present in control urine; U.V. light of 254 nm.

noted that two of the metabolites are found as glucuronides. The results of feeding 400 mg of carbamazepine to humans are also shown in Table 1. The extraction procedure was the same as for the urine from rats.

3. Identification of the Metabolites

For reasons which are explained in Section A(5), gas chromatography was not found to be useful for separating and identifying the metabolites of carbamazepine due to the chemical lability of the compounds using this technique. This of course renders gas chromatography–mass spectrometry unuseful in identifying the metabolites. Instead the best procedure was found to be separation of the metabolites by thin-layer chromatography and then identification by direct injection mass spectrometry. As a final check each separated material was compared with the spectra and R_f's of authentic synthesized samples. In general the amounts of material present in rat urine are much less than in human urine and so the latter is used for identification purposes. The thin-layer chromatography of the human urine showed three materials not present in the control and the investigation of each is shown below. Iminostilbene has so far been found only in rat urine.

i. Iminostilbene (III). This compound was identified from its R_f and mass spectrum, both of which are identical to those of authentic material. The mass spectrum showed a molecular ion at m/e 193 and very little further fragmentation. This metabolite was obtained in greatest quantity after β-glucuronidase incubation at acid pH and as a proof that iminostilbene did not arise from hydrolysis of carbamazepine at the acid pH used (4 to 5.5); labeled carbamazepine was placed in urine and subjected to the same treatment as for the metabolites. Neither before nor after β-glucuronidase incubation was any iminostilbene found as a decomposition product of carbamazepine and hence iminostilbene was demonstrated as a true metabolite of carbamazepine. The glucuronide of iminostilbene (III) would be much more susceptible to hydrolysis than an oxygen glucuronide and this would account for the finding of this material in the organic extracts before β-glucuronidase incubation.

ii. Carbamazepine (I). This spot as shown in Table 1 corresponds in R_f to that of authentic material. The mass spectrometric fragmentation is rationalized in Fig. 2; the base peak at m/e 193 is produced from a loss of HNCO from the molecular ion. This transition is confirmed by the presence of a metastable ion.

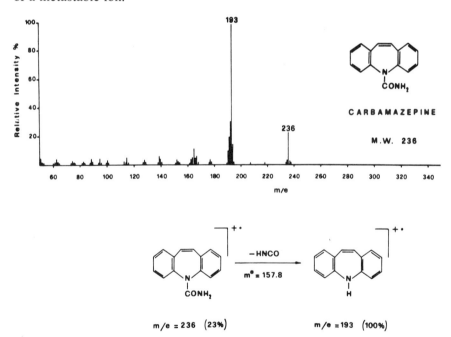

FIG. 2. Mass spectrum obtained by direct injection system and fragmentation pathway for carbamazepine (I).

iii. 10,11-Dihydro-10,11-epoxy-5H-dibenz[b,f]azepine-5-carboxamide (carbamazepine-10,11-epoxide) (II). This metabolite accounts for 24% of the metabolites extracted and has a mass spectrum identical to that of authentic material (Frigerio et al., 1972). The mass spectrum could only be obtained by direct injection, gas chromatography giving a single peak due to a decomposition product identified as 9-acridinecarboxaldehyde (V) [see Section A(5)]. The mass spectrum obtained by direct injection shows a molecular ion at m/e 252 with a loss of 29 mass units to give a peak at m/e 223 which subsequently loses 43 mass units (HNCO) to give the base peak at m/e 180 corresponding to the very stable protonated acridinium ion [see Section A(4)]. A rationalization of this spectrum is shown in Fig. 3.

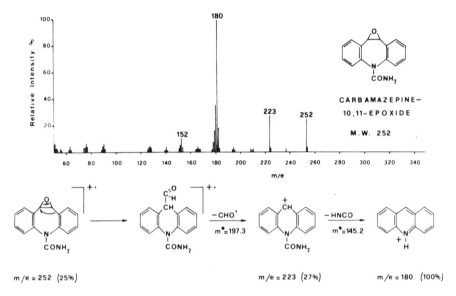

FIG. 3. Mass spectrum obtained by direct injection system and fragmentation pathway for carbamazepine-10,11-epoxide (II).

iv. 10,11-Dihydro-10,11-dihydroxy-5H-dibenz[b,f]azepine-5-carboxamide (IV). This metabolite has recently been found by Goenechea and Hecke-Seibicke (1972) and the structure has been confirmed by isolation and comparison with an authentic sample in this laboratory. The compound IV exists for the major part as the glucuronide in urine because it is extracted after β-glucuronidase incubation. The gas chromatographic behavior is described in Section A(5). The structure of the material can be determined from its mass spectrum which showed a molecular ion at m/e 270 with fragment ions at m/e 253 and 252 (M–OH,M–H$_2$O). The spectrum showed a base peak at m/e 180 due to the protonated acridinium ion. An authentic sample prepared by osmium tetroxide oxidation of carbamazepine showed the same fragmentation pathway and same thin-layer chromatographic behavior as the material isolated from urine. The osmium tetroxide product would be expected to have a *cis* configuration for the hydroxyl groups and therefore it is probable that metabolite IV does also.

4. Mass Spectral Characteristics of Carbamazepine and Derivatives

Carbamazepine (I) and its metabolites show fragmentation patterns very characteristic of their structures. The parent compound carbamazepine (I) and its metabolite iminostilbene (III) each give intense ions at m/e 193, with little further fragmentation. This ion, m/e 193, is the very stable, completely conjugated system, the iminostilbene radical ion (VI). The conjugation and hence charge delocalization accounts for the extreme stability of this ion. The other two metabolites carbamazepine-10,11-epoxide (II) and 10,11-dihydro-10,11-dihydroxy-5H-dibenz[b,f]azepine-5-carboxamide (IV) each show base peaks at m/e 180 in their mass spectra, in this case due to the completely conjugated protonated acridinium ion (VII). In these two cases a rearrangement of the parent ion takes place to form the conjugated system.

These stable ions VI and VII undergo several fragmentations worthy of note. They each lose hydrogen cyanide, and also lose the dihydrogen cyanide radical in a previously unknown concerted process (Baker and Frigerio, 1973). It is proposed that the loss of the dihydrogen cyanide radical occurs after extensive rearrangement of the parent ions as illustrated in Figs. 4 and 5 for the iminostilbene and protonated acridinium ions respectively.

Each of the illustrated mechanisms proposes an intermediate tropylium or azatropylium species of the same type as has been proposed for the loss

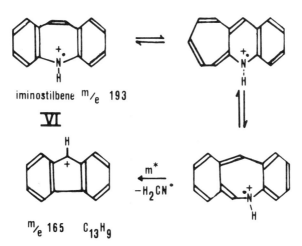

FIG. 4. Mechanism for the loss of the dihydrogen cyanide radical from the iminostilbene radical ion (VI).

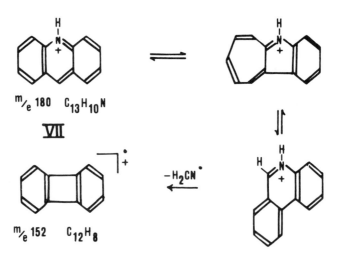

FIG. 5. Mechanism for the loss of the dihydrogen cyanide radical from the protonated acridinium ion (VII).

of hydrogen cyanide from less complicated aromatic nitrogen heterocycles (Spiteller, 1966; Loader, Palmer, and Timmons, 1967; Powers, 1968; Safe, Jamieson, and Hutzinger, 1972). These mechanisms resemble that proposed for quinoline (Porter and Baldas, 1971) in involving substituted cyclobuta-dienes as product ions.

The extensive rearrangements proposed may explain why the loss of the dihydrogen cyanide radical in carbamazepine-10,11-epoxide (II) and hydro-gen cyanide from acridine is less facile (9% and 7% respectively) than the loss of hydrogen cyanide from quinoline (20%) although other factors affect the abundance of ions.

5. Gas Chromatographic Degradations

Work in our laboratories has shown that carbamazepine (I) and several of its metabolites are unstable under the conditions of gas chromatography and undergo acid catalyzed rearrangements or degradations on the column at the elevated temperatures used. Carbamazepine itself when injected as a solution in methanol gives besides a major peak due to carbamazepine, two less-polar components of low abundance. Injection as an ethanol or acetone solution does not give this degradation reaction. Gas chromatog-raphy–mass spectrometry shows that these two are iminostilbene (III) and 9-methylacridine (VIII). The methanol present in the system amplifies the acidity of the column which produces the hydrolysis reaction to form iminostilbene and the extensive rearrangement to give 9-methylacridine.

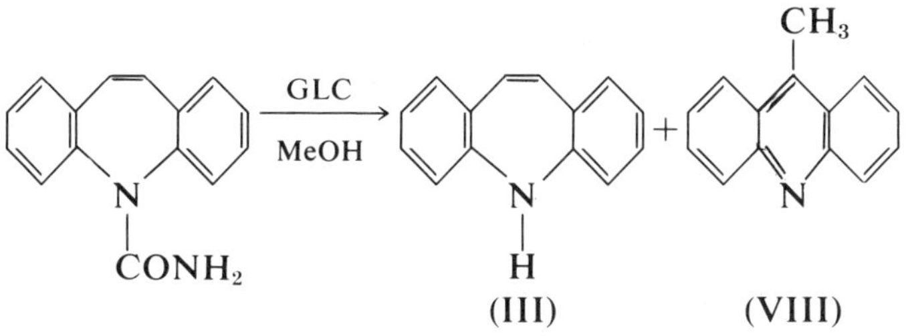

The epoxide (II) formed as a metabolite of carbamazepine is extremely unstable to the conditions of gas chromatographic analysis, the acidity and the high-temperature column being enough to cause a complete rearrangement to 9-acridinecarboxaldehyde (V). The structure of this material was deduced from the mass spectrum obtained by gas chromatography–mass spectrometry, preparative gas chromatography of the aldehyde, and comparison with an authentic sample (Baker, Frigerio, Morselli, and Pifferi, 1973). A rationalization of the rearrangement mechanism is shown in Fig. 6. The dihydroxylated metabolite 10,11-dihydro-10,11-dihydroxy-5H-dibenz-[b,f]azepine-5-carboxamide (IV) also undergoes a similar, but less facile, rearrangement under the conditions of gas chromatography. Injection as a methanol solution on to an OV-17 column, gives a single nonpolar material identical again with 9-acridinecarboxaldehyde. The diol (IV) undergoes a pinacol type rearrangement and a rationalization is shown in Fig. 6. The

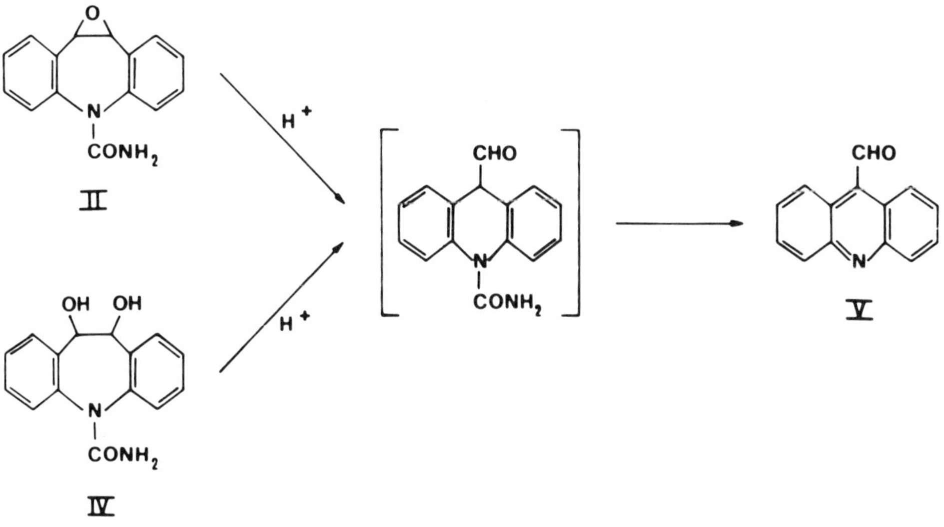

FIG. 6. Mechanism for the acid-catalyzed rearrangements of carbamazepine-10,11-epoxide (II) and 10,11-dihydro-10,11-dihydroxy-5H-dibenze[b,f]azepine-5-carboxamide (IV).

rearrangement of the diol (IV) is not so facile as that of the epoxide as it does not take place on SE-30 column which is much less acidic than an OV-17 (Kruppa, 1972). The rearrangements emphasize the need when studying drug metabolism, and indeed any mixture of compounds, by gas chromatography to be sure that the material obtained in the effluent is that injected, and no rearrangement or degradations have taken place. Where possible the results should be checked by other physical methods, e.g., synthesis of the material or separation by other chromatographic means.

B. Imipramine

The metabolic pathway of imipramine is already well known and we will deal only with the levels of unchanged imipramine in plasma after administration of the drug.

Imipramine is one of the group of tricyclic antidepressant drugs whose metabolism has been extensively studied (Bickel and Baggiolini, 1966). Recent reports suggest that for these tricyclic drugs, the assessment of their optimum plasma level may be an important factor in determining therapeutic efficacy (Åsberg, Cronholm, Sjöqvist, and Tuck, 1970, 1971). Some information is available on the plasma concentrations of imipramine or amitriptyline in depressed patients undergoing chronic treatment with these drugs (Moody, Tait, and Todrick, 1967; Rafaelsen, and Christiansen, 1969; Braithwaite and Widdopp, 1971). Fluorometric and radiochemical methods are available for imipramine determination in biological fluids (Herrmann and Pulver, 1960; Dingell, Sulser, and Gillette, 1964; Kuntzman and Tsai, 1967; Weder and Bickel, 1968; Harris, Gaudette, Efron, and Manian, 1970), but they suffer from various disadvantages such as a lack of specificity or sensitivity or they require large amounts of blood. For these reasons they are not ideal for routine assay to titrate the dose required by various patients.

1. Methods

Conventional gas chromatographic analysis of imipramine in plasma will assay drug concentrations of 100 to 150 ng/ml, or higher; in many instances this sensitivity is not sufficient. An adequate sensitivity and specificity is provided by mass fragmentography (Hammar et al., 1968, 1969, 1971; Borgä, et al., 1971; Hammar, 1972; Frigerio, Belvedere, De Nadai, Fanelli, Pantarotto, Riva, and Morselli, 1972; Gordon and Frigerio, 1972).

In mass fragmentography the mass spectrometer is used as a gas chromatographic detector: its sensitivity is 1,000 to 10,000 times greater than that of the usual detectors. In brief, one prominent ion of the imipramine mass spectrum is selected and one near ion from an internal standard. Using the accelerating voltage alternator, the instrument is set to monitor the characteristic ions during the elution time of the imipramine and internal standard from the gas chromatograph. The intensity of the ion current is read by the electron multiplier and recorded by the U.V. recorder. The actual quantitation of the material can be carried out by using the internal standard. There are a number of things which must be stipulated about the internal standard. It must have a prominent fragment ion with its mass number within 10% of that of the fragment ion of the substance under determination, and it must have similar gas chromatographic properties to the material being quantitated. In quantifying imipramine, promazine can be used as an internal standard because of its suitable gas chromatographic properties (Fig. 7) and mass spectra (Fig. 8). The ion m/e 235 is used to measure

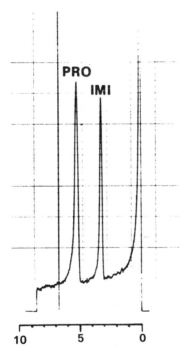

FIG. 7. Gas chromatogram of imipramine (IMI) and promazine (PRO). Conditions: 3% OV-17 glass column 2 m at 240°C, helium flow rate 30 ml/min.

imipramine and m/e 238 to measure promazine. The fragmentation process involved in the formation of these ions is explained in Fig. 9.

Further details about the mass spectra are discussed in a paper by Frigerio et al. (1972). The LKB 9000 was set at an electron energy of 30 eV because at this energy the ionization potential of the electron beam has reached maximum intensity. Maximal amplification of the ion currents detected by the LKB 9000 can be obtained by connecting the mass spectrometer amplifier to the peak matcher amplifier and then to the pen recorder.

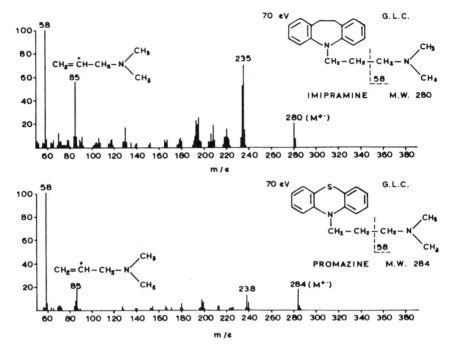

FIG. 8. Mass spectra of imipramine (upper) and promazine (lower).

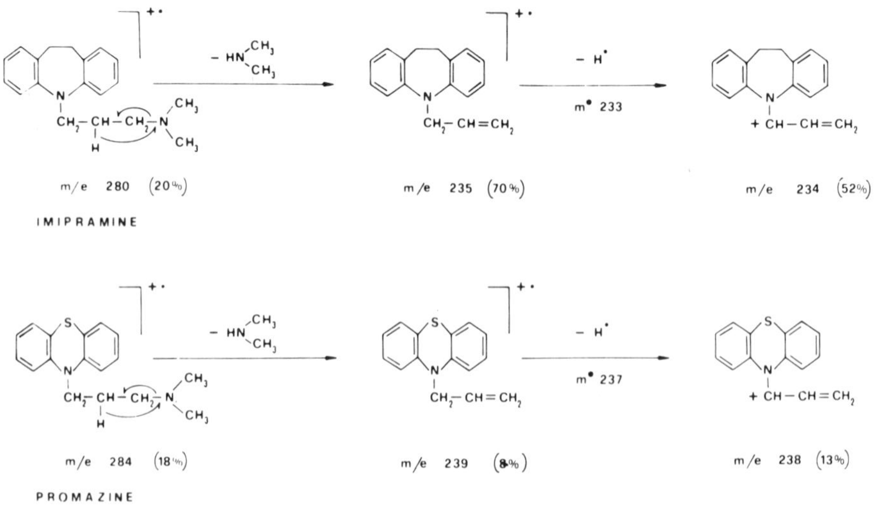

FIG. 9. Comparison of the fragmentation pathways of imipramine and promazine.

2. Results

The imipramine is extracted from plasma at pH 7 into n-hexane which is then evaporated under nitrogen (about 81% recovery). This is estimated by dissolving the hexane residue in a solution containing a known amount of the internal marker, i.e., promazine. It is possible to construct a calibration curve by plotting the ratio of imipramine and promazine peak areas against the imipramine concentration. The minimum detectable amount of imipramine by this method is 50 pg. Mass fragmentograms obtained from plasma of a patient receiving imipramine are shown in Fig. 10.

Plasma obtained from different patients receiving imipramine concurrently with various other drugs (barbiturates or benzodiazepines) can be analyzed without any interference by these drugs. The method appears to be sensitive enough to cover the drug concentration normally present during

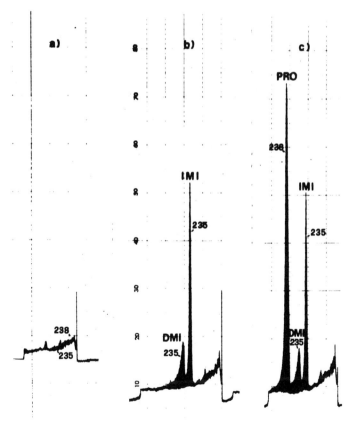

FIG. 10. Mass fragmentograms obtained from plasma of a patient undergoing chronic treatment with imipramine (125 mg/day). (a) before treatment, (b) after treatment with imipramine (IMI), without internal marker, (c) the same as (b) after addition of the internal marker promazine (PRO). DMI is desmethylimipramine.

chronic treatment with imipramine. The imipramine plasma levels and the therapeutic effects obtained over a 28-day period in two patients receiving imipramine are shown in Fig. 11.

From the reported data, even if preliminary and representative of only two cases, it can be seen that for imipramine, as for other tricyclic antidepressants (Åsberg et al., 1971), the plasma levels are not related to the dosage. Even these two patients show wide variations in plasma levels. In fact case VC had plasma levels of about 180 ng/ml while receiving 150 mg/day, and case BS had plasma levels of about 75 ng/ml while receiving 175 mg/day. It is also interesting to note that case VC, who had higher plasma levels of imipramine, responded favorably to the therapy, but also presented severe side effects (cardiocirculatory collapse) when imipramine plasma levels approached 180 ng/ml. Case BS with very low plasma levels of both imipramine and desmethylimipramine did not show any improvement or side effects.

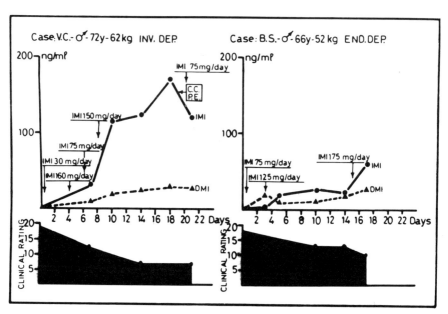

FIG. 11. Plasma levels of imipramine (IMI) and desmethylimipramine (DMI) during chronic treatment in two depressive patients. Case V.C. – involutive depression, cc = cardiocirculatory collapse, pe = pulmonary edema. Case B.S. – endogenous depression. The DMI levels were measured according to the method of Hammer and Brodie (1967).

3. Discussion

Several studies have been devoted to the measurement of tricyclic antidepressant agents present in plasma during a therapeutic treatment of depressed patients. However, such determinations have been limited to the secondary amines (Sjöqvist, Hammer, Borgå, and Azarnoff, 1969) taking advantage of the simple and rapid radioisotope method developed by Hammer and Brodie (1967). It is becoming more evident, however, that tertiary and secondary amines attached to the same tricyclic structure show different biochemical effects. In fact although desipramine is more effective than imipramine in inhibiting noradrenaline uptake at the nerve endings (Carlsson, Corrodi, Fuxe, and Hökfelt, 1969b), imipramine is more active than desipramine in blocking serotonin uptake (Carlsson, Corrodi, Fuxe, and Hökfelt, 1969a; Samanin, Ghezzi, and Garattini, 1972). This different biochemical effect may in turn be responsible for different therapeutic and/or side effects for tertiary and secondary amines. Since tertiary amines (e.g., imipramine) are transformed in the body into secondary amines (e.g., desipramine), the need to follow both molecular species in the plasma is evident, in order to adapt the dose of the drug to individual patients. The variability of plasma levels for both imipramine and desipramine observed in the two cases studied supports the need to perform additional studies in the hope that the monitoring of plasma levels may represent a useful tool for providing a rational therapy with tricyclic antidepressant drugs.

ACKNOWLEDGMENTS

The authors wish to thank Professor S. Garattini for his encouragement and helpful comments. This work was supported by National Institutes of Health grant No. 1P01 GM18376–02 PTR.

REFERENCES

Åsberg, M., Cronholm, B., Sjöqvist, F., and Tuck, D. (1970): Correlation of subjective side effects with plasma concentrations of nortriptyline. *British Medical Journal*, 4:18–21.
Åsberg, M., Cronholm, B., Sjöqvist, F., and Tuck, D. (1971): Relationship between plasma level and therapeutic effect of nortriptyline. *British Medical Journal*, 3:331–334.
Baker, K. M., and Frigerio, A. (1973): The loss of hydrogen cyanide and the dihydrogen cyanide radical from aromatic nitrogen heterocycles on electron impact. *Journal of the Chemical Society, Perkin II (in press).*
Baker, K. M., Frigerio, A., Morselli, P. L., and Pifferi, G. (1973): Identification of a rearranged degradation product from carbamazepine-10,11-epoxide. *Journal of Pharmaceutical Sciences*, 62:475–476.
Bickel, M. H., and Baggiolini, M. (1966): The metabolism of imipramine and its metabolites by rat liver microsomes. *Biochemical Pharmacology*, 15:1155–1169.
Blom, S. (1962): Trigeminal neuralgia: Its treatment with a new anticonvulsant drug (G-32883). *Lancet*, 1:839–840.
Bonduelle, M., Bouygues, P., Sallon, C., and Chemaly, R. (1964): Bilan de l'expérimentation clinique de l'anti-épileptique. In: *Neuropsychopharmacology*, Vol. 3, pp. 312–316. Elsevier, Amsterdam.
Borgå, O., Palmér, L., Linnarsson, A., and Holmstedt, B. (1971): Quantitative determination of nortriptyline and desmethylnortriptyline in human plasma by combined gas chromatography–mass spectrometry. *Analytical Letters*, 4:837–249.
Braithwaite, R. A., and Widdopp, B. (1971): A specific gas-chromatographic method for the measurement of "steady-state" plasma levels of amitriptyline and nortriptyline in patients. *Clinica Chimica Acta*, 35:461–472.

Brooks, C. J. W. (1971): Gas chromatography–mass spectrometry. In: *Mass Spectrometry*, Vol. 1, edited by D. H. Williams, pp. 288–307. Chemical Society, London.

Brooks, C. J. W., Thawley, A. R., Rocher, P., Middleditch, B. S., Anthony, G. M., and Stillwell, W. G. (1970): Characterization of steroidal drug metabolites by combined gas chromatography–mass spectrometry. In: *Advances in Chromatography*, edited by A. Zlatkis, pp. 262–272. University of Houston Press, Houston.

Burlingame, A. L., and Johanson, G. A. (1972): Mass spectrometry. *Analytical Chemistry*, 44:337R–378R.

Carlsson, A., Corrodi, H., Fuxe, K., and Hökfelt, T. (1969a): Effect of antidepressant drugs on the depletion of intraneuronal brain 5-hydroxytryptamine stores caused by 4-methyl-α-ethyl-meta-tyramine. *European Journal of Pharmacology*, 5:357–366.

Carlsson, A., Corrodi, H., Fuxe, K., and Hökfelt, T. (1969b): Effects of some antidepressant drugs on the depletion of intraneuronal brain catecholamine stores caused by 4α-dimethyl-meta-tyramine. *European Journal of Pharmacology*, 5:367–373.

Cram, S. P., and Juvet, R. S. (1972): Gas chromatography. *Analytical Chemistry*, 44:213R–241R.

Dalessio, D. J., and Abbott, K. H. (1966): A new agent in the treatment of tic douloureux: A preliminary report. *Headache*, 5:103–107.

Davis, E. (1964): Clinical evaluation of a new anti-convulsant, G.32883. *Medical Journal of Australia*, 1:150–152.

Dingell, J. V., Sulser, F., and Gillette, J. R. (1964): Species differences in the metabolism of imipramine and desmethylimipramine (DMI). *Journal of Pharmacology and Experimental Therapeutics*, 143:14–22.

Fichsel, H., and Heyer, R. (1970): Carbamazepin in der behandlung kindlicher epilepsien. *Deutsch Medizinische Wochenschrift*, 95:2367–2370.

Frigerio, A., Belvedere, G., De Nadai, F., Fanelli, R., Pantarotto, C., Riva, E., and Morselli, P. L. (1972). A method for the determination of imipramine in human plasma by gas-liquid chromatography–mass fragmentography. *Journal of Chromatography*, 74:201–208.

Frigerio, A., Biandrate, P., Fanelli, R., Baker, K. M., and Morselli, P. L. (1972): Carbamazepine-10,11-epoxide: A metabolite of carbamazepine isolated from human urine and identified by GLC–MS. In: *Proceedings of the International Symposium on Gas Chromatography–Mass Spectrometry*, edited by A. Frigerio, pp. 389–402. Tamburini, Publ., Milan.

Frigerio, A., Fanelli, R., Biandrate, P., Passerini, G., Morselli, P. L., and Garattini, S. (1972): Mass spectrometric characterization of carbamazepine-10,11-epoxide, a carbamazepine metabolite isolated from human urine. *Journal of Pharmaceutical Sciences*, 61:1144–1147.

Goenechea, S., and Hecke-Seibicke, E. (1972): Beitrag zum stoffwechsel von carbamazepin. *Zeirschrift für Klinische Chemie und Klinische Biochemie*, 10:112–113.

Goldbaum, L. R., Johnston, E. H., and Blumberg, J. M. (1963): The practice of identification in analytical toxicology. *Journal of Forensic Sciences*, 8:286–294.

Gordon, A. E., and Frigerio, A. (1972): Mass fragmentography as an application of gas liquid chromatography–mass spectrometry in biological research. *Chromatographic Reviews*, 73:401–417.

Gudzinowicz, B. J. (1967): *Gas Chromatographic Analysis of Drugs and Pesticides*. Dekker, New York.

Hammar, C.-G. (1972): Qualitative and quantitative analyses of drugs in body fluids by means of mass fragmentography and a novel peak matching technique. In: *Proceedings of the International Symposium on Gas Chromatography–Mass Spectrometry*, edited by A. Frigerio, pp. 1–18. Tamburini Publ., Milan.

Hammar, C.-G., Alexanderson, B., Holmstedt, B., and Sjöqvist, F. (1971): Gas chromatography–mass spectrometry of nortriptyline in body fluids of man. *Clinical Pharmacology and Therapeutics*, 12:496–505.

Hammar, C.-G., Holmstedt, B., and Ryhage, R. (1968): Mass fragmentography. Identification of chlorpromazine and its metabolites in human blood by a new method. *Analytical Biochemistry*, 25:532–548.

Hammar, C.-G., Holmstedt, B., Lindgren, J. E., and Tham, R. (1969): The combination of gas-chromatography and mass spectrometry in the identification of drugs and metabolites. In: *Advances in Pharmacology and Chemotherapy*, Vol. 7, edited by S. Garattini, A. Goldin, F. Hawking, and I. J. Kopin, pp. 53–89. Academic Press, New York.

Hammer, W. M., and Brodie, B. B. (1967): Application of isotope derivative technique to assay of secondary amines: Estimation of desipramine by acetylation with H³-acetic anhydride. *Journal of Pharmacology and Experimental Therapeutics,* 157:503–508.

Harris, S. R., Gaudette, L. E., Efron, D. H., and Manian, A. A. (1970): A method for the measurement of plasma imipramine and desmethylimipramine concentrations. *Life Sciences,* 9:781–788.

Herrmann, B., and Pulver, R. (1960): Der stoffwechsel des psychopharmakons tofranil. *Archives Internationales de Pharmacodynamie et de Thérapie,* 126:454–469.

Horning, E. C., and Horning, M. G. (1971): Human metabolic profiles obtained by GC and GC/MS. *Journal of Chromatographic Science,* 9:129–140.

Jongmans, J. W. M. (1964): Report on the anti-epileptic action of Tegretol. *Epilepsia,* 5:74–82.

Junk, G. A. (1972): Gas chromatograph–mass spectrometer combinations and their applications. *International Journal of Mass Spectrometry and Ion Physics,* 8:1–71.

Kroman, H. S., and Bender, S. R., editors (1968): *Theory and Application of Gas Chromatography in Industry and Medicine.* Grune and Stratton, New York.

Kruppa, R. F. (1972): *personal communication.*

Kuntzman, R., and Tsai, I. (1967): A sensitive radiochemical method for the assay of some tertiary amines. *Pharmacologist,* 9:240.

Livingston, S. (1966): *Drug Therapy for Epilepsy,* p. 125. Charles C Thomas, Springfield, Ill.

Loader, C. E., Palmer, T. F., and Timmons, C. J. (1967): Mass spectra of some heterocyclic ring systems. In: *Some Newer Physical Methods in Structural Chemistry,* edited by R. Bennett and J. G. Davis, pp. 80–83. United Trade Press Ltd., London.

Meinardi, H. (1971): Other antiepileptic drugs. Carbamazepine. In: *Antiepileptic Drugs,* edited by D. M. Woodbury, J. K. Penry, and R. P. Schmidt, pp. 487–496. Raven Press, New York.

Merritt, C., Jr. (1970): The combination of gas chromatography with mass spectrometry. In: *Applied Spectroscopy Reviews,* Vol. 3, pp. 263–326. M. Dekker Inc., New York.

Moody, J. P., Tait, A. C., and Todrick, A. (1967): Plasma levels of imipramine and desmethylimipramine during therapy. *British Journal of Psychiatry,* 113:183–193.

Morselli, P. L., Biandrate, P., Frigerio, A., and Garattini, S. (1973): Pharmacokinetics of carbamazepine in rats and humans. In: *Proceedings of the Sixth Annual Meeting: European Society for Clinical Investigation.* Abstract No. 114, p. 88. *European Journal of Clinical Investigation,* in press.

Morselli, P. L., Gerna, M., Frigerio, A., Zanda, G., and De Nadai, F. (1971): Some observations on the metabolism of carbamazepine in animals and man. In: *Fifth World Congress of Psychiatry,* Abstract No. 883. La Prensa Medica Mexicana.

Morselli, P. L., Gerna, M., and Garattini, S. (1971): Carbamazepine plasma and tissue levels in the rat. *Biochemical Pharmacology,* 20:2043–2047.

Powers, J. C. (1968): Mass spectrometry of simple indoles. *Journal of Organic Chemistry,* 33:2044–2050.

Porter, Q. N., and Baldas, J. (1971): *Mass Spectrometry of Heterocyclic Compounds.* Wiley-Interscience, New York.

Rafaelsen, O. J., and Christiansen, J. (1969): Imipramine metabolism in man. In: *The Present Status of Psychotropic Drugs,* edited by A. Cerletti and F. J. Bové, pp. 118–119. Excerpta Medica Foundation, Amsterdam.

Safe, S., Jamieson, W. D., and Hutzinger, O. (1972): The ion kinetic energy and mass spectra of isomeric methyl indoles. *Organic Mass Spectrometry,* 6:33–37.

Samanin, R., Ghezzi, D., and Garattini, S. (1972): Effect of imipramine and desipramine on the metabolism of serotonin in midbrain raphe stimulated rat. *European Journal of Pharmacology,* 20:281–283.

Sisenwine, S. F., Tio, C. O., Shrader, S. R., and Ruelius, H. W. (1970): The biotransformation of protriptyline in man, pig and dog. *Journal of Pharmacology and Experimental Therapeutics,* 175:51–59.

Sjöqvist, F., Hammer, W., Borgå, O., and Azarnoff, D. L. (1969): Pharmacological significance of the plasma level of monomethylated tricyclic antidepressants. In: *The Present Status of Psychotropic Drugs,* edited by A. Cerletti and F. J. Bové, pp. 128–136. Excerpta Medica Foundation, Amsterdam.

Spiteller, G. (1966): Mass spectrometry of heterocyclic compounds. In: *Advances in Hetero-cyclic Chemistry,* Vol. 7, edited by A. R. Katritzky and A. J. Boulton, pp. 301–376. Academic Press, New York.

Sutherland, J. W., Williamson, D. E., and Theivagt, J. G. (1971): Pharmaceuticals and related drugs. *Analytical Chemistry Annual Reviews,* 43:206R–266R.

Taylor, J. F. (1971): Methods of chemical analysis. In: *Narcotic Drugs — Biochemical Pharmacology,* edited by D. H. Clouet, pp. 17–88. Plenum Press, New York.

Weder, H. J., and Bickel, M. H. (1968): Separation and determination of imipramine and its metabolites from biological samples by gas-liquid chromatography. *Journal of Chromatography,* 37:181–189.

Weist, F., and Zicha, L. (1967): Dünnscichtchromatographische untersuchungen über 5-carbamyl-5H-dibenzo [b, f] azepin in harn und liquor bei neuen indikationsgebieten. *Arznei-mittel-Forschung,* 17:874–875.

Mass Spectrometry in Biochemistry and Medicine,
edited by A. Frigerio and N. Castagnoli.
Raven Press, New York © 1974

Gas Chromatography–Mass Spectrometry in the Distribution and Metabolism of Phentermine

Arthur K. Cho, Björn Lindeke, and Donald J. Jenden

Department of Pharmacology, University of California School of Medicine, Los Angeles, California 90024

I. INTRODUCTION

This is a summary of studies carried out in our laboratory on the pharmacokinetics of phentermine (2-methyl-1-phenylisopropylamine, I), a compound closely related to amphetamine in its pharmacology and chemistry.

Both compounds are anorexic agents but phentermine is thought to have reduced central nervous system (CNS) stimulatory activity (Yelnosky, Panasevich, Borrelli, and Lawlor, 1969). Amphetamine-like compounds can be metabolized along several different pathways. In order to examine the effect of an α-methyl group on metabolism and distribution, the kinetics of phentermine metabolism and elimination were examined in detail (Cho, Lindeke, and Hodshon, 1972; Cho, Lindeke, and Sum, 1973*b*; Cho, Lindeke, Hodshon, and Miwa, 1973*a*).

GC–MS was used throughout this study as a quantitative and qualitative tool. In quantitative studies it was used to estimate levels of phentermine in biological tissue with a stable isotope-labeled internal standard and specific ion detection. In qualitative procedures GC–MS was used to identify the major *in vitro* metabolite of phentermine, the N-hydroxy derivative.

II. METHODS OF PROCEDURE

A. Instrumentation

The GC–MS used was an EAI Quad 300 quadrupole mass spectrometer interfaced with a Varian 1400 Gas Chromatograph with a glass frit separator. The mass spectrometer has been modified (Jenden and Silverman, 1973) by the use of an eight-channel multiplexing device that permits simultaneous monitoring of the ion current at as many as eight different mass numbers at a switching frequency of about 100 Hz, so that throughout a single GC peak, all eight masses can be continuously monitored. The GC employed was a Varian 2100 equipped with flame ionization detectors. The GC columns used were 2 mm × 2 m all glass columns and contained either a 3% OV 17 on

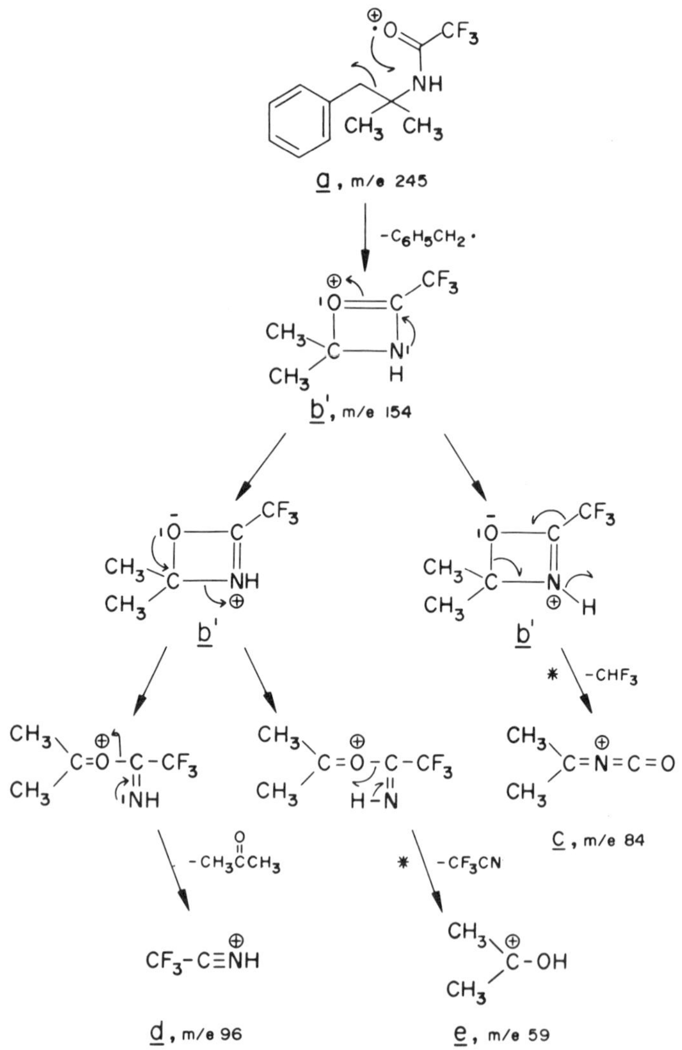

FIG. 1. Fragmentation pathway for N-trifluoroacetyl phentermine.

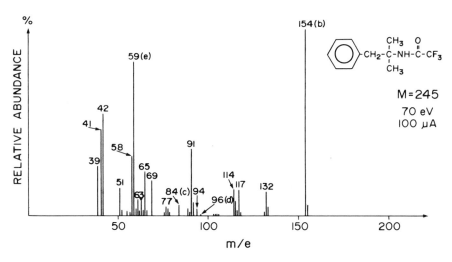

FIG. 2. Mass spectrum of N-trifluoroacetyl phentermine.

Gas Chrom Q or a 1% OV 101/1% DDTS (Jenden, Roch, and Booth, 1972) on Gas Chrom Q.

When used in the specific ion detection mode for quantitative analysis, the mass spectrometer was focused on two m/e values, one for a key fragment of the compound to be analyzed and the other the m/e value for the corresponding fragment of the internal standard. The appropriate GC peak was monitored at these settings and the ratio of spectrometer response to the two masses was compared to a standard curve. The fragments used for phentermine analysis were the 154 and 157 peaks, which are the major fragments of the trifluoroacetyl derivatives of phentermine and its deuterium-substituted variant after loss of the benzyl radical (Figs. 1b' and 2). For the qualitative identification of the N-hydroxy metabolite, the specific ion detector was focused on four masses to obtain more accurate estimates of relative peak intensities (Table 1).

TABLE 1. *Identification of N-hydroxyphentermine by specific ion analysis*

Source	Relative abundance (m/e)			
	56	91	130	146
Metabolite	100	30.1	13.0	20.6
N-Hydroxyphentermine	100	29.8	12.3	20.3

The mass spectrometer was programmed to monitor the mass numbers indicated throughout the gas chromatographic peaks for the compounds. The relative abundance represents the integrated response of the ion detector at the mass indicated.

B. Analytical Methods

The procedure developed for the quantitative analysis of phentermine (Cho et al., 1973*a*) was a modification of an assay for amphetamine (Cho, Lindeke, Hodshon, and Jenden, 1973) utilizing the N-trifluoroacetyl derivative for gas chromatographic separation and trideuterated phentermine as the internal standard. Before introduction into the GC–MS, the plasma or brain extracts to be analyzed were prepurified by extraction of the phentermine into an organic solvent, back extracted into HCl, and reextracted into benzene for the trifluoroacetylation. The *in vitro* incubation mixtures were treated in the same way when analyzed for phentermine.

The metabolism of phentermine was examined *in vitro* using microsomes from rabbit, rat, and guinea pig. The initial studies were carried out with rabbit liver preparations and incubation with phentermine and an NADPH-generating system. The result was a metabolite that was identified as the N-hydroxy compound (II). In this procedure a microsomal incubation mix-

$$\langle O \rangle\!\!-\!\!CH_2\!\!-\!\!\underset{\underset{CH_3}{|}}{\overset{\overset{CH_3}{|}}{C}}\!\!-\!\!NHOH \qquad (II)$$

ture was extracted with an isopropanol petroleum ether mixture and the organic extract evaporated to purify and concentrate the metabolite. The metabolite was converted to its O-trimethylsilyl derivatives and identified by GC–MS. In subsequent studies the metabolite was estimated by a GC procedure with phenylacetone oxime as the internal standard (Cho et al., 1973).

C. Chemical Synthesis

The deuterium-substituted internal standard used in this study was synthesized from the alcohol obtained by treatment of phenylacetone with methylmagnesium iodide-$^2H_3$. This alcohol was converted to its methane sulfonate ester which was then treated sequentially with sodium azide and lithium aluminum hydride. The isolated 2-methyl-1-phenyl(3,3,3-$^2H_3$)isopropylamine had an isotopic purity of 99.8% (Lindeke, Cho, and Fedorchuk, 1972). N-Hydroxyphentermine was prepared by reduction of the corresponding nitro compound with aluminum amalgam (Lindeke, Cho, Thomas, and Michelson, 1973).

III. RESULTS AND DISCUSSION

The use of stable isotope-substituted internal standards for quantitative GC–MS was first described by Samuelsson, Hamberg, and Sweeley (1970), who used the technique to quantitate levels of prostaglandins. The technique has had numerous applications (Gordon and Frigerio, 1972; Jenden and Cho, 1972), and we have used it to measure levels of phentermine in brain and plasma of rats (Cho et al., 1973*a*). The isotopically substituted internal standard functions in the same way that conventional internal standards do but has the added advantage of having almost identical physical properties, so that it also acts as a carrier and hence minimizes the losses due to surface adsorption throughout the procedure.

The m/e 154 and 157 peaks were chosen for the quantitative analysis because they are the base peaks and because they are relatively isolated in the spectrum (Fig. 2). The choice of the base peak maximizes sensitivity and the isolation of the peak in the spectrum is important for minimal overlap between the two peaks used for analysis. Since quantitation is based on the ratios of the spectrometer response to these peaks, minimal overlap is desirable. In this analysis the relative abundance of the m/e 154 peak was only 0.2% in the spectrum of the N-trifluoroacetyl derivative of the internal standard. Therefore, even with a 100-fold excess of internal standard, levels of phentermine can be accurately estimated.

The results of a pharmacokinetic study in rats are shown in Fig. 3. The animals were given i.v. doses of 0.5 mg/kg of phentermine, and brain and plasma levels were determined at the indicated times. The decline followed the kinetics of the two compartment system as described by Riegelman, Loo, and Rowland (1968) and the equations for the curves drawn are shown in the figure. The results indicate that phentermine equilibrates rapidly between brain and plasma and the two levels decline at approximately the same rate. These data are very similar to data collected for amphetamine, although phentermine is more localized in the brain. Therefore, the lower CNS stimulatory activity of phentermine cannot be explained simply in terms of its distribution since it reaches higher levels in the brain than does amphetamine.

In qualitative GC–MS studies the *in vitro* metabolism of phentermine was examined with microsomal preparations from rabbit, guinea pig, and rat. GC monitoring of an organic extract of the incubation mixture revealed the appearance of a new peak whose height increased with time. The metabolite was unstable when chromatographed directly so it was converted to the TMS derivative. This derivative gave a symmetrical GC peak and was identified as the TMS derivative of N-hydroxyphentermine by its mass spectrum. The presence of a peak at m/e = 146 corresponding to M-(91) and the absence of peaks such as 90 + O–TMS reflecting phenolic TMS derivatives suggested that the metabolite was the N-hydroxy compound. This was verified by comparison of the mass spectra with that obtained with authentic material (Cho et al., 1972). A quantitative comparison of the three compounds was made with specific ion detection techniques in which the relative abundances of four different mass peaks were determined. The results are shown in Table 1.

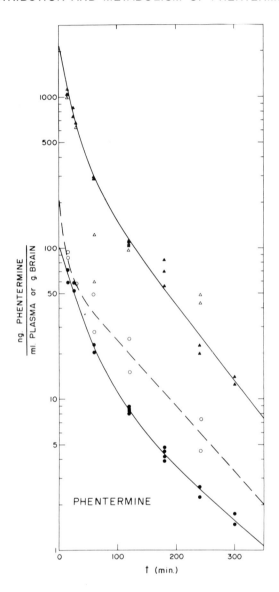

FIG. 3. Pharmacokinetics of phentermine in the rat. Brain and plasma levels of phentermine were determined in animals given i.v. doses of 0.5 mg/kg of phentermine (●, plasma; ▲, brain) or N-hydroxyphentermine (○, plasma; △, brain) at the times indicated. The data were fitted to a biexponential equation of the form $C_t = Ae^{-\alpha t} + Be^{-\beta t}$ where C_t is the tissue level at time t and A, α, B, β are the parameters determined (Riegelman, Loo, and Rowland, 1968).

This metabolite appears to be the major metabolite formed *in vitro* based on preliminary material balance experiments in which the phentermine lost was compared to the N-hydroxy compound formed. The reaction is sensitive to SKF 525A (1 mM) and carbon monoxide, indicating the involvement of cytochrome P_{450}. A GC estimation procedure was developed, and, when the kinetics of the reaction were examined, the results indicated that more than one enzyme may be involved (Cho et al., 1973b). Details of this study will be published elsewhere.

Additional substrates examined for this metabolic transformation included phenethylamine, amphetamine, and p-chlorophentermine. The mass spectra obtained after trimethylsilylation indicated that all three compounds formed the N-hydroxy derivatives (Lindeke et al., 1973). Because Beckett (1971) has previously reported N-hydroxy metabolites for other arylalkylamines, this metabolic pathway appears to be a general one. However, the hydroxylamines formed are not always stable and those amines with α-hydrogen atoms are labile and decompose to the corresponding oximes (Beckett, 1971). Phentermine gives a more stable metabolite because it does not have α-hydrogens; therefore, it should be a better substrate for studying this reaction.

The pharmacological properties of the N-hydroxy derivative of amphetamine have been reported to be very similar to amphetamine (Benington, Morin, and Clark, 1965). The pharmacology of N-hydroxyphentermine (O. Hinsvark and C. Beattie, *private communication*) is quantitatively and qualitatively like that of the parent compound. This similarity suggested that the N-hydroxy compound may be reduced *in vivo* to phentermine and this possibility was investigated. Animals were given i.v. doses of 0.5 mg/kg of N-hydroxyphentermine, and plasma levels were measured at different times after administration. The results are shown in Fig. 3 and indicate a rapid reduction of the hydroxylamine to phentermine. Thus, it appears that this N-hydroxy metabolite if formed *in vivo* would be rapidly reduced. A similar explanation could account for the observations made for N-hydroxy-amphetamine.

ACKNOWLEDGMENTS

Portions of this research were supported by U.S. Public Health Service grants MH-20473, MH-17691, GM-02040, and by a grant from the California Institute for Cancer Research. We gratefully acknowledge the collaboration of Barbara Hodshon, Gerald Miwa, Joseph Steinborn, Check Sum, and Robert Silverman in various phases of this work and the Pennwalt Corporation for supplies of phentermine and N-hydroxyphentermine.

REFERENCES

Beckett, A. H. (1971): Metabolic oxidation of aliphatic basic nitrogen atoms and their α-carbon atoms. *Xenobiotica*, 1:365–383.

Benington, F., Morin, R. D., and Clark, L. C., Jr. (1965): Behavioral and neuropharmacological actions of N-aralkylhydroxylamines and their O-methyl ethers. *Journal of Medicinal Chemistry*, 8:100–104.

Cho, A. K., Lindeke, B., and Hodshon, B. J. (1972): The N-hydroxylation of phentermine (2-methyl-1-phenylisopropylamine) by rabbit liver microsomes. *Research Communications in Chemical Pathology and Pharmacology*, 4:519–527.

Cho, A. K., Lindeke, B., Hodshon, B. J., and Jenden, D. J. (1973): Deuterium substituted amphetamine as an internal standard in a gas chromatographic/mass spectrometric (GC/MS) assay for amphetamine. *Analytical Chemistry*, 45:570–574.

Cho, A. K., Lindeke, B., Hodshon, B. J., and Miwa, G. (1973a): Application of quantitative gas chromatography/mass spectrometry (GC/MS) to a study of the pharmacokinetics of amphetamine and phentermine *Journal of Pharmaceutical Sciences*, 62:1491–1494.

Cho, A. K., Lindeke, B., and Sum, C. (1973b): A gas chromatographic method for the quantitative estimation of N-hydroxyphentermine. *Federation Proceedings*, 32:734.

Gordon, A. E., and Frigerio, A. (1972): Mass fragmentography as an application of gas-liquid chromatography–mass spectrometry in biological research. *Journal of Chromatography*, 73:401–417.

Jenden, D. J., and Cho, A. K. (1972): Application of integrated gas chromatography/mass spectrometry in pharmacology and toxicology. In: *Annual Reviews of Pharmacology, Vol. 13*, edited by H. W. Elliott, p. 371. Annual Reviews, Palo Alto.

Jenden, D. J., Roch, M., and Booth, R. (1972): A new liquid phase for the separation of amines and alkaloids. *Journal of Chromatographic Sciences*, 10:151–153.

Jenden, D. J., and Silverman, R. W. (1973): A multiple specific ion detector and analog data processor for a gas chromatograph/quadrupole mass spectrometer system. *Journal of Chromatographic Sciences*, 11:601–606.

Lindeke, B., Cho, A. K., and Fedorchuk, M. (1972): Specifically deuterated 1-phenylisopropylamines. II. Synthesis of deuterium labelled phentermine. *Acta Pharmaceutica Suecica*, 9:605–608.

Lindeke, B., Cho, A. K., Thomas, T. L., and Michelson, L. (1973): Microsomal N-hydroxylation of phenylalkylamines: Identification of N-hydroxylated phenylalkylamines as their trimethylsilyl derivatives by GC/MS. *Acta Pharmaceutica Suecica*, 10:493–503.

Riegelman, S., Loo, J. C. K., and Rowland, M. (1968): Shortcomings in pharmacokinetic analysis by conceiving the body to exhibit properties of a single compartment. *Journal of Pharmaceutical Sciences*, 57:117–123.

Samuelsson, B., Hamberg, M., and Sweeley, C. C. (1970): Quantitative gas chromatography of prostaglandin E_1 at the nanogram level: Use of deuterated carrier and multiple-ion analyzer. *Analytical Biochemistry*, 38:301–304.

Yelnosky, J., Panasevich, R. E., Borrelli, A. R., and Lawlor, R. B. (1969): Pharmacology of phentermine. *Archives Internationales de Pharmacodynamie et de Therapie*, 178:62–76.

Mass Spectrometry in Biochemistry and Medicine,
edited by A. Frigerio and N. Castagnoli.
Raven Press, New York © 1974

Mass Spectrometric Identification of Metabolites from Microbial Dehydrogenation of Tomatidine and Derivatives

V. Kramer, I. Belič[a], and H. Sočič[b]

J. Stefan Institute, University of Ljubljana, Ljubljana, Yugoslavia, [a]Biochemical Institute of the Medical Faculty, University of Ljubljana, Ljubljana, Yugoslavia, and [b]Chemical Institute Boris Kidric, Ljubljana, Yugoslavia

I. INTRODUCTION

During the last decade many publications have appeared demonstrating the extensive use of mass spectrometry in steroid research. Although the breakdown of steroids under electron impact is rather complex, present knowledge about fragmentation processes of this class of compound enables a considerable number of identification problems to be successfully resolved. Hetero atoms and/or double bonds in steroid molecules greatly simplify the mass spectra and make identification more easy and rapid. Tomatidine and its derivatives are steroidal alkaloids and produce many typical fragment ions which facilitate the characterization of these compounds.

II. RESULTS AND DISCUSSION

In the present report we wish to describe the mass spectra of products of the microbial dehydrogenation of tomatidine and dihydrotomatidines. When tomatidine was incubated under appropriate conditions with *Nocardia restrictus*, besides some unchanged tomatidine, four new products were isolated (Belič and Sočič, 1972). The metabolites were separated by thin-layer chromatography (TLC), crystallized from a methanol or chloroform-methanol mixture, and their R_f-values, m.p., IR, UV, and mass spectra determined.

The low ends of the mass spectra of all metabolites were similar to one another and to that of tomatidine (Fig. 1). They exhibited very intense peaks at m/e 114 and m/e 138, which are typical for the unchanged structure of rings E and F (Budzikiewicz, 1964). Thus, the alteration must take place in the remaining steroidal part of the molecules.

The first metabolite (10% of the total) showed a molecular ion at m/e 413 and ions at m/e 385 (M^+–CO) and m/e 271 in the upper part of the mass

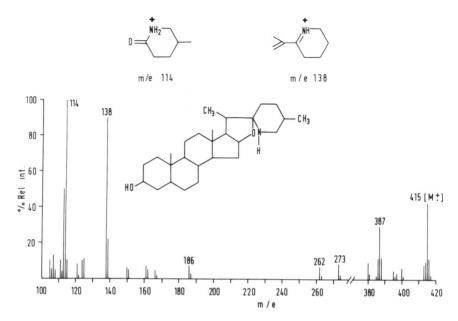

FIG. 1. Mass spectrum of tomatidine.

spectrum. The latter is the fragment ion including rings A, B, C, and D formed by rupture of ring E and hydrogen rearrangement. These three peaks are always 2 mass units lower than those found in tomatidine. This finding, as well as the absence of a M^+-H_2O peak, suggests a 3-keto structure in ring A. The mass spectrum and other spectral data are actually identical with those of synthesized *tomatanin-3-one* (Belič and Sočič, 1972).

The next two metabolites exhibited almost identical mass spectra on the whole and could not be distinguished from each other. Their molecular ions at m/e 411 are 4 mass units lower than tomatidine. The same holds for the other ions in the M^+-region. Because the steroidal ion of tomatidine at m/e 273 is also shifted down to m/e 269, a double bond in addition to the keto-structure of ring A (no M^+-H_2O peak was found) must be formed. Although the cluster of ions in the m/e 120–125 mass region allows some presumption that the double bond is located in ring A, an unquestionable structure could not be determined from only the mass spectra. The exact structures of these metabolites were determined in combination with other spectral methods as *1-tomatenin-3-one* (3%) and *4-tomatenin-3-one* (7%). The results agree well with those obtained on synthesized products prepared by a known method (Toldy, 1958).

The main metabolite (80%) gave a molecular ion at m/e 409 and ions at m/e 394 (M^+-CH_3) and m/e 267 (formed by ring E fission). These ions are 6 mass units lower than those found in tomatidine. Further, appearance of m/e 121 and 288 (M^+-121) points to the 1,4-dien-3-one structure of ring A

(Budzikiewicz, Djerassi, and Williams, 1964). The structure of the metabolite must therefore be 1,4-tomatadien-3-one. The above structures for the metabolites show that tomatidine is transformed first to the keto and then to keto-ene and keto-diene structures. It is well known from the literature (e.g., review by Hörhold, Böhme, and Schubert, 1969) that such a pathway is common for the steroids. However, the sequence normally does not stop at the 1,4-dien-3-one structure, but continues to lower molecular species. To determine whether rings E and F or the nitrogen atom in ring F are responsible for stopping the reaction at the diene-one structure, it seemed desirable to study compounds where rings E and F (or both) are open, i.e., the dihydrotomatidines. For this reason three compounds were synthesized:

A. *22,26-Epimino-5α-cholestane-3β,16β-diol* was prepared from tomatidine according to the procedure of Adam and Schreiber (1969). Actually two isomeric diols with differing configurations at C-22 were isolated (22S:25S) and (22R:25S). However, their mass spectra did not differ and therefore in further discussion this differentiation will be omitted.

B. *26-Acetylamino-5α-furostan-3β-ol* was prepared from tomatidine diacetate (Sato and Latham, 1956).

C. *26-Amino-5α-cholestane-3β,16β,22ξ-triol* was prepared by refluxing the acetylaminofurostanol (B) with 6 N HCl for 7 hr. After HCl was evaporated in vacuum the product was purified by TLC. Actually, we expected the deacetylated product only, but as the spectra show, the ring E was also opened.

The mass spectrum of the synthesized diol (A) (Fig. 2) showed a molecular ion at m/e 417 (2 mass units higher than tomatidine), a relatively intense M^+-1, and a cluster of ions in the M^+–H_2O region. The most characteristic ion at m/e 98, which is the base peak, is formed by the expected cleavage α- to the nitrogen atom (i.e., the C-20–22 bond) and confirms an open ring E.

With *Nocardia restrictus,* (A) metabolized completely producing only one metabolite. This metabolite gave a molecular ion at m/e 411, corresponding to the loss of six hydrogen atoms (Fig. 3). The M^+ decomposed further losing alternatively a CH_3 radical and carbon monoxide. The elimination of H_2O giving rise to a small peak at m/e 393 is appreciably reduced in comparison to the original compound. Again by far the most abundant fragment ion at m/e 98 confirms the presence of the unchanged ring F. The ion at m/e 121 originates from ring B fission and indicates a keto-diene structure of ring A. Thus, the structure of the metabolite should be *22,26-epimino-1,4-cholestadien-3-one.*

The mass spectrum of the compound (B) is shown in Fig. 4. The M^+ of medium intensity was formed at m/e 459. The base peak at m/e 185 corresponds to ring E fission, namely C-17–20 and C-16–0 bonds, and the peak at m/e 158, on the other hand, to the rupture of C-20–22 and C-16–0 bonds. Further evidence for the first mode of ring E fission is the second most prominent peak at m/e 273, which represents the remaining steroidal part

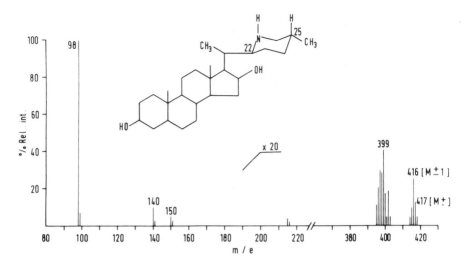

FIG. 2. Mass spectrum of 22,26-epimino-5α-cholestane-3β,16β-diol.

of the molecule. Here, a hydrogen atom was transferred to the lost side chain. Another cleavage near the oxygen atom (i.e., C-22–23 bond) gave rise to the m/e 331 (M+-side chain). Peaks at m/e 444, 441, and 416 represent the loss of CH_3, H_2O and CH_3CO from the M+, respectively. Thus the mass spectrum agrees with the structure given in Fig. 4.

When the product (B), having an open ring F, was incubated for 100 hr, a new metabolite, besides some unchanged starting material, was chromatographically isolated. Its molecular ion at m/e 453 was the base peak and

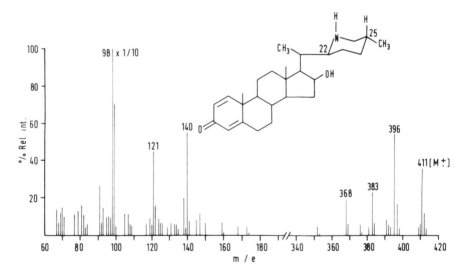

FIG. 3. Mass spectrum of 22,26-epimino-1,4-cholestadien-3-one.

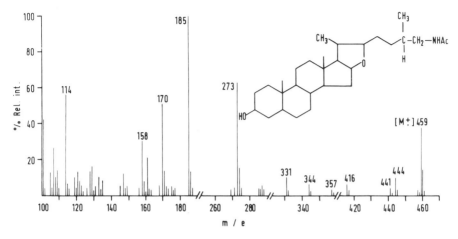

FIG. 4. Mass spectrum of 26-acetylamino-5α-furostan-3β-ol.

immediately suggests that the loss of six hydrogen atoms took place. No elimination of water was observed in the spectrum indicating that no free OH-groups are present. Ions arising from ring E fission, i.e., m/e 185 and m/e 158, were still present in the spectrum. The steroidal ion at m/e 273 for starting compound (B) is now shifted to m/e 267 as expected. This fact and the peak at m/e 332 (M^+-121), the peak at m/e 121 as well as m/e 147 (267–120), are diagnostic for the keto-diene form of ring A and support the given structure of *26-acetylamino-1,4-furostadien-3-one* (Fig. 5).

The mass spectrum of (C) (Fig. 6) showed a parent ion at m/e 435. Successive loss of three H_2O from the M^+ shows clearly that at least three hydroxyl groups are present. The base peak at m/e 143 resulted from splitting the C-17–20 bond as a consequence of the McLafferty rearrange-

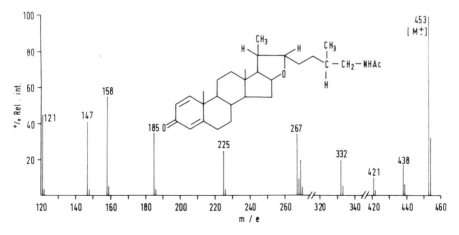

FIG. 5. Mass spectrum of 26-acetylamino-1,4-furostadien-3-one.

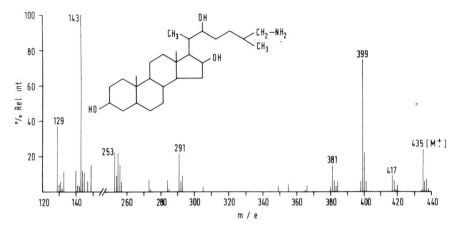

FIG. 6. Mass spectrum of 26-amino-5α-cholestane-3β,16β,22ξ-triol.

ment where a neutral water molecule is lost. The hydrogen transfer most likely takes place from C-22 to hydroxyl at C-16. Simple C-17–20 bond fission, leaving a positive charge on the steroid portion, led to the m/e 291. This latter ion decomposed further, expelling two water molecules.

After incubation of (C) with *Nocardia restrictus*, again only one metabolite was found. Its fragmentation pattern (Fig. 7) is very similar to that of the starting compound. The molecular ion increased 42 mass units from m/e 435 to m/e 477. The loss of water molecules and/or OH-radicals from M⁺ indicates again that three OH-groups must still be present. Fragment ions at m/e 254 and m/e 291 demonstrate the unchanged steroid structure. From the C-17–20 bond cleavage an ion at m/e 185 appears. This confirms that the alteration by an acetyl-group on the aliphatic side chain was performed.

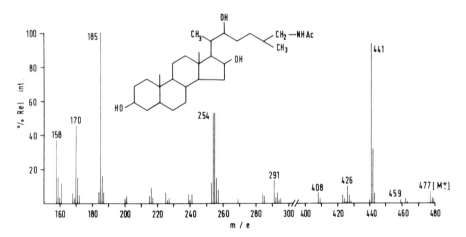

FIG. 7. Mass spectrum of 26-acetylamino-5α-cholestane-3β,16β,22ξ-triol.

The mass spectrum corresponds to the structure drawn in Fig. 7. Therefore, that metabolite is *26-acetylamino-5α-cholestane-3β,16β-22ξ-triol.* This assumption was supported also by other spectroscopic methods.

III. CONCLUSION

Summarizing the above findings on microbial transformation of tomatidine and related compounds, we observe that the reaction always [except for the compound (C)] stops at the 1,4-dien-3-one structure, even though different experimental conditions were applied. The incubation of (C) produces a N-acetylated product and no dehydrogenation was detected. From these observations, it seems that the nitrogen atom in steroidal alkaloids exhibits a decisive effect on the course of microbial transformations. Because structural alterations on these transformed compounds do not fundamentally affect the main fragmentation mode, it was relatively convenient to follow the reactions by mass spectrometry.

IV. EXPERIMENTAL

Mass spectra were carried out on a CEC-21–110 C mass spectrometer at 6-kV accelerating voltage, 70 eV and 100 μA ionizing current. Several fragmentation pathways were confirmed by first field-free metastable technique. The samples were introduced by direct inlet probe. The probe temperature was maintained as low as possible.

ACKNOWLEDGMENT

We wish to thank the "Boris Kidrič" Fund for financial support.

REFERENCES

Adam, G., and Schreiber, K. (1969): Reduktion von Spirosolan-Alkaloiden mit Natriumboranat. *Zeitschrift für Chemie*, 9:227–228.

Belić, I., and Sočič, H. (1972): Microbial dehydrogenation of tomatidine. *Journal of Steroid Biochemistry*, 3:843–846.

Budzikiewicz, H. (1964): Zum Massenspektrometrischen Fragmentierungsverhalten von Steroidalkaloiden. *Tetrahedron*, 20:2267–2278.

Budzikiewicz, H., Djerassi, C., and Williams, D. H. (1964): *Structure Elucidation of Natural Products by Mass Spectrometry*, p. 91. Holden-Day, San Francisco.

Hörhold, C., Böhme, K. H., and Schubert, K. (1969): Über den Abbau von Steroiden durch Mikroorganismen. *Zeitschrift für Allgemeine Mikrobiologie*, 9:235–246.

Sato, Y., and Latham, G. H. (1956): Chemistry of dihydrotomatidines. *Journal of the American Chemical Society*, 78:3146–3250.

Toldy, L. (1958): Untersuchungen mit Tomatidin, I. *Acta Chimica Academiae Scientiarum Hungaricae*, 16:401–409.

Mass Spectrometry in Biochemistry and Medicine,
edited by A. Frigerio and N. Castagnoli.
Raven Press, New York © 1974

The Identification and Quantitation of Tetrahydropapaveroline in Rat Brain by Mass Fragmentography

A. J. Turner, K. M. Baker, S. Algeri, and A. Frigerio

Istituto di Ricerche Farmacologiche "Mario Negri," Via Eritrea 62, 20157 Milano, Italy

I. INTRODUCTION

Mass fragmentography, a term originally introduced by Hammar, Holmstedt, and Ryhage (1968), is a technique that involves the use of the mass spectrometer as a detector for the effluent from the gas chromatograph. The mass spectrometer can be set to detect one or more characteristic fragment ion of the compound under study, thereby introducing a further parameter of identification in addition to the retention time in the gas chromatograph. Thus the method combines the high resolving power of the gas chromatograph with the high sensitivity and specificity of identification provided by the mass spectrometer. The technique and its application to studies on drug metabolism have been reviewed (Hammar, Holmstedt, Lindgren, and Tham, 1969; Gordon and Frigerio, 1972), and additional examples are given in this volume.

Recently, mass fragmentography has become a valuable analytical technique for the identification and measurement of "biogenic amines" and their metabolites in the central nervous system as well as for the detection of possible new neurotransmitter compounds. The occurrence of acetylcholine was confirmed in rat brain although the higher homologues propionylcholine and butyrylcholine were undetectable (Hammar, Hanin, Holmstedt, Kitz, Jenden, and Karlén, 1968). The simultaneous measurement of norepinephrine and dopamine in discrete brain regions by mass fragmentography has been reported to have a sensitivity in the picogram range (Koslow, Cattabeni, and Costa, 1972). The method has also been applied to the measurement of indoleamines and some of their metabolites in cerebrospinal fluid, pineal gland, and hypothalamus (Bertilsson and Palmér, 1972; Cattabeni, Koslow, and Costa, 1972; Green, Koslow, and Costa, 1973). We have become interested in the application of mass fragmentography to the detection in brain of tetrahydropapaveroline (THP; norlaudanosoline), a postulated metabolite of the biogenic amine dopamine (Holtz, Stock, and Westermann, 1964a; Walsh, Davis, and Yamanaka, 1970). Our interest in this compound lies in its possible involvement in some of the effects of L-DOPA therapy in Parkinson's disease (Sourkes, 1971).

II. ASPECTS OF DOPAMINE METABOLISM

Figure 1 illustrates some aspects of the biosynthesis and metabolism of dopamine. The amine is derived from the amino acid L-tyrosine by hydroxylation to L-DOPA followed by decarboxylation. An important pathway of metabolism available to dopamine is oxidative deamination by the enzyme monoamine oxidase (MAO) to 3,4-dihydroxyphenylacetaldehyde. This aldehyde is then normally rapidly metabolized by aldehyde dehydrogenase to the corresponding acid (Breese, Chase, and Kopin, 1969). A small proportion of the aldehyde may also be reduced enzymatically to the alcohol by aldehyde reductase (Turner and Tipton, 1972). This divergence of aldehyde metabolism is reminiscent of one of the possible pathways of metabolism of the drug propranolol (Walle, Saelens, Privitera, and Gaffney, 1973). Deamination of propranolol leads to the formation of an aldehyde which can then be oxidized *in vivo* to an acid or reduced to an alcohol, presumably by similar enzyme systems to those involved in biogenic amine metabolism.

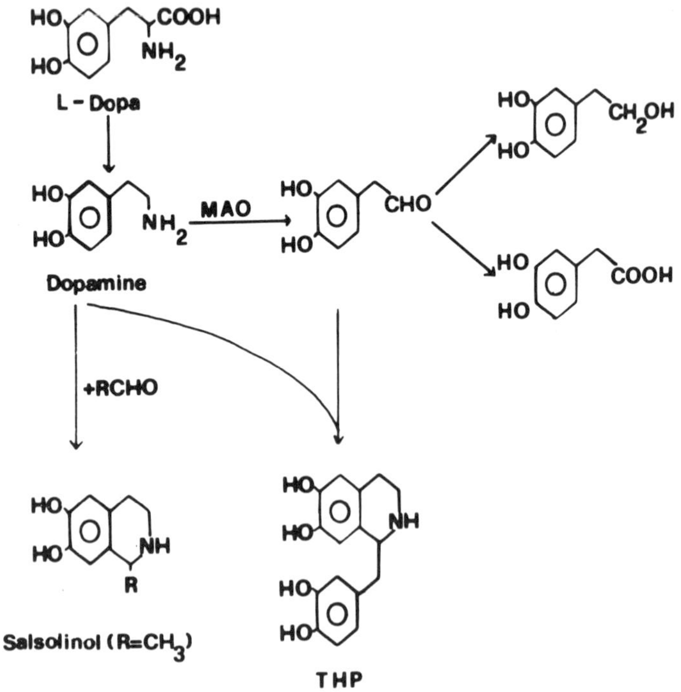

FIG. 1. The biosynthesis of dopamine and its metabolism via the monoamine oxidase (MAO) pathway.

Another possible metabolite of dopamine was suggested in 1964 by Holtz et al. (1964a), who proposed that the aldehyde derived from dopamine could react nonenzymically with the parent amine by Schiff's base formation and then cyclize to form the tetrahydroisoquinoline alkaloid, THP. This reaction, the Pictet-Spengler condensation (Whaley and Govindachari, 1951), can occur at physiological pH and is known to occur in the opium poppy since morphine biosynthesis proceeds from tyrosine via dopamine and tetrahydropapaveroline (see review by Kirby, 1967). It is also possible that the parent amine, dopamine, could condense with other available aldehydes. For example, alcohol ingestion leads to the production of acetaldehyde which can condense with dopamine to produce salsolinol (Fig. 1) (Yamanaka, Walsh, and Davis, 1970). This reaction is analogous to the use of formaldehyde for the histofluorescent detection of catecholamines in brain as quinoidal dihydroisoquinolines (Falck, 1962). The possibility that THP may be a metabolite of dopamine *in vivo* has led to speculation that THP may be a mediator of certain of the effects of administered L-DOPA (Sourkes, 1971). It now appears that dopamine alone cannot account for all the observed effects of DOPA administration to parkinsonian patients (Sandler, Bonham Carter, Hunter, and Stern, 1973).

III. SOME PROPERTIES OF THP

THP has pharmacological activities in its own right that are distinct from those of dopamine. It has many of the properties of a β-adrenergic agonist and has a blood pressure lowering effect. This hypotensive activity may be related to that observed as a side effect of L-DOPA therapy in Parkinson's disease (for discussion, see Sandler et al., 1973). Several recent articles have reviewed at length the pharmacological properties of THP, particularly with regard to its peripheral effects (Holtz, Stock, and Westermann, 1964b; Santi, Ferrari, Toth, Contessa, Fassina, Bruni, and Luciani, 1967; Simon, Goujet, Chermat, and Boissier, 1971).

Tetrahydroisoquinoline (THIQ) alkaloids can probably compete for the same uptake and storage mechanisms as the catecholamines and can be released following stimulation (Cohen, Mytilineou, and Barrett, 1972; Greenberg and Cohen, 1973). Available evidence therefore supports the hypothesis that THIQ alkaloids may be capable of acting as false transmitters (Greenberg and Cohen, 1973; Tennyson, Cohen, Mytilineou, and Heikkila, 1973). THP, if formed *in vivo*, may be rapidly metabolized. For example, several authors have pointed out that oxidation and rearrangement of THP could lead to compounds resembling apomorphine or bulbocapnine (Simon et al., 1971; Sourkes, 1971; Sandler et al., 1973), and the structures of these compounds are compared in Fig. 2. In connection with the above, apomorphine itself has been reported to have a DOPA-like action in Parkinson's disease, although of short duration (see Sourkes, 1971).

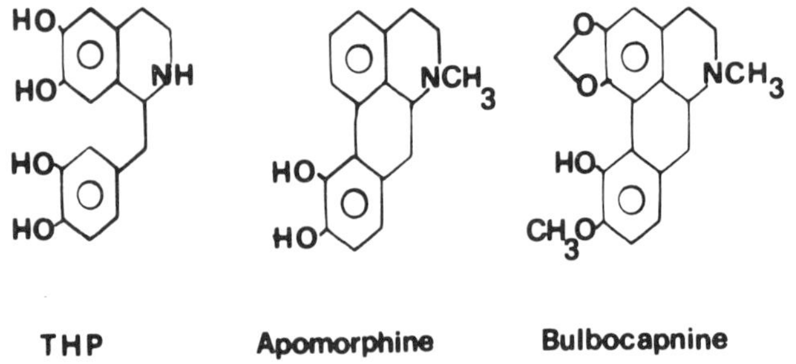

FIG. 2. Comparison of the structures of tetrahydropapaveroline (THP), apomorphine, and bulbocapnine.

IV. THP AS A METABOLITE OF DOPAMINE *in vivo* AND *in vitro*

The possible importance of THP as a metabolite of dopamine in animals has been supported to a certain extent by studies carried out *in vitro* by various groups of workers. Some 10 years ago, Holtz, Stock, and Wester-mann (1963, 1964a) showed that incubation of dopamine in millimolar concentrations with liver homogenates resulted in the formation of signifi-cant amounts of a compound with the same R_f as THP on thin-layer chroma-tography. More recently Walsh et al. (1970) have revived interest in this *in vitro* work, and they confirmed that THP could be formed from dopamine by incubation with tissue homogenates. Thus it appeared that it was not only the opium poppy that could synthesize such isoquinoline alkaloids.

The formation of THP in the tissue homogenates was prevented by pre-incubation with the MAO inhibitor pargyline (Holtz et al., 1964a) and in-creased in the presence of ethanol (Davis, Walsh, and Yamanaka, 1970). The effect of ethanol has been explained on the basis of its conversion to acetaldehyde which will competitively inhibit the oxidation of 3,4-dihy-droxyphenylacetaldehyde by aldehyde dehydrogenase. The metabolism of the latter aldehyde may therefore be diverted to condensation with the parent amine to give an increasing amount of THP. Davis and Walsh (1970) have proposed that this effect of ethanol to promote THP formation may be related to the processes involved in alcohol addiction.

Recently Halushka and Hoffmann (1970) have cast some doubt on the criteria used by Davis and Walsh for the identification of THP. Halushka and Hoffmann (1968) were unable to detect significant quantities of THP formed *in vivo* in guinea pig liver or kidney after the i.v. administration of dopamine-[14]C. However, they did not examine brain tissue. *In vitro* work by these authors showed that THP was an important metabolite of dopamine only when tissue homogenates were incubated with millimolar concentra-tions of the amine. Such quantities are greatly above physiological concen-trations of dopamine available to MAO.

In view of the above work, we became interested in developing a method for THP determination involving mass fragmentography in order to obtain a more positive identification of THP formed *in vitro* from either L-DOPA or dopamine. Furthermore, we wanted to see whether it was possible to detect THP formed *in vivo* in brain tissue, particularly after the chronic administration of L-DOPA and/or ethanol.

Very recently, Sandler et al. (1973) have reported the identification of small quantities of THP in the urine of parkinsonian patients receiving L-DOPA, strengthening the possible relationship between DOPA therapy in parkinsonism and THP formation. However, urinary THP levels may well not reflect THP formation in brain, and, although these authors used GC–MS for the identification of THP, they were unable to obtain sufficient THP to provide a complete mass spectrum for confirmation of its structure.

V. EXPERIMENTAL

In our work we have used the penta-trifluoroacetate derivative of THP for GC–MS. The mass spectrum of this compound is shown in Fig. 3. We used an L.K.B. 9000 GC–MS system and the gas chromatography was carried out with a 1-m glass column packed with 1% OV-17 on Gas Chrom Q operating at a temperature of 205°C. The ionization energy was chosen to be 50 eV.

The mass spectrum of the THP trifluoroacetylated derivative has a molecular ion at m/e = 767 and shows a base peak at m/e = 452, which was used

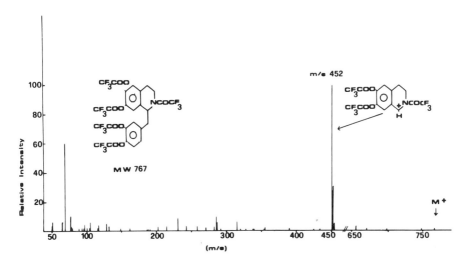

FIG. 3. Mass spectrum of the trifluoroacetylated derivative of THP. An LKB 9000 GC–MS was used with a 1-m glass column packed with 1% OV-17 on Gas Chrom Q, operating at a temperature of 205°C. The ionization energy was chosen to be 50 eV as this gave the highest yield of m/e = 452 from the THP derivative. MW = molecular weight; m/e = mass to charge ratio.

as the peak for the mass fragmentographic monitoring. The fragment ion corresponding to the base peak is illustrated in Fig. 3. The mass spectrum is fairly simple and no other suitable ion was available for multiple-ion detection. The trifluoroacetate was selected as the derivative of choice because of its ease of formation; in contrast the trimethylsilyl derivative was difficult to form and often a mixture of the penta- and tetra-substituted compounds was present. The penta-trifluoroacetate was also least easily adsorbed on the gas chromatographic column. The minimum detectable amount of derivatized THP by mass fragmentography was about 50 pg. Above this level the response was linear with respect to the amount injected into the gas chromatograph. For the extraction of THP from tissue, procedures involving organic solvent extractions were found to be unsuccessful because of the great instability of THP as the free base. Thus, tissue was homogenized in 4 volumes of 0.4 M $HClO_4$ containing 0.1% (w/v) of ascorbate and of E.D.T.A. to stabilize the catechol compounds and to protect against oxidation. The homogenate was centrifuged at 15,000 g for 10 min at 4°C and the precipitate was rehomogenized with a further 4 volumes of 0.4 M $HClO_4$ and recentrifuged. The combined supernatants were used for further purification. $KClO_4$ was precipitated at pH 4 by the careful addition of 1 M KOH and catechol compounds, including THP, were then adsorbed on alumina at pH 8.3. After washing the alumina with water, the THP was eluted with 2 × 2 ml of 1 M acetic acid and the eluate was lyophilized. The freeze-dried powder was then allowed to react with a mixture of trifluoroacetic anhydride and ethyl acetate (1:3) at 60°C for 3 hr and the derivative was used for GC–MS. Rather than evaporating the solvents to dryness and redissolving the residue in ethyl acetate, 1 to 2 μl of the reaction mixture was injected directly into the gas chromatograph.

The recovery of THP was routinely calculated by taking 25 ng of authentic THP through the procedure. The recovery was in the range 40 to 60%, and was not affected by the presence of brain tissue. Unfortunately, the lack of an ideally suitable internal standard for mass fragmentography has so far hindered the accurate quantitation of THP extracted from tissue.

For *in vitro* studies on the formation of THP, rat brainstems were homogenized in 10 volumes of 0.1 M sodium phosphate buffer, pH 7.4, and incubated with L-DOPA or dopamine for 60 min at 37°C in a total volume of 1 ml. The incubation mixture contained 10 μg/ml of ascorbic acid to prevent air oxidation of dopamine. The enzymic reaction was stopped by the addition of $HClO_4$ to give a final concentration of 0.4 M and THP was then isolated and derivatized as described above. Control incubations were carried out with boiled homogenates or homogenates preincubated for 20 min with 0.4 mM pargyline to inhibit MAO. In this way aldehyde and hence THP formation should be prevented.

The extracts were subjected to mass fragmentography after derivatization and typical mass fragmentograms are shown in Fig. 4. The peaks rep-

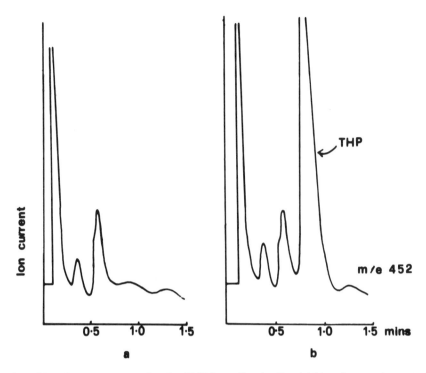

FIG. 4. Mass fragmentogram showing THP formation *in vitro*. (a) Mass fragmentogram of an extract prepared from an incubation of dopamine with a brain homogenate pretreated with 0.4 mM pargyline. The mass spectrometer was set to detect m/e = 452, with instrumentation conditions as described in the legend to Fig. 3. (b) Mass fragmentogram prepared from an incubation of 5mM dopamine with an untreated brain homogenate. Instrumentation conditions as above. The peak corresponding to THP is marked.

resent compounds eluting from the GC that give specific fragment ions of m/e = 452. These include compounds other than THP, but THP is identified by its specific retention time in the GC.

Figure 4b shows a mass fragmentogram from an extract prepared after incubation of a brain homogenate with 5 mM dopamine. Sufficient THP was formed in this case (about 50 μg) to obtain a complete mass spectrum of the derivative which was identical to that of the standard. Thus we have positive confirmation that THP can be a metabolite of dopamine *in vitro*. Figure 4a illustrates the results of a control experiment in which the brain homogenate had been preincubated with pargyline before the addition of dopamine. No peak with the retention time of THP is seen. Thus, THP formation from dopamine is indeed dependent upon the activity of MAO.

Incubation of homogenates with millimolar concentrations of L-DOPA also resulted in THP formation. Similar quantities of THP were formed as in the incubations with dopamine, probably because of the rapid decarboxylation of L-DOPA to dopamine. The concentrations of THP precursor used in these experiments are well above the normal concentrations of do-pamine available to MAO in the neuron *in vivo* (Halushka and Hoffmann, 1970). When the incubations were carried out with 50 μM DOPA or do-pamine, THP formation could again be detected by mass fragmentography but the amount of THP synthesized represented a conversion of only about 1% of the added dopamine. Thus, under these conditions, THP appears to be a minor metabolite of dopamine *in vitro,* as previously shown by Halushka and Hoffman (1968).

Having obtained positive confirmation of THP formation in brain homog-enate from either L-DOPA or dopamine, we turned our attention to the de-tection of THP formed *in vivo* in rat brain and our findings are reported here.

For the oral administration of compounds, female rats were divided into four groups of three rats each, which were supplied with the following solu-

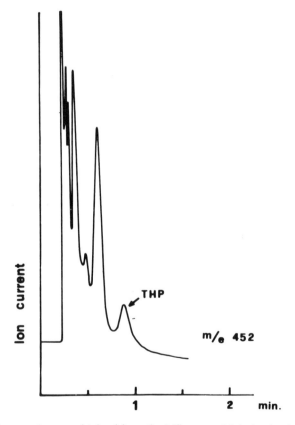

FIG. 5. Mass fragmentogram obtained from the trifluoroacetylated extract from rat brain tissue after the oral administration of L-DOPA to the animals (see text for details). The presence of THP was confirmed by coinjecting the brain sample with a sample of authentic THP derivative. The mass fragmentogram corresponds to about 400 pg of injected THP. Instrumentation conditions as above.

tions as sole source of drinking water: (a) tap water alone; (b) 10% ethanol (v/v); (c) 4 mg/ml L-DOPA + 1 mg/ml Ro 4–4602 (to inhibit the peripheral decarboxylation of DOPA); (d) 4 mg/ml L-DOPA + 1 mg/ml Ro 4–4602 + 10% ethanol (w/v). All solutions were supplemented with 0.1% (w/v) ascorbic acid, and solutions were changed once daily. After 8 days the rats were killed, the brains removed, frozen on dry ice, and stored at −20°C until the extraction was carried out as described above. The derivatized brain extracts were then subjected to mass fragmentography. In untreated rats and rats treated with ethanol alone we were unable to detect any THP. However, in rats treated with L-DOPA, small quantities of THP could be detected.

The mass fragmentogram corresponding to such an extract is shown in Fig. 5. Again we observe several compounds eluting from the gas chromatograph that show specific fragment ions at m/e = 452. The identity of THP was confirmed by reinjecting a mixture of the extract together with authentic THP trifluoroacetate. In Fig. 5 the peak corresponding to THP is equivalent to approximately 400 pg of injected THP with a total of about 8 ng/g brain. After treatment with DOPA + ethanol, levels of THP in the range of 10 to 25 ng/g brain were detected.

VI. DISCUSSION

We have shown that, after the chronic administration of L-DOPA to rats, THP can be detected in brain, although only in nanogram quantities. The administration of DOPA together with ethanol resulted in only a small increase in THP levels, possibly because of the occurrence of the alternative condensation of dopamine with acetaldehyde to form salsolinol. The low levels of THP detected in our experiments may be a reflection of rapid metabolism of this compound, e.g., by methylation of the phenolic hydroxyl groups or of the amino group. In fact, THP has been shown to be a substrate for the methylating enzyme catechol O-methyl transferase and can competitively inhibit the methylation of other catecholamines (V. E. Davis, *personal communication*).

Another possibility for metabolism is the oxidation and rearrangement of THP to a morphine-type alkaloid (Fig. 2). Such reactions occur in certain plants, and also the penta-methylated derivative of THP (laudanosine) can be chemically oxidized in high yield to morphinandienone O-methyl flavinantine, a morphine derivative (Miller, Stermitz, and Falck, 1971). Thus, in certain pharmacological situations, THP may be an intermediate in the conversion of dopamine to a variety of previously uninvestigated metabolites in animals and man. Our results provide the first demonstration of the occurrence of THP in brain *in vivo*. The involvement of THP or related alkaloids in the therapeutic actions of L-DOPA in Parkinson's disease is therefore not an unreasonable proposition.

We are at present investigating the effects of acute administration of

L-DOPA on THP formation and we are trying to determine whether THP is localized in discrete brain regions. The limitations of the method, which we are at present trying to overcome, are the lack of an ideally suitable internal standard for mass fragmentography and some problems of instability of the derivative. Taken together, these limitations have restricted the routine applicability of the method for quantitative studies.

ACKNOWLEDGMENTS

We would like to thank Professor S. Garattini for much helpful advice and criticism. A. J. T. was supported by a Royal Society European Fellowship and K. M. B. by a fellowship from the Wellcome Trust.

REFERENCES

Bertilsson, L., and Palmér, L. (1972): Indole-3-acetic acid in human cerebrospinal fluid: Identification and quantification by mass fragmentography. *Science,* 177:74–76.

Breese, G. R., Chase, T. N., and Kopin, I. J. (1969): Metabolism of some phenylethylamines and their β-hydroxylated analogs in brain. *Journal of Pharmacology and Experimental Therapeutics,* 165:9–13.

Cattabeni, F., Koslow, S. H., and Costa, E. (1972): Gas chromatographic–mass spectrometry assay of four indole alkylamines of rat pineal. *Science,* 178:166–168.

Cohen, G., Mytilineou, C., and Barrett, R. E. (1972): 6,7-Dihydroxytetrahydroisoquinoline: Uptake and storage by peripheral sympathetic nerve of the rat. *Science,* 175:1269–1272.

Davis, V. E., and Walsh, M. J. (1970): Alcohols, amines, and alkaloids: A possible biochemical basis for alcohol addiction. *Science,* 167:1005–1007.

Davis, V. E., Walsh, M. J., and Yamanaka, Y. (1970): Augmentation of alkaloid formation from dopamine by alcohol and acetaldehyde in vitro. *Journal of Pharmacology and Experimental Therapeutics,* 174:401–412.

Falck, B. (1962): Observations on the possibilities of the cellular localization of monoamines by a fluorescence method. *Acta Physiologica Scandinavica,* 56 (Suppl. 197):1–25.

Gordon, A. E., and Frigerio, A. (1972): Mass fragmentography as an application of gas-liquid chromatography–mass spectrometry in biological research. *Journal of Chromatography,* 73:401–417.

Green, A. R., Koslow, S. H., and Costa, E. (1973): Identification and quantitation of a new indolealkylamine in rat hypothalamus. *Brain Research,* 51:371–374.

Greenberg, R. S., and Cohen, G. (1973): Tetrahydroisoquinoline alkaloids: Stimulated secretion from the adrenal medulla. *Journal of Pharmacology and Experimental Therapeutics,* 184:119–128.

Halushka, P. V., and Hoffmann, P. C. (1968): Does tetrahydropapaveroline contribute to the cardiovascular actions of dopamine? *Biochemical Pharmacology,* 17:1873–1880.

Halushka, P. V., and Hoffmann, P. C. (1970): Alcohol addiction and tetrahydropapaveroline. *Science,* 169:1104–1105.

Hammar, C.-G., Hanin, I., Holmstedt, B., Kitz, R. J., Jenden, D. J., and Karlén, B. (1968): Identification of acetylcholine in fresh rat brain by combined gas chromatography–mass spectrometry. *Nature,* 220:915–917.

Hammar, C.-G., Holmstedt, B., Lindgren, J.-E. and Tham, R. (1969): The combination of gas chromatography and mass spectrometry in the identification of drugs and metabolites. In: *Advances in Pharmacology and Chemotherapy,* Vol. 7, edited by S. Garattini, A. Goldin, F. Hawking, and I. J. Kopin, pp. 53–89. Academic Press, New York.

Hammar, C.-G., Holmstedt, B., and Ryhage, R. (1968): Mass fragmentography. Identification of chlorpromazine and its metabolites in human blood by a new method. *Analytical Biochemistry,* 25:532–548.

Holtz, P., Stock, K., and Westermann, E. (1963): Uber die blutdruckwirkung des dopamines.

Naunyn-Schmiedeberg's Archiv für Experimentelle Pathologie und Pharmakologie, 246: 133–146.

Holtz, P., Stock, K., and Westermann, E. (1964a): Formation of tetrahydropapaveroline from dopamine *in vitro. Nature,* 203:656–658.

Holtz, P., Stock, K., and Westermann, E. (1964b): Pharmakologie des tetrahydropapaverolins und seine entstehung aus dopamin. *Naunyn-Schmiedeberg's Archiv für Experimentelle Pathologie und Pharmakologie,* 248:387–405.

Kirby, G. W. (1967): Biosynthesis of the morphine alkaloids. *Science,* 155:170–173.

Koslow, S. H., Cattabeni, F., and Costa, E. (1972): Norepinephrine and dopamine: Assay by mass fragmentography in the picomole range. *Science,* 176:177–180.

Miller, L. L., Stermitz, F. R., and Falck, J. R. (1971): Electrooxidative cyclization of laudanosine. A novel nonphenolic coupling reaction. *Journal of the American Chemical Society,* 93:5941–5942.

Sandler, M., Bonham Carter, S., Hunter, K. R., and Stern, G. M. (1973): Tetrahydroisoquinoline alkaloids: *In vivo* metabolites of L-DOPA in man. *Nature,* 241:439–443.

Santi, R., Ferrari, M., Toth, C. E., Contessa, A. R., Fassina, G., Bruni, A., and Luciani, S. (1967): Pharmacological properties of tetrahydropapaveroline. *Journal of Pharmacy and Pharmacology,* 19:45–51.

Simon, P., Goujet, M. A., Chermat, R., and Boissier, J. R. (1971): Etude pharmacologique d'un métabolite présumé de la dopamine, la tétrahydropapavéroline. *Thérapie,* 26:1175–1192.

Sourkes, T. L. (1971): Possible new metabolites mediating the actions of L-DOPA. *Nature,* 229:413–414.

Tennyson, V. M., Cohen, G., Mytilineou, C., and Heikkila, R. (1973): 6,7-Dihydroxytetrahydroisoquinoline: Electron microscopic evidence for uptake into the amine-binding vesicles in sympathetic nerves of rat iris and pineal gland. *Brain Research,* 51:161–169.

Turner, A. J., and Tipton, K. F. (1972): The characterization of two reduced nicotinamide-adenine dinucleotide phosphate-linked aldehyde reductases from pig brain. *Biochemical Journal,* 130:765–772.

Walle, T., Saelens, D. A., Privitera, P. J., and Gaffney, T. E. (1973): Applications of gas chromatography–mass spectrometry to studies of the metabolism of the β-blocking drug propranolol. Identification and characterization of pharmacologically active metabolites. *This volume.*

Walsh, M. J., Davis, V. E., and Yamanaka, Y. (1970): Tetrahydropapaveroline: An alkaloid metabolite of dopamine *in vitro. Journal of Pharmacology and Experimental Therapeutics,* 174:388–400.

Whaley, W. M., and Govindachari, T. R. (1951): The Pictet-Spengler synthesis of tetrahydroisoquinolines and related compounds. In: *Organic Reactions,* Vol. 6, edited by R. Adams, pp. 151–190. Wiley & Sons, Inc., New York.

Yamanaka, Y., Walsh, M. J., and Davis, V. E. (1970): Salsolinol, an alkaloid derivative of dopamine formed *in vitro* during alcohol metabolism. *Nature,* 227:1143–1144.

Mass Spectrometry in Biochemistry and Medicine,
edited by A. Frigerio and N. Castagnoli.
Raven Press, New York © 1974

Quantitative Analysis of Catecholamines in Picomole Range by an Improved Gas-Liquid Chromatographic Method. Compared Performance of Electron Capture Detection and Mass Fragmentography

J.-C. Lhuguenot and B. F. Maume

Centre de Biochimie de la Différenciation Cellulaire de l'Université de Dijon, Laboratoire d'Application en Chromatographie Gazeuse et en Spectrometrie de Masse, Faculté de Médecine, 7 bd Jeanne d'Arc, 21033 Dijon, France

I. INTRODUCTION

In order to analyze low amounts of catecholamines in animal tissues, several studies have been made on the use of gas-liquid chromatography (GLC) with electron capture detection (ECD) (Clarke, Wilk, and Gitlow, 1966; Horning, Moss, Boucher, and Horning, 1968; Änggärd and Sedvall, 1969; Wilkinson, 1970; Edwards and Blau, 1972). At nanogram levels, homogeneity of GLC effluents cannot be investigated by conventional GLC mass spectrometry (MS) because of the minuteness of the samples. Mass fragmentography is able to perform specific analyses of small amounts of substances present in biological materials (Hammar, Holmstedt, and Rhyage, 1968; Gordon and Frigerio, 1972; Maume, Bournot, Lhuguenot, Baron, Barbier, Maume, Prost, and Padieu, 1973). This method appears to be suitable for biological amines (Koslow, Cattabeni, and Costa, 1972; Karoum, Cattabeni, Costa, Ruthven, and Sandler, 1972; Lhuguenot and Maume, 1974).

In this work, we have optimized the conditions of formation and stability of perfluorobenzylimine-trimethylsilyl ether (PFB–TMS) derivatives for quantitative analysis. This method was described by Moffat and Horning (1970) for qualitative analysis only. The catecholamine derivatives used allow two types of detection. The excellent "electron capturing" properties bring a high sensitivity with ECD. On the other hand, the MS behavior of this derivative leads to a good sensitivity with a large specificity when a mass fragmentographic method is used.

II. MATERIALS AND METHODS

A. GLC and MS

A Packard model 7400 dual-column instrument was used. Either a 4-V direct current (DC) or a 45-V pulsed voltage (pulse period: 100 μsec, pulse width: 1 μsec) was applied to the 15-μC ^{63}Ni electron capture detector. The columns were 4 m × 3 mm i.d. silanized glass tubes. The packings were 1% SE-30 Ultraphase and 1% OV-17 (Supelco Inc.) on 100 to 120 mesh acid washed and silanized Gas Chrom P (Applied Science Laboratories Inc.). All columns were packed according to Horning, Van den Heuvel, and Creech (1963). The temperature of the separation was 190°C with argon-methane (90:10) as carrier gas (1.3 bar). The injector block temperature was 250°C and the detector oven temperature was 270°C.

An LKB 9000 GLC–MS instrument with accelerating voltage alternator was used. The silanized glass columns (3 m × 3 mm i.d.) were packed with 1% OV-I or 1% OV-17 on 100 to 120 mesh Gas Chrom P in the usual way. The temperatures were flash heater 260°C, molecular separator 270°C, and ion source 290°C. Temperature-programmed separations were carried out at 1°C/min from 170°C with helium as carrier gas (35 ml/mm). The accelerating voltage was 3.5 kV, trap current 60 μA, and electron energy 70 eV. The accelerating voltage alternator allowed detection of three fragment ions up to 20% higher than the lower mass. In this case, a 28-eV electron energy was used.

B. Sample Preparations

A modified procedure for quantitative formation of PFB–TMS derivatives at microgram and nanogram levels (Lhuguenot and Maume, 1974), derived from Moffat and Horning's method (1970), was set up. The pentafluoro-benzaldehyde was purified before use (Barbour, Buxton, Coe, Stephens, and Tatlow, 1966) and redistilled. The silylation was carried out with *bis*-trimethylsilylacetamide (BSA). The total time of PFB-TMS derivative formation was about 20 min. The derivatives were extracted from the reaction mixture by hexane. The derivative stability in hexane solution was excellent and made it possible to store them for several days before analysis if necessary.

Catecholamines were extracted from rat adrenals according to Imaï, Sugiura, and Tamura (1971). The 2-amino-2-phenylpropanol (Decombe and Patouraux, 1968) used as internal standard was added to the adrenal extract and the mixture was submitted to derivative formation reaction. The equivalent of 60 μg of adrenal gland was injected into the GLC–ECD instrument, and the equivalent of 300 μg of gland in the GLC–MS instrument.

III. RESULTS AND DISCUSSION

A. Sensitivity

Theoretically, ECD greatly improves the sensitivity of detection of GLC effluents, provided that the derivative group has high "electron capturing" capability. As far as we know, among the different types of derivatives the PFB–TMS have the best sensitivity. Figure 1 shows that 10 pg can be detected and quantitated without any doubt by this method.

On the other hand, in mass spectral analysis the catecholamine PFB–TMS derivatives have a major fragmentation pattern: the cleavage of the α and β bonds associated with the retention of the positive charge on the catechol fragment. Figure 2 shows mass spectra recorded with two electron

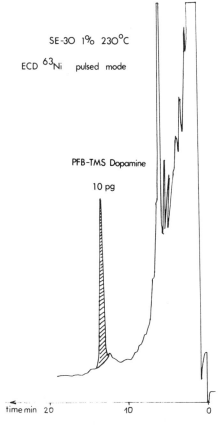

SE-30 1% 230°C

ECD ^{63}Ni   pulsed mode

PFB-TMS Dopamine

10 pg

time min 20   10   0

FIG. 1. Chromatogram with electron capture detection using pulsed current. The same chromatograph run with faster chart speed allows quantitative analysis with good reliability at picogram level.

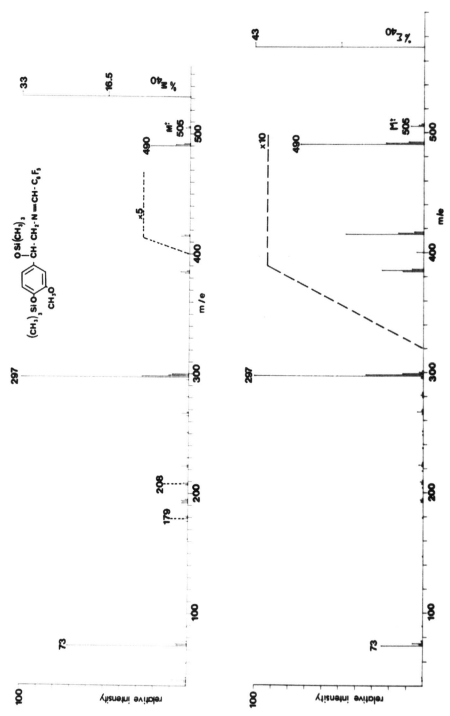

FIG. 2. Mass spectra of PFB–TMS normetanephrine. Above, spectrum obtained with an electron energy of 70 eV; below, spectrum scanned with an electron energy of 28 eV. The base peak corresponding to M–CH$_2$—N=CH—C$_6$F$_5$ was obtained for all primary catecholamines as PFB–TMS derivatives.

energies (28 and 70 eV) for PFB–TMS normetanephrine. This preferential fragmentation allows a reproducible sensitivity at nanogram levels when the corresponding fragment ion is monitored on the mass spectrometer.

B. Quantitative Analysis

As known, with ECD the response is linear in a limited range. Figure 3 shows that the responses of different catecholamines as PFB–TMS derivatives are linear from 0.1 to 1 ng when the DC mode is used. For more than 1 ng, the response reaches a maximum which does not allow further quantification. The response coefficient per gram is inversely proportional to the amine molecular weight, that is to say that molar response coefficients are identical for the different primary catecholamines. We have used an internal standard having a similar structure to those of catecholamines in order to get reproducible quantitative results. This standard was submitted to the same reactions of PFB–TMS derivative formation.

The response coefficients are multiplied by more than three when the pulsed mode is used; furthermore the pulsed mode enhances reliability and accuracy of the quantitative results. For example, response in coulombs per picomole of dopamine is increased from 1.65×10^{-8} (DC mode) up to 5.23×10^{-8} (pulsed mode). This pulsed mode gives rise to a new linear response range between 10 and 400 pg.

A larger linear response range down to 1 ng is obtained when the mass spectrometer is used as a detector, but it was observed that some losses in the GLC–MS system may occur for subnanogram amounts of these perfluoro derivatives. This phenomenon requires further attention and could be avoided by the use of labeled amine as a carrier.

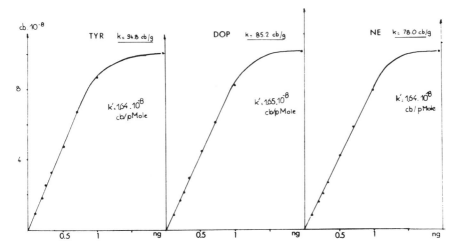

FIG. 3. Response curves of some PFB–TMS amines with electron capture detection (DC mode). Molar response coefficients k' are identical for the three catecholamines.

C. Specificity

The specificity of gas chromatography depends on the resolution of the injected mixture by the GLC column. Combined GC–MS offers a well-known advantage. But, even if the compounds are in too small amounts to record interpretable mass spectra and are incompletely resolved by the column, it is possible to use mass fragmentography. Several works were published in this field and results obtained in the catecholamine series were recently given by Koslow et al. (1972), Karoum et al. (1972), and Maume et al. (1973). As we know, PFB–TMS derivatives have mass spectra showing a preferential fragmentation. The very important value of the base peak intensity leads to high sensitivity with good specificity when the mass spectrometer is focused on the corresponding m/e value. Table 1 indicates the $\Sigma_{40}\%$ values of the fragment ion $M-CH_2-N=CH-C_6F_5$ which is the base peak for the amines studied except for the internal standard. But the absence of other fragmentations in these mass spectra prevents the detection of several specific fragment ions with the same sensitivity. However, the new multiple-ion detector–peak matching device (Hammar and Hessling, 1971) allows a simultaneous detection of ions very different in relative intensities by adjusting the gain of each channel separately. The possibility of a greater dynamic range allows detection in the same recording of the base peak and another ion of low intensity such as the M-15. In Fig. 4 the mass fragmentogram obtained from an adrenal extract indicates the high specificity of mass fragmentography for norepinephrine analysis.

TABLE 1. *Values of $\Sigma_{40}\%$ of base peak of PFB–TMS catecholamines*

Amine	base peak (m/e value)	$\Sigma_{40}\%$ value of base peak	
		70 eV	28 eV
2-Phenyl-2-aminopropanol (Internal standard)	298	14	29
Tyramine	179	28	48
3-O-Methyldopamine	209	27	38
Octopamine	267	13	44
Dopamine	267	20	31
Normetanephrine	297	33	43
Norepinephrine	355	23	40

These values are higher than those generally obtained and led to good specificity when mass fragmentography was used for catecholamine analysis.

IV. CONCLUSION

The PFB–TMS derivatives permit one to obtain a high sensitivity of electron capture and a very specific mass fragmentation. We think that the

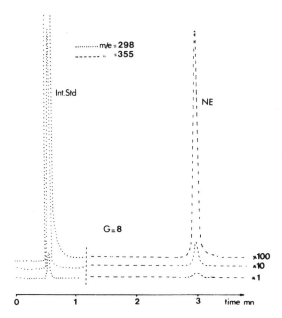

FIG. 4. Mass fragmentogram of adrenal extract corresponding to 300 µg of adrenal gland. The norepinephrine (NE) peak was recorded by use of the m/e value of the base peak: 355.

two described methods, namely GLC with ECD and mass fragmentography, are complementary for catecholamine analysis. It appears, also, that the reliability of electron capture measurements is considerably improved by mass fragmentographic assays run on the same sample.

ACKNOWLEDGMENTS

Financial help is gratefully acknowledged through research grants from: Délégation Générale à la Recherche Scientifique et Technique, Action Complémentaire Coordonnée, Développement Périnatal; Institut National de la Santé et de la Recherche Médicale, Action Thématique Programmée no. 6; Centre National de la Recherche Scientifique, ERA 267; Enseignement Supérieur, Vème et VIème Plan; and Fondation pour la Recherche Médicale Française.

REFERENCES

Änggärd, E., and Sedvall, G. (1969): Gas chromatography of catecholamine metabolites using electron capture detection and mass spectrometry. *Analytical Chemistry,* 41:1250–1256.

Barbour, A. K., Buxton, M. W., Coe, P. L., Stephens, R., and Tatlow, J. C. (1966): Aromatic polyfluoro compounds. Part VIII. Pentafluoro-benzaldehyde and related pentafluorophenyl ketones and carboxylic acids. *Journal of the Chemical Society,* C:808–817.

Clarke, D. D., Wilk, S., and Gitlow, S. E. (1966): Electron capture properties of halogenated amine derivatives. *Journal of Gas Chromatography,* 4:310–313.

Decombe, J., and Patouraux, D. (1968): Sur la synthèse d'un isomère de l'éphédrine à fonction alcool primaire. *Comptes Rendus de l'Académie des Sciences (Paris)*, 266:473–474.

Edwards, D. J., and Blau, K. (1972): Analysis of phenylethylamines in biological tissues by gas-liquid chromatography with electron capture detection. *Analytical Biochemistry*, 45:387–401.

Gordon, R. E., and Frigerio, A. (1972): Mass fragmentography as an application of gas liquid chromatography–mass spectrometry in biological research. *Journal of Chromatography*, 73:401–417.

Hammar, C. G., and Hessling, R. (1971): Novel peak matching technique by means of a new and combined multiple ion detector peak matcher device. Elemental analyses of compounds in submicrogram quantities without prior isolation. *Analytical Chemistry*, 43:298–306.

Hammar, C. G., Holmstedt, B., and Rhyage, R. (1968): Mass fragmentography. Identification of chlorpromazine and its metabolites in human blood by a new method. *Analytical Biochemistry*, 25:532–548.

Horning, E. C., Van den Heuvel, W. J. A., and Creech, B. G. (1963): Separation and determination of steroids by gas chromatography. In: *Methods of Biochemical Analysis*, edited by D. Glick. Interscience, New York.

Horning, M. G., Moss, A. M., Boucher, E. A., and Horning, E. C. (1968): The GLC separation of hydroxyl-substituted amines of biological importance including the catecholamines. *Analytical Letters*, 1:311–321.

Imaï, K., Sugiura, M., and Tamura, Z. (1971): Catecholamines in rat tissues and serum determined by gas chromatography. *Chemical and Pharmaceutical Bulletin (Tokyo)*, 19:409–411.

Karoum, F., Cattabeni, F., Costa, E., Ruthven, C. R. J., and Sandler, M. (1972): Gas chromatographic assay of picomole concentrations of biogenic amines. *Analytical Biochemistry*, 47:550–561.

Koslow, S. H., Cattabeni, F., and Costa, E. (1972): Norepinephrine and dopamine. Assay by mass fragmentography in the picomole range. *Nature*, 176:177–180.

Lhuguenot, J.-C., and Maume, B. F. (1974): Improvements in quantitative gas phase analysis of catecholamines in the picomole range by electron capture detection and mass fragmentography of their pentafluorobenzylimine-trimethylsilyl derivatives. *Journal of Chromatographic Science, accepted for publication.*

Maume, B. F., Bournot, P., Lhuguenot, J.-C., Baron, C., Barbier, F., Maume, G., Prost, M., and Padieu, P. (1973): Mass fragmentographic analysis of steroids, catecholamines and amino acids in biological materials. *Analytical Chemistry*, 45:1073–1082.

Moffat, A. C., and Horning, E. C. (1970): A new derivative for gas liquid chromatography of picogram quantities of primary amines of the catecholamine series. *Biochimica et Biophysica Acta*, 222:248–250.

Wilkinson, G. R. (1970): The GLC separation of amphetamine and ephedrine as pentafluorobenzamide derivatives and their determination by electron capture detection. *Analytical Letters*, 3:289–298.

Mass Spectrometry in Biochemistry and Medicine,
edited by A. Frigerio and N. Castagnoli.
Raven Press, New York © 1974

Mass Fragmentography Assay of Known N-Methylated Amino Acids Occurring in Acto-Myosin of Heart Cell Cultures

F. Barbier, B. F. Maume, and P. Padieu

Centre de Biochimie de la Différenciation Cellulaire de l'Université de Dijon, Laboratoire de Culture Tissulaire et Laboratoire d'Application en Chromatographie Gazeuse et en Spectrométrie de Masse, Faculté de Médecine, 7 bd Jeanne d'Arc, 21033 Dijon, France

I. INTRODUCTION

In an attempt to investigate the regulation of cardiac contractile protein biosynthesis, cardiac muscle cells explanted from postnatal rat hearts were grown in Petri dishes. A way to follow the maturation of cells *in vitro* in comparison with the organ is to study the genesis into myoblasts through the onset of specific proteins of the myofilament, namely, myosin and actin, which are the most abundant among the contractile proteins. The specific location of some rare methylated amino acids in these proteins is useful since the methylation of the amino acid residues may be related to the maturation of the cells from the myoblast to myocyte stage. ϵ-N-monomethyllysine (MML), ϵ-N-dimethyllysine (DML), and ϵ-N-trimethyllysine (TML) are encountered in the cardiac myosin. They are not found in actin. Conversely, actin contains 3-methylhistidine (3-CH_3His) which is absent in cardiac myosin. The amount of methylated lysine may vary with different types of muscles and with the age of animal samples. On the other hand, actin from a wide variety of species consistently contains one 3-CH_3His residue per molecule. The function of protein methylation is unclear. Methylated lysine is also encountered in the primary structure of an acidic protein from ribosomes of E. coli (Terhorst, Moller, Laursen, Wittmann, and Liebold, 1973) as well as histones (Paik and Kim, 1971) for which specific methylation seem to be a factor that causes blocking off the DNA function before mitosis (Paik and Kim, 1967). Cytochrome C of yeast and *Neurospora* contains trimethyllysine, but this amino acid does not seem to be found in higher animals (Delange, Glazer, and Smith, 1969). These methylated amino acids have not yet been described in other myofibrillar proteins (Johnson, Harris, and Perry, 1967; Hardy, Harris, Perry, and Stone, 1970). For this reason, we have tried to set up an assay of these amino acids during the maturation of the cardiac cells.

II. METHOD

The problem was to establish a micromethod for the assay of these rare amino acids: effectively, the ratio of MML to lysine is 1 to 1,000 and we have to work on small quantities such as a few micrograms of proteins. In one molecule of myosin having a molecular weight of 500,000 there are 630 residues of lysine, one residue of MML, and five residues of TML (Hardy et al., 1970). In actin there is one residue of 3-CH_3His per molecule (Johnson et al., 1967). Using gas chromatography, it is impossible to detect simultaneously on the same channel amino acids in such a concentration ratio. Moreover, if we work with 500 mg of cultured cardiac cells, the use of a flame ionization detector does not allow the detection of one residue of MML per molecule of myosin. For this reason, we set up an assay of these methylated amino acids by coupling gas chromatography to mass fragmentography. The general methodology of mass fragmentography is described elsewhere (Maume, Bournot, Lhuguenot, Baron, Barbier, Maume, Prost, and Padieu, 1973).

III. RESULTS AND DISCUSSION

Derivatives were prepared according to Gehrke, Zumwalt, and Wall (1968), Gehrke, Roach, Zumwalt, Stalling, and Wall (1968), and Gehrke, Kuo, and Zumwalt (1971). The amino acids were studied as their n-butyl-ester N-trifluoroacetamide derivatives. The complete separation of acid and neutral amino acids was achieved on a two phase bed column composed of one-third 2% XE-60 for the beginning of the column and two-thirds 1% OV-17 for the remainder of the column (Barbier, 1972). The chromatogram shown in Fig. 1 was obtained by a temperature programming of 2°C/min from 80°C to 220°C. Hydroxyproline and C_{16} are two internal standards (IS). Arginine, histidine, and cystine are not recorded since they are destroyed on the column. The elution temperature of these amino acids is incompatible with the XE-60 phase, which is stable only until 220°C. These three amino acids (Fig. 2) are resolved on a second column of 1% OV-17 but on this phase leucine, isoleucine, tryptophan, and 3-CH_3His are not separated.

This two-column procedure could be used to study protein hydrolysates. First, to the mixed column is applied myosin and actin hydrolysates (Fig. 3a,b). Then, using an OV-17 column for a myosin hydrolyzed by 6 N HCl, we get good resolution, particularly of lysine. The MML and DML are not recorded since they occur below the limit of detection.

We thus determined the retention times of standard methylated amino acids encountered in myosin and actin (Fig. 5) and we took the total mass spectra of each of them. We obtained the bar spectrogram for lysine (Fig. 6). We choose the fragment m/e 320 because of its low abundance to balance

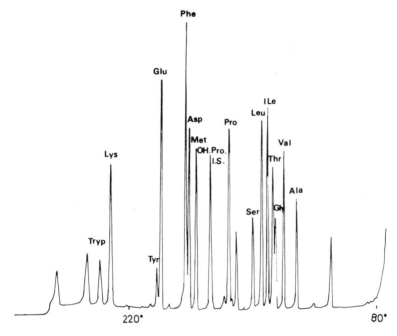

FIG. 1. Separation of acidic and neutral amino acids (10 μM of each) as N–TFA n butyl esters on a 4-m long column (φ 3 mm) composed of ⅓ 2% XE-60 and ⅔ 1% OV-17. Temperature programming: 2°C/min from 80 to 220°C. Internal standard (IS) hydroxyproline.

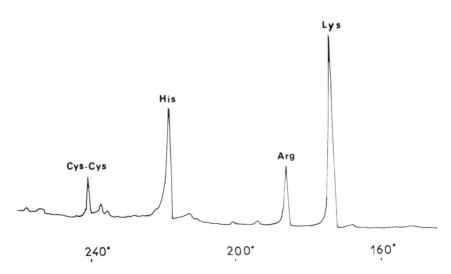

FIG. 2. Separation of histidine, arginine, lysine, and cystine (10 μM of each) on a 3-m long column (φ 3 mm) composed of 1% OV-17. Temperature programming: 2°C/min from 120 to 250°C.

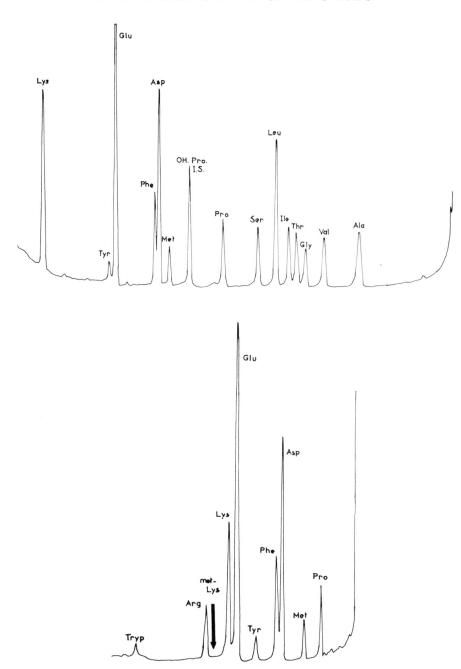

FIG. 3. Chromatogram obtained on GLC of rat myosin amino acids N–TFA, n butyl ester. (Top) 4,000 × 3 mm column of ¹/₃ XE-60 ²/₃ OV-17 1%. Temperature programmed at a rate of 2°C/min from 90 to 220°C. Internal standard (IS) hydroxyproline. (Bottom) 3,000 × 3 mm column of OV-17 1%. Temperature programmed at a rate of 2°C/min from 150 to 240°C.

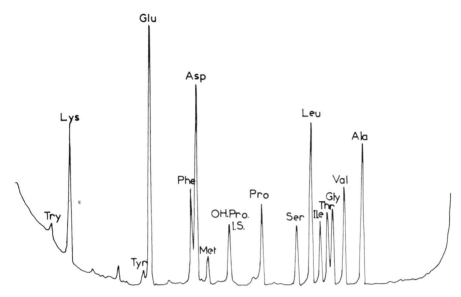

FIG. 4. Chromatogram obtained on GLC of rat actin amino acids N–TFA butyl ester on a 4,000 × 3 mm column of ⅓ XE-60 2% and ⅔ OV-17 1%. Temperature programmed at a rate of 2°C/min from 80 to 20°C. Internal standard hydroxyproline.

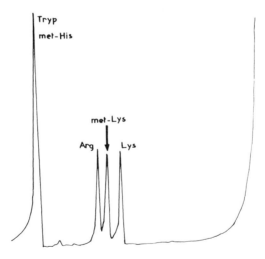

FIG. 5. Detection of standard methylated amino acids encountered in myosin and actin. 3,200 × 3 mm column of OV-17 1%. Temperature programmed at a rate of 2°C/min from 150 to 210°C. Met-His: 3-methylhistidine, met-Lys: ε-N-monomethyllysine.

the high amount of lysine occurring in the hydrolysate and because of its high value to avoid interferences with other substances having the same retention time and identical fragments.

The spectra of MML and DML and 3-CH₃His are shown in Figs. 7–9.

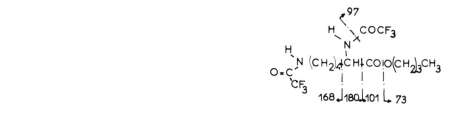

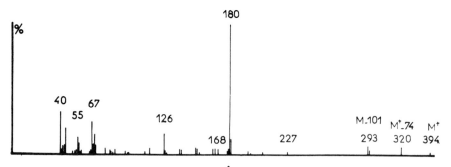

FIG. 6. Mass spectrogram of lysine. MW = 394, 70 eV.

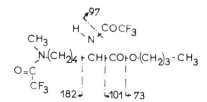

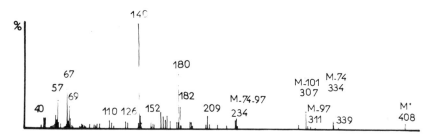

Fig. 7. Mass spectrogram of ε-N-monomethyllysine. MW = 408, 70 eV.

The identification of the MML in hydrolysates was achieved by using two specific fragments: 334 and 307. The ratio of these two fragments provides a third proof of the identity of the MML if it stays constant. For 3-CH₃His and DML, molecular ions are taken as specific fragments.

With the acceleration voltage alternator, three or four fragments can be detected simultaneously if their masses do not differ by more than 20%. Generally, we assay MML, lysine, and DML under isothermal condition (160°C) and lysine and 3-CH₃His at 200°C.

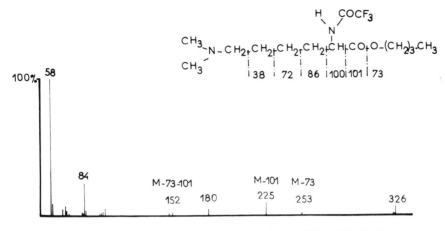

FIG. 8. Mass spectrogram of ε-N-dimethyllysine. MW = 326, 70 eV.

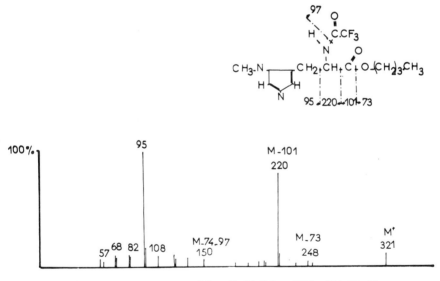

FIG. 9. Mass spectrogram of 3-methylhistidine. MW = 321, 70 eV.

Therefore, for each injection, we obtain two recordings, one for total ionization current with a sensitivity approximately identical to that of a flame ionization detector and the mass fragmentometric recording corresponding to the selected fragments. Figure 10 shows the detection of standard methylated amino acids according to this method. The detection of lysine, MML, and DML is linear and no adsorption is observed on the gas chromatographic column (Figs. 11 and 12). By contrast, adsorption of $3\text{-}CH_3His$ is considerable and below 0.46 nM it is impossible to detect this amino acid (Fig. 13). But this is a gas chromatographic problem. In the case of TML it is impossible to detect it by gas chromatography because of the

LYSINE 2nM

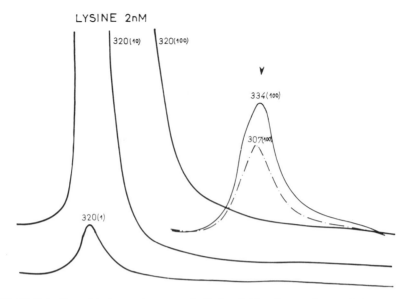

FIG. 10. Detection of lysine and monomethyllysine (0.02 μM) according to their respective fragment m/e 320, 307, and 334.

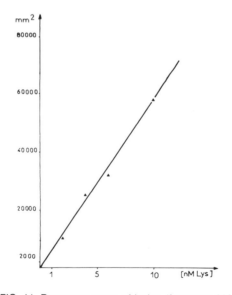

FIG. 11. Response curve of lysine (fragment 320).

presence of a quaternary ammonium moiety which prevents sublimation of the molecule and induces its breakdown in the injector.

Figure 14 displays the detection of MML in cardiac myosin. MML was not recorded using the total ionization current but lysine was measurable. On the other side, we see the recording of the two selected fragments of the

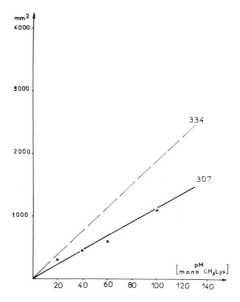

FIG. 12. Response curve of ε-monomethyllysine (fragment 334).

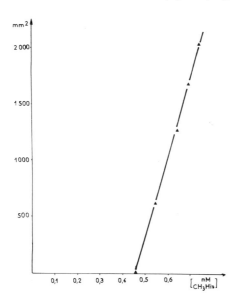

FIG. 13. Response curve of methylhistidine (fragment 321).

MML. As expected we found the normal amount of 3-CH₃His in the actin (Fig. 15) and the absence of MML in the same actin (Fig. 16). But 3-CH₃His was always found at a low level in our cardiac myosin preparations suggesting a contamination by actin. This has been confirmed by acrylamide gel electrophoresis.

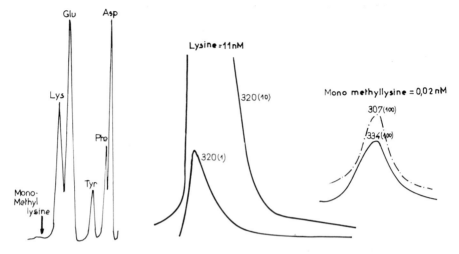

FIG. 14. Detection of MML in cardiac myosin in the adult rat. On the left, recording total ionization current on a 10-foot column, isothermic temperature 160°C. On the right, recording of the specific fragment of lysine (320) and MML (307 and 334).

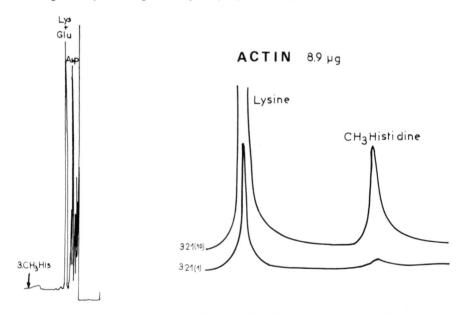

FIG. 15. Detection of methylhistidine in actin: specific fragment 321.

From myosin extracted from 500 mg of cardiac cells grown in Petri dishes, we obtained the results shown in Fig. 17.

This method provides us an adequate tool to study the methylation of myofibrillar protein *in vitro* which could be considered as a maturation process of the organ.

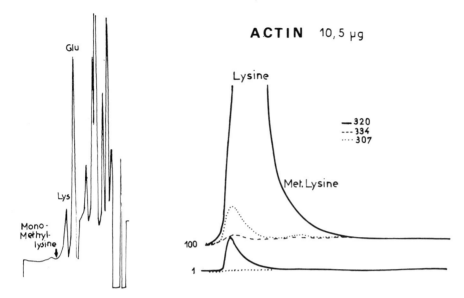

FIG. 16. Absence of MML in actin. No detection of the specific fragment: 320 for lysine, 334 and 307 for MML.

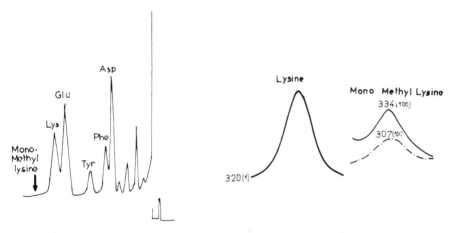

FIG. 17. Detection of MML in cardiac cells. Specific fragment: lysine = 320 and MML = 307 and 334.

ACKNOWLEDGMENTS

Financial help is gratefully acknowledged through research grants from: Institut National de la Recherche et de la Santé Médicale, Commission no. 6, Action Thématique Programmée no. 6, Action Thématique Programmée no. 2; Centre National de la Recherche Scientifique, ERA 267;

Enseignement Supérieur, Vème et VIème Plan; Fondation pour la Recherche Médicale Française.

REFERENCES

Barbier, F. (1972): Etude des méthodes de purification de la myosine cardiaque en vue d'un dosage par immunoprécipitation et par séparation quantitative des acides aminés méthylés par le couplage chromatographie gazeuse–fragmentométrie de masse. *Thèse de 3ème cycle,* Faculté des Sciences, Dijon.

Delange, R. J., Glazer, A. N., and Smith, E. L. (1969): Presence and location of an unusual amino acid N. trimethyllysine in cytochrome C of wheat germ and *neurospora. Journal of Biological Chemistry,* 244:1385.

Gehrke, C. W., Zumwalt, R. W., and Wall, L. L. (1968): Gas liquid chromatography of proteins amino acids. Separation factors. *Journal of Chromatography,* 37:398.

Gehrke, C. W., Roach, D., Zumwalt, R. W., Stalling, D. L., and Wall, L. L. (1968): Quantitative gas liquid chromatography of amino acids in proteins and biological substances. Macro-semimacro-micro methods. In: *Analytical Biochemistry Laboratories Incorporation,* Columbia, Missouri 65201, P.O. Box 1097.

Gehrke, C. W., Kuo, K., and Zumwalt, R. W. (1971): The complete gas liquid chromatographic separation of the twenty protein amino acids. *Journal of Chromatography,* 57:209.

Hardy, M. F., Harris, C. I., Perry, S. V., and Stone, D. (1970): Occurrence and formation of the N-ϵ-methyllysines in myosin and fibrillar proteins. *Biochemical Journal,* 120:653–660.

Johnson, P., Harris, C. I., and Perry, S. V. (1967): 3 CH_3His in actin and other muscle proteins. *Biochemical Journal,* 105:361.

Maume, B. F., Bournot, P., Lhuguenot, J.-C., Baron, C., Barbier, F., Maume, G., Prost, M., and Padieu, P. (1973): Mass fragmentographic analysis of steroids, catecholamines and amino acids in biological material. *Analytical Chemistry,* 45:1073.

Paik, W. K., and Kim, S. (1967): ϵ-N-dimethyllysine in histones. *Biochemical and Biophysical Research Communications,* 27:479.

Paik, W. K., and Kim, S. (1971): Protein methylation. *Science,* 174:114–118.

Terhorst, C., Moller, W., Laursen, R., Wittmann, B., and Liebold, D. (1973): The primary structure of an acidic protein from SDS ribosomes of E. coli which is involved in GTP hydrolysis dependent on elongation factors G and T. *European Journal of Biochemistry,* 34:138–152.

Mass Spectrometry in Biochemistry and Medicine,
edited by A. Frigerio and N. Castagnoli.
Raven Press, New York © 1974

The Use of Dimethylsilyl Ethers in the GC–MS Analysis of Steroids

D. H. Hunneman

Varian Mat, 28 Bremen 10, Postfach 4062, West Germany

I. INTRODUCTION

Gas chromatographic (GC) and gas chromatographic–mass spectrometric (GC–MS) analyses often require or, at least, are facilitated by derivitization of the compounds to be examined. This derivitization is undertaken to:
1. Improve GC behavior.
   a. Reduce tailing or adsorptive losses, e.g., methylation of fatty acids.
   b. Achieve better separation, e.g., benzyloxime derivitization of keto-steroids.
   c. Increase volatility to reduce or eliminate decomposition upon injection, e.g., silylation or acetylation of sugars.
2. Improve MS characteristics.
   a. Achieve a more readily interpretable spectrum.
   b. Obtain a derivative whose spectrum is not subject to thermal or catalytic decomposition.
An acceptable derivative must fulfill certain conditions:
1. Easy to make, require a minimum of manipulations.
2. Rapidly formed.
3. Formed in 100% yield.
4. Result in only one derivative.
5. No by-products.
6. Stable.
For the analysis of steroids, trimethylsilyl (TMS) ethers have become among the most widely used derivatives (Eneroth, Hellström, and Ryhage, 1964; Diekman and Djerassi, 1967; Pierce, 1968) since they fulfill the above requirements fairly well. The success with which TMS derivatives are employed in GC and GC–MS suggested that an examination of dimethylsilyl (DMS) ethers might be rewarding. Examination of the TMS and DMS ethers of hydroxy acids (Hunneman and Richter, 1972) and alcohols (Richter and Hunneman, 1973) showed that the mass spectral fragmentation of the DMS derivatives parallels very closely that of the TMS compounds. We present here a comparison of the fragmentation of the TMS and DMS ethers of selected steroids.

II. RESULTS AND DISCUSSION

A. Cholesterol

In Fig. 1 the mass spectra of the DMS and TMS ethers of cholesterol are shown. Many of the same or analogous fragments are evident; for example: M, M-15, M-90 (M-76 for the dimethylsilyl ether; the loss of trimethylsilanol or dimethylsilanol); M-90–15 (M-76–15); M-129 (M-115); m/e 129 $[(CH_3)_3SiO^+\!\!=\!\!CH\!-\!CH\!=\!CH_2]$ (m/e 115) $[(CH_3)_2HSiO^+\!\!=\!\!CH\!-\!CH\!=\!CH_2]$. However, the general impression of the TMS spectrum is much cleaner than that of the DMS compound. The nonspecific fragmentation in the lower range of the spectrum is relatively more intense in the DMS compound. Noteworthy is the absence of an M-1 peak in the DMS ether. Although one might expect this fragmentation, it does not occur, at least in the compounds here examined.

The molecular peak of the DMS is both relatively (60% compared to

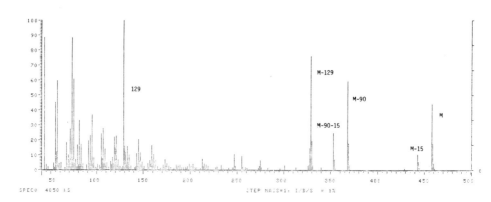

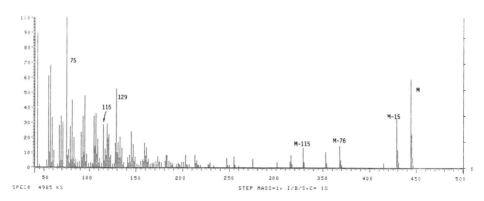

FIG. 1. Mass spectra of cholesterol trimethylsilyl ether (top) and cholesterol dimethylsilyl ether (bottom).

43%) and absolutely more intense than that (% Σ_{40} 3.4% compared to 2.6%) of the TMS ether. These facts suggest that one of the most important driving forces for the fragmentation of TMS and DMS ethers is release of steric hindrance which is, of course, not so great in the DMS, hence the more stable molecular ion, nor so well relieved by the loss of H˙ as by the loss of ˙CH_3, hence the absence of an M-1 peak in the DMS compounds.

The fragment m/e 129 in cholesterol TMS has been identified as $(CH_3)_3$-SiO=CH—CH=CH_2 (Diekman and Djerassi, 1967) and the presence of a significant m/e 129 in cholesterol DMS was totally unexpected and in disagreement with the postulated origin for this ion. However, in the DMS ether, m/e 129 was found, by peak matching on the emerging GC peak, to be $C_{10}H_9$ and not $C_6H_{13}OSi$. Work is continuing on the origin of this ion.

B. Epiandrosterone

The mass spectra of the TMS and DMS ethers of epiandrosterone (Fig. 2) reflect further many of the aspects mentioned with respect to cholesterol.

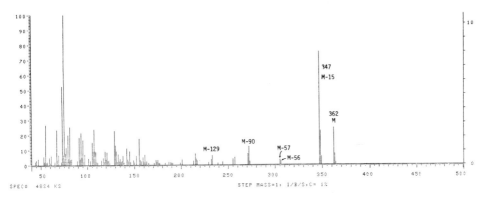

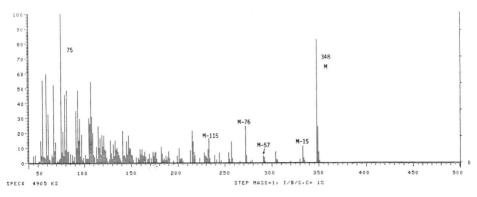

FIG. 2. Mass spectra of epiandrosterone trimethylsilyl ether (top) and epiandrosterone dimethylsilyl ether (bottom).

The molecular ion of the DMS compound is again both absolutely and relatively more intense than the molecular ion of the TMS ether. Analogous fragmentation is also fairly clear here.

C. Pregnanediol

Pregnanediol TMS (Fig. 3) gives a relatively unattractive spectrum for the purpose of identification in complex mixtures. Except for the usual fragments at m/e 73 and 75 and the dominating m/e 117 ion all other peaks are fairly small and may be easily lost in a complex background. The fragmentations observable are, however, all of a readily understandable origin. Particularly here, any increase in the molecular ion intensity would be most welcome for a derivative for analytical purposes. The DMS ether (Fig. 4) does indeed give a more intense molecular ion than the TMS, albeit still not of useful intensity. In the DMS ether an M-1 ion is also observed which probably does not come from the silyl group but rather from the tertiary carbon attached to the silyloxy group. In our previous studies (Hunneman

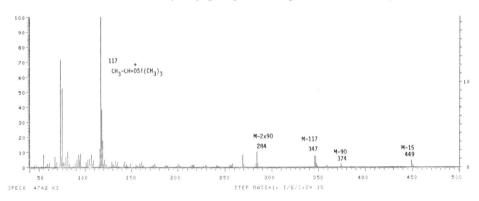

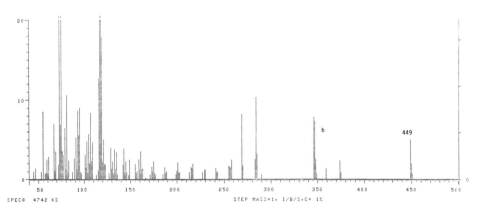

FIG. 3. Mass spectrum of pregnanediol trimethylsilyl ether. The top spectrum is the same as the bottom but with an expanded vertical scale.

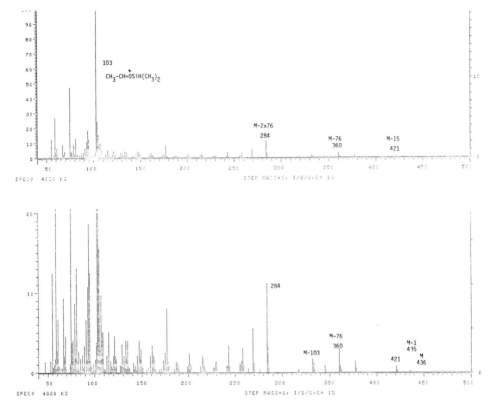

FIG. 4. Mass spectrum of pregnanediol dimethylsilyl ether. The top spectrum is the same as the bottom but with an expanded vertical scale.

and Richter, 1972; Richter and Hunneman, 1973) of the DMS ethers of simple molecules both the TMS and DMS ethers gave M-1 peaks when the silyloxy group was attached to a tertiary carbon, suggesting that the M-1 fragmentation does not usually come from the dimethylsilyl group. The slight increase in the intensity of the molecular ion of the pregnanediol DMS over the TMS does not fulfill the wish for a more useful derivative for the analysis of pregnanediol.

D. Estriol

Again the molecular peak is both relatively and absolutely more intense in the DMS than in the TMS derivative (Fig. 5). In the lower mass range the m/e 59 ion in the DMS is significantly smaller than the analogous m/e 73 in the TMS, reflecting the lowered driving force to form this ion in order to relieve steric hindrance. The peak at m/e 133 in the DMS is obviously analogous to the well-known m/e 147 of poly TMS compounds:

$$(CH_3)_2Si \overset{+}{=} O—Si(CH_3)_3 \qquad (CH_3)_2Si \overset{+}{=} O—SiH(CH_3)_2$$
$$m/e\ 147 \qquad\qquad\qquad m/e\ 133$$

E. Temperature Dependence of Molecular Ion Intensities

The dependence of the mass spectrum of cholesterol on inlet conditions was noted some years ago (Diekman and Djerassi, 1968) but in general most workers in GC–MS pay little attention to this aspect of mass spectrometry. In Fig. 6 the temperature dependence of the molecular ion intensities of the DMS and TMS derivatives of cholesterol and estriol is given. Only the ion source temperature is varied here; all other parameters remain constant. These results are the average of a number of spectra scanned cyclicly over the emerging GC peak. The spectra were corrected for changing total ion current, and the background subtracted. At all ion source temperatures the DMS ethers gave more intense molecular peaks than the

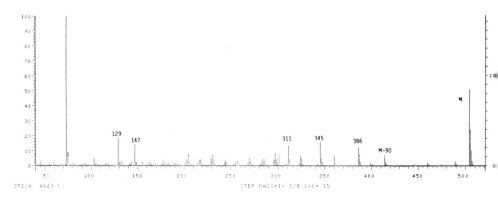

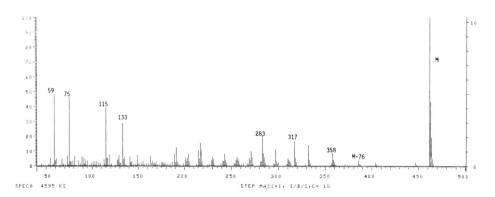

FIG. 5. Mass spectra of estriol trimethylsilyl ether (top) and estriol dimethylsilyl ether (bottom).

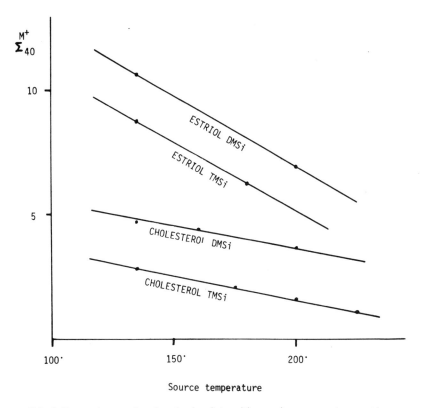

Source temperature

FIG. 6. Dependence of molecular ion intensities on ion source temperature.

corresponding TMS ether. Particularly for mass fragmentography where the molecular peak is used for quantitation this means that the DMS would be the preferred derivative since the molecular ion carries more of the total ion current and thus should show a better detection limit. Further, whichever derivative is used, the ion source should be kept as cool as possible.

III. CONCLUSION

DMS ethers of steroids give generally more intense molecular peaks than the corresponding TMS ethers. Fragmentation of both derivatives is fairly analogous although the lower mass range silyl fragments are usually much less intense in the DMS compounds. Conclusions reached in the literature concerning TMS ethers can generally be applied to the DMS derivatives as well.

Ion source temperature has an adverse effect on the molecular ion intensity of both the TMS and DMS ethers and an attempt should always be made to keep the ion source as cool as possible.

IV. EXPERIMENTAL

All spectra were determined on a Varian MAT CH 7 GC–MS using a two stage Biemann-Watson separator. Spectra were normalized, background subtracted and corrected for total ion current changes by means of a Varian MAT SS-100 data system and were plotted on a Varian Statos 21 recorder. Instrumental conditions were as follows: GC injection block: 240°C; column, 3% SE-30, 1.5 m × 2 mm i.d. glass, 230°C, separator: 230°C; ion source: 137°C (except in the measurements when the ion source temperature was varied); emission: 300 μA; electron energy: 70 eV, accelerating voltage: 3 kV.

TMS ethers were formed by adding 20 μl of *bis*(TMS) acetamide (Supelco, Inc., Bellefonte, Pa.) to a solution of 2 mg of the sterol in 1-ml ethyl acetate. The solution was then warmed at ~ 50°C for 1 hr. The DMS ethers were formed similarly with *bis*(DMS) acetamide (Supelco, Inc., Bellefonte, Pa.).

REFERENCES

Diekman, J., and Djerassi, C. (1967): Mass spectrometry in structural and stereochemical problems. CXXV. Mass spectrometry of some steroid trimethylsilyl ethers. *Journal of Organic Chemistry*, 32:1005–1012.

Eneroth, P., Helström, K., and Ryhage, R. (1964): Identification and quantification of neutral fecal steroids by gas liquid chromatography and mass spectrometry: Studies of human excretion during two dietary regimens. *Journal of Lipid Research*, 5:245–262.

Hunneman, D. H., and Richter, W. J. (1972): Migration of dimethylsilyl substituents upon electron impact: The fragmentation of methyl 12-dimethylsiloxyoctadecanoate. *Organic Mass Spectrometry*, 6:909–916.

Pierce, A. E. (1968): *Silylation of Organic Compounds*. Pierce Chemical Company, Rockford, Illinois.

Richter, W. J., and Hunneman, D. H. (1973): Use of dimethylsilyl ethers for characterizing primary aliphatic alcohols: A comparison of mass spectrometric fragmentation of di- and trimethylsilyl derivatives. *Submitted for publication*.

Mass Spectrometry in Biochemistry and Medicine,
edited by A. Frigerio and N. Castagnoli.
Raven Press, New York © 1974

Hormonal Steroids in Biological Materials: A Study by Mass Fragmentography

M. Prost and B. F. Maume

Centre de Biochimie de la Différenciation Cellulaire de l'Université de Dijon, Laboratoire d'Application en Chromatographie Gazeuse et en Spectrométrie de Masse, Faculté de Médecine de Dijon, 7 bd Jeanne d'Arc, 21033 Dijon, France

I. INTRODUCTION

A fascinating way to approach the study of steroidogenesis is to use isolated systems where cells keep their differentiated functions for a sufficient time to allow the study of biochemical events and their evolution under the action of effectors. It was demonstrated that normal cells from rat liver in culture and subculture (Padieu, Barbier, Chessebeuf, Cordier, Gerique, Lallemant, and Olsson, 1971*a;* Padieu, Lallemant, Barbier, and Chessebeuf, 1971*b*) keep their differentiated state especially in metabolizing steroid hormones (Bournot, Chessebeuf, Maume, Olsson, Maume, and Padieu, 1974). On the other hand, we have recently found in this laboratory that normal isolated adrenal cells from newborn rats can grow in primary culture for several weeks (Maume and Prost, 1974). This offers an excellent model system for the study of steroidogenesis in terms of mechanism and regulation in the postnatal life.

The enzyme activities of the steroid biosynthetic pathway can be studied through the quantitation of the steroid intermediates from cholesterol to corticosterone and/or aldosterone, with and without addition of steroidogenesis effectors to the culture medium. The need of a quantitative and specific analysis of this large range of steroids, prompted us to develop a *separative* method by gas-liquid chromatography (GLC) and gas-liquid chromatography–mass spectrometry (GLC–MS). But the limited size of our biological samples containing minute amounts of biosynthetized steroids obliged us to use the sensitive technique of mass fragmentography (Hammar, Holmstedt, and Ryhage, 1968).

In a pioneer work on the use of a mass spectrometer as a detector of the GLC effluents, Sweeley, Elliott, Fries, and Ryhage (1966) described the separation of two mixed steroids, epiandrosterone and dehydroepiandrosterone, from human plasma. Recently, Baillie, Brooks, and Middleditch (1972) have investigated the sensitivity of detection for several 11-deoxy-cortisol derivatives by monitoring single ions on the mass spectrometer.

They concluded that the O-methyloxime fully silylated (MO-perTMS) derivative (Thenot and Horning, 1972) is the most suitable for detection of this corticosteroid at the nanogram level. In our hands (Maume, Bournot, Lhuguenot, Baron, Barbier, Maume, Prost, and Padieu, 1973a; Maume, Maume, Prost, Bournot, Lhuguenot, Durand and Padieu, 1973b; Padieu, Barbier, Bègue, Bournot, Desgrès, Durand, Maume, Lhuguenot, Prost, and Maume, 1973) these MO-perTMS derivatives have also appeared to be the derivative of choice for adrenal corticosterone with respect to convenience of preparation, GLC separation, nanogram detection with a multiple-ion detector, and quantitation with deuterium-labeled corticosterone as a carrier and as an internal standard.

In this chapter we describe the use of mass fragmentography for the quantitative analysis of a large range of corticosteroids: 17-deoxycorticosteroids, 17-hydroxycorticosteroids, 18-hydroxycorticosteroids, and aldosterone.

Application is given for the quantitative evaluation of the evolution of endogenous corticosteroid amounts and ratios in the adrenal of the developing postnatal rat. Special attention will be focused on the neonatal rats from which cells of different organs are explanted for tissue culture.

II. MATERIALS AND METHODS

A. GLC and MS

A Packard model 7400 and Packard-Becker model 420 equipped with flame ionization detectors were employed. The columns were 4 m × 3 mm i.d. silanized glass tubes. The packings were 1% SE-30 Ultraphase, 1% OV-I, 1% OV-17, and 1% Dexsil-300 (Supelco Inc.) on 100 to 120 mesh acid washed and silanized Gas Chrom P (Applied Science Laboratories Inc.). All column packings were prepared according to Horning, Vanden-Heuvel, and Creech (1963). The separations were carried out at programming temperature from 180°C to 300°C at 1°C/min with nitrogen (1.2 bar). The injector block temperature was 250°C and the detector oven temperature was 300°C.

An LKB 9000 GLC–MS instrument with accelerating voltage alternator was employed. The silanized glass columns (4 m × 3 mm i.d.) were packed with 1% OV-I on 100 to 120 mesh Gas Chrom P in the usual way. The temperatures were flash heater 260°C, molecular separator 285°C, and ion source 310°C. Temperature-programmed separations were carried out at 1°C/min from 180 to 200°C with helium as the carrier gas (30 ml/min). The accelerating voltage was 3.5 kV, trap current 60 μA, and electron energy 70 eV. An accelerating voltage alternator allowed detection of three fragment ions up to 20% higher in mass than the lower mass. In this case, a 20 or 28 eV electron energy was used. A new multiple ion detector-peak matching device (Hammar and Hessling, 1971) has allowed simultaneous recordings at four different masses within a mass range of 25%.

B. Sample Preparations

The adrenals were removed from Wistar/US/Commentry (Dr. Causeret, Institut National de la Recherche Agronomique, Dijon) and Sprague-Dawley, OFA rats after decapitation. Adrenals (20 to 50 newborn glands and 16 adult glands) from male and female rats were pooled separately. The rats were 1, 4, 7, 15, and 45 days old. Each pool of adrenal glands was homogenized in methanol-water (70:30). Then 10,000 dpm 4-^{14}C-corticosterone (specific activity 50 mC/m M) was added in order to evaluate the recovery of corticosterone after extraction. After centrifugation the supernatant was kept at −18°C for 24 hr to precipitate the lipids; the methanol-water phase was removed from the lipid precipitate by centrifugation at −18°C. The methanol was evaporated under a nitrogen stream at 60°C and steroids were extracted from the aqueous solution three times with ethyl acetate, then three times with methylene chloride. After the evaporation of solvents, the steroid extract was submitted to derivative formation. In separate experiments when total mass spectra were needed and for the analysis of aldosterone, a fractionation by thin-layer chromatography (TLC) was carried out before derivative formation (a first run with isopropyl ether; then a second run with isopropyl ether-acetone, 60:40 as eluents). O-Methyloximes were formed by using O-methyloxyamine hydrochloride (Pierce Chemical Co.) in pyridine (16 mg/ml) at 65°C for 3 hr (or at room temperature for 2 hr for aldosterone analysis); then reaction with BSA + TMCS (10:1) (Pierce Chemical Co.) was performed at 60°C for 16 hr in order to obtain a complete silylation of all hydroxyl groups present in the corticosteroids. About $^{1}/_{10}$ of the initial sample was introduced in the GLC–MS system for each mass fragmentographic analysis. MO–*d9*–TMS derivatives of reference steroids (Ikapharm, Israel and Searle, Mexico) were prepared in the same way but with *d18*–BSA and in some instances *d9*–TMCS (Merck Sharpe and Dohme, Montreal, Canada).

III. RESULTS AND DISCUSSION

The GC methods previously used for quantitative analysis of the corticosteroids in biological materials have essentially employed a preliminary oxidation of the side chain. For instance, methods have been developed by Kittinger (1964) for analysis of microgram quantities of corticosteroids produced by rat adrenal glands *in vitro,* by Palem, Lapière, Coninx, and Margoulies (1970) for 18-hydroxy-11-deoxycorticosterone in adrenal incubates, and by Fabre, Fenimore, Farmer, Davis, and Farell (1969) for aldosterone in blood. The MO–TMS derivatives allow one to retain the parent corticosteroid structure in the course of the GLC analysis. This leads to the separation of hormonal steroids closely related in their chemical and stereochemical structure. Furthermore, mass fragmentography enables us to quantitate nanogram amounts of these steroids in adrenal tissue or blood.

A. 17-Deoxycorticosteroids

The following steroid intermediates in the biosynthetic pathway of corticosterone have been separated by GLC as their MO–TMS derivatives on 1% SE-30 with temperature programming: pregnenolone, progesterone, 11-deoxycorticosterone, corticosterone, and 11-dehydrocorticosterone. In the steroid extract of adrenal glands, the presence of large amount of cholesterol does not affect the GLC resolution of the steroid peaks as is shown in Fig. 1.

The fragmentation pattern of 17-deoxycorticosteroid MO–TMS under electron impact includes well-known losses of $O-CH_3$ from O-methyloxime

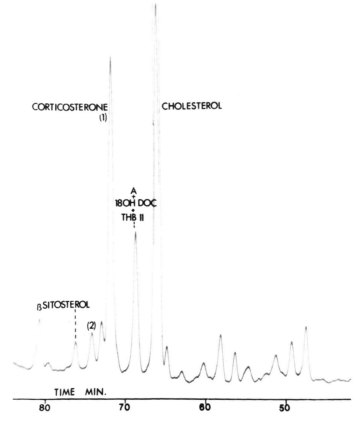

FIG. 1. Gas chromatographic separation of a total extract of adrenal corticosteroids from the adult male rat: cholesterol, 11-dehydrocorticosterone (A), 18-hydroxy-11-deoxycorticosterone (18-OH DOC), $3\beta,11\beta,21$-trihydroxy-5α-pregnan-3,20-dione (THB II), corticosterone, and β-sitosterol. The mixed steroids A, 18-OH DOC, and THB II are separated and assayed by mass fragmentography. The β-sitosterol has a dietary origin. The analysis was performed with a combined GLC–MS instrument: detection by the total ionization current, temperature programming at 1°C/min from 180°C, 1% SE-30 column.

groups and of silanol from TMS groups. Other characteristic cleavages are shown on the following corticosterone formula as an example:

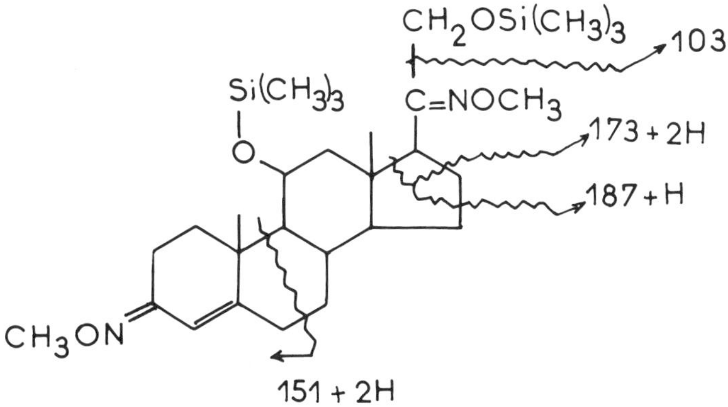

The total mass spectra of corticosterone as MO–TMS and MO–*d18*–TMS derivatives are shown in Fig. 2. The high values of the ion intensities in the high-mass range (m/e = 548 or 517) allow sensitive and specific detection of corticosterone at the nanogram level in blood and adrenals. The MO–*d18*–TMS derivative of corticosterone added to the biological sample before the analysis in the 100- to 200-ng range for each analysis plays the role of carrier in the GLC–MS system. Additionally, it can be used as an internal standard by detection of its m/e = 535 ion, which is of negligible intensity in the unlabeled corticosterone (Maume et al., 1973*a* and also Fig. 3). The m/e = 517 ion detected simultaneously for unlabeled corticosterone is not present in the mass spectrum of the deuterated standard. These ions and some others in the high-mass range have been used for detection in the adrenal glands of endogenous corticosterone, 11-deoxycorticosterone m/e = 460 (M), with a relative intensity of 51%, progesterone m/e = 372 (M), 100%; m/e = 341 (M-31), 52% and pregnenolone m/e = 417 (M), 9%; m/e = 402 (M-15), 19%; m/e = 386 (M-31), 24%. MO-deuterated TMS steroids can be used in each case as internal standard and carrier except for progesterone for which a deuterated MO could be employed for this purpose.

B. 17α-Hydroxycorticosteroids

It is generally believed that there is a very limited 17α-hydroxylase activity in rat adrenals; but Kalavsky (1971) has recently detected by TLC and protein-binding 11-deoxycortisol and cortisol in fetal adrenal glands. In man and in other animal species including cattle, sheep, dogs, hamsters, and guinea pigs, cortisol is the main secreted adrenal steroid.

On the other hand 17α-hydroxycorticosteroids must also be present in the culture media since, at the present time, it is necessary for a normal

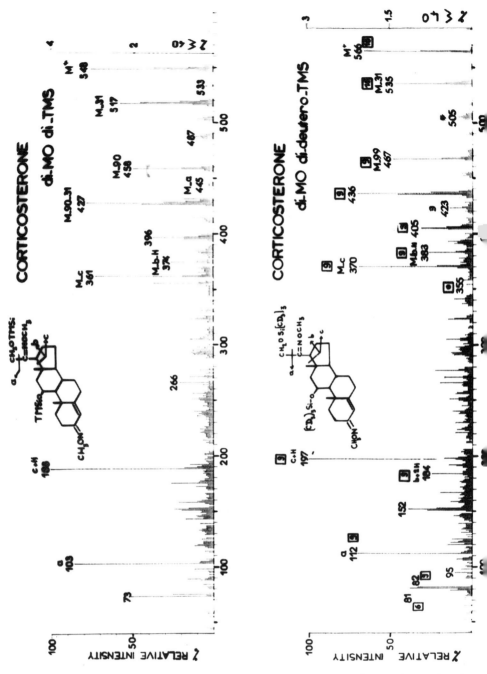

FIG. 2. Mass spectra of corticosterone as 3,20-di-MO-11β,21-di-TMS derivative (top) and of corticosterone as 3,20-di-MO-11β,21-di-perdeutero-TMS derivative (bottom). The mass shifts of 18, 9, or 0 correspond to 2, 1, or 0 TMS groups respectively in the considered fragments. The fragment ions at m/e 517 (protium derivative) and 535 (deuterium derivative) are used for the mass fragmentographic assay. Bombarding electron energy is 70 eV.

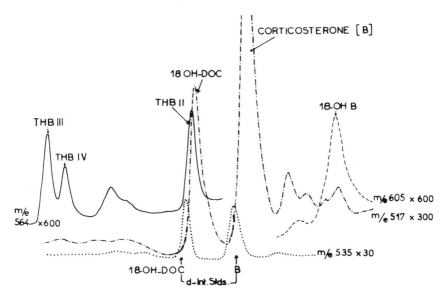

FIG. 3. Mass fragmentogram of total steroid extract from 4-day-old male rats. Corticosterone (B) and 18-hydroxy-11-deoxycorticosterone (18-OH–DOC) are detected by the m/e 517 ion. Each reference hormone (200 ng) as their MO-perdeuterated-TMS derivatives are injected with the sample. They are detected by the m/e 535 fragment ion and used as internal standards. Three THB isomers are detected by the fragment ion at m/e 564: 3α,11β,21-trihydroxy-5β-pregnane-3,20-dione (THB III), 3β,11β,21-trihydroxy-5β-pregnane-3,20-dione (THB IV), and 3β,11β,21-trihydroxy-5α-pregnane-3,20-dione (THB II). The 18-hydroxycorticosterone (18-OH B) is recorded by the m/e 605 fragment ion.

growth of cells in culture to add human serum, fetal calf serum, and/or horse serum to the synthetic medium.

For these reasons we have investigated the GLC–MS properties of 17α-hydroxypregnenolone, 17α-hydroxyprogesterone, 11-deoxycortisol, cortisol, and cortisone as MO–TMS derivatives. These steroid derivatives are well separated on a 4-m 1% OV-I column. Characteristic mass spectra have been obtained by GLC–MS.

11-Deoxycortisol can be assayed by mass fragmentography with the same fragment ions as those of corticosterone: m/e 517 (M-31) and 548 = (M) but the 548/517 intensity ratio is 0.2 instead of 1 for corticosterone at 28 eV. On the other hand the same ions are used for cortisol and 18-hydroxycorticosterone: for instance m/e = 636 (M), 29.4% and m/e = 605 (M-31), 100%.

In the rat adrenal glands very small amounts of cortisol and 11-deoxycortisol have been detected by this method only in the neonatal period; the largest amount has been found in the 1-day-old rat.

C. 18-Hydroxycorticosteroids

The 18-hydroxy-11-deoxycorticosterone is the second most prominent steroid secreted by the rat adrenals; the 18-hydroxycorticosterone is also synthetized by this animal but not in such large amount as 18-hydroxy-11-deoxycorticosterone. Furthermore the former steroid is the precursor of aldosterone in the human adrenal and adrenal tumor (Pasqualini, 1964).

We have developed a GLC and GLC–MS method of analysis for these 18-hydroxycorticosteroids plus 18-hydroxyprogesterone as MO–TMS derivatives (Prost and Maume, 1973). The formation of these derivatives leads to 20-MO-18-TMS form and allows good separation from other corticosteroids except between 18-hydroxy-11-deoxycorticosterone and 11-dehydrocorticosterone (see Fig. 1). But the mass fragmentographic method, by the use of m/e 517 and 548 ions, provides a specific assay of 18-hydroxy-11-deoxycorticosterone since the molecular ion of 11-dehydrocorticosterone has a lower value (M = 474). The m/e 517 ion intensity of 18-hydroxy-11-deoxycorticosterone has a high value but the intensity ratio of the ions 548 and 517 is 0.2 instead of 1 for corticosterone. An example of the quantitative determination of the 18-hydroxycorticosteroids is given in the paragraph concerning newborn rat steroids (see Fig. 3).

D. Aldosterone

It was shown in a previous paper that aldosterone can be converted to suitable derivatives which keep the parent aldosterone structure for gas phase analysis (Horning and Maume, 1969). The 3,20-di-MO-18,21-di-TMS derivatives led to two reproducible and characteristic gas chromatographic peaks of equal size, corresponding very likely to the $18\alpha/\beta$-trimethylsilyloxy isomers. These aldosterone derivatives are detected at the microgram level with the flame ionization detector; but as aldosterone levels in the human or animals are usually low, a more sensitive detection is needed. Our studies on mass fragmentography gave us the opportunity to reinvestigate this problem. With multiple-ion detection it is possible to quantitate down to 1 ng of aldosterone-MO–TMS. The chosen fragment ion at m/e 459 (M-103) is by far the most intense fragment ion in the aldosterone MO–TMS spectrum ($\Sigma_{40}\% = 23$); this is favorable for a sensitive measurement. The isotopic ions at m/e 460 and 461 are simultaneously recorded and the isotopic abundances are measured in order to increase the specificity of the assay.

Retention times of aldosterone peaks are very close to those of 11-dehydrocorticosterone and of corticosterone respectively. A satisfactory separation is obtained on a 20-m glass capillary column.

But on a 1% SE-30 packed column, peaks of aldosterone I (MU value = 31.73) and 11-dehydrocorticosterone (MU = 31.67) and of aldosterone II (MU = 32.14) and corticosterone (MU = 32.27) interfere. When mass fragmentography of aldosterone is performed with such a packed column, it is necessary to submit the biological sample to a TLC separation before the gas phase analysis. Figure 4 shows the mass fragmentogram of adrenal aldosterone. The injected sample corresponds to aldosterone extracted

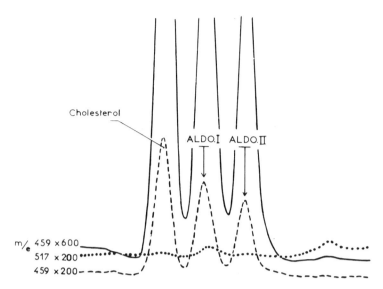

FIG. 4. Mass fragmentogram of a TLC fraction containing aldosterone from adrenals of adult female rats. A small amount of cholesterol-TMS is a contaminant from the TLC fraction. The m/e 459 fragment ion corresponds to M-103 from aldosterone and to M+1 from cholesterol. Only traces of 18-hydroxydeoxycorticosterone are detected by the m/e 517 fragment (slightly to the right of aldosterone I peak) and no corticosterone. Aldo I and Aldo II are the two $18\alpha/\beta$ isomers of the di-MO–di-TMS derivative (hemiacetal form). SE-30 column, temperature programmed from 200°C at 1°C/min. The injected sample is equivalent to three rat adrenal glands.

from three adrenal glands. Only one TLC separation is necessary to eliminate corticosterone interference since no response is found for the m/e 517 ion in the sample.

This proves that aldosterone can be assayed by this method with a high level of specificity and sensitivity. With regard to the corticosterone assay in the rat with m/e 517 and 548 fragment ions, the contribution of aldosterone to the corticosterone peak is lower than one per thousand because of the very low intensity of these ions in the aldosterone spectrum.

E. Sterols and Other Steroids of the Rat Adrenal

In addition to the already described steroids of the adrenal, the other components of the adrenal steroid extract have been studied by TLC fractionation and GLC–MS. The sterol TLC fraction submitted to GLC separation shows more than 10 peaks besides the cholesterol. Sterols such as β-sitosterol, campesterol, stigmasterol, 5β-cholestan-3β-ol and 3β-hydroxy-5-cholesten-7-one have been identified by GLC–MS (Prost, Maume, and Padieu, *to be published*). The three phytosterols identified in the rat foods are concentrated in relatively large amounts in the adrenals. Their biological effects on this organ are unknown.

Additionally, we have identified by GLC–MS the tetrahydrogenated derivatives of corticosterone (THB) in the total extract of the rat adrenal and in TLC fractions: the $3\beta,11\beta,21$-trihydroxy-5α-pregnan-3,20-dione was identified by its total mass spectrum and by mass fragmentography (ions at m/e 595 = M, 564 = M-31, and 505 = M-90). Furthermore, the mass fragmentographic data and the retention data on a glass capillary column are consistent with the presence of the two 5β isomers of THB.

The mass fragmentographic method enables us to quantitate specifically the THB isomers from the total adrenal steroid extract. It shows that the nature of the isomers and their relative amounts change considerably with age (from 1 day old to adulthood), sex, and strain of rat. However, the $3\beta,5\alpha$ isomer is always present.

The dramatic effect of sex hormones on the level of the liver steroid reductase in the rat is discussed elsewhere (Bournot, Maume and Padieu, 1974). The action of the sex hormones on the adrenal appears to be also at the level of reductases as is shown by our results on the newborn rat and by the results of Colby and Kitay (1972) on the gonadectomized adult rat.

F. Neonatal Corticosteroids

Figure 3 shows the separation and the quantitation of some steroids from the adrenals of a 4-day-old male rat (Sprague-Dawley strain). The injected sample corresponds to 10 mg of fresh tissue (about six adrenals). On the mass fragmentogram six steroids can be detected: corticosterone and 18-hydroxy-11-deoxycorticosterone (fragmentation at m/e 517), three of the four isomers of tetrahydrocorticosterone (m/e 564), and 18-hydroxy-corticosterone (m/e 605). Two hundred ng amounts of deuterated corticosterone and of 18-hydroxy-11-deoxycorticosterone injected with the sample are detected at m/e = 535 and are used as internal standards for the quantitation.

The amounts of corticosterone and of 18-hydroxy-11-deoxycorticosterone in the rat are reported in Fig. 5 for ages from 1 to 15 days. The levels of these two hormones are relatively high at birth in the female and on the 4th day in the male. This is followed by a decrease at day 4 and 7 respectively. In the 15-day-old animal, the amount goes up around the adult level: 30 ng and 45 ng/mg of fresh tissue for the male and the female respectively. As is shown in Fig. 5 the formation of the THB isomers is strikingly enhanced at day 4. At day 1 and 15 the THB level is low in absolute amount and in percent of the total secreted corticosterone.

As known, the neonatal testicle secretes testosterone at birth, then reaches a latency period until puberty. The low level of THB at birth and in adulthood suggests that its high level on day 4 could result from an inhibitory effect of testosterone on the reductase which is released around day 4.

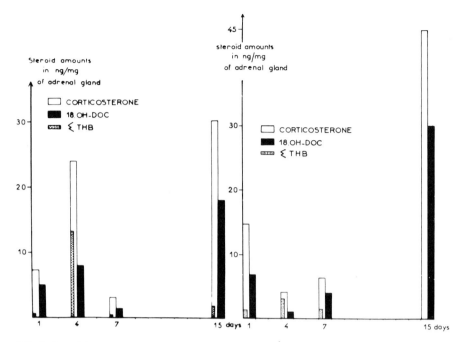

FIG. 5. Steroid amounts in ng/mg of adrenal gland against the age of the male (left) and female (right) rat. Corticosterone, 18-hydroxy-11-deoxycorticosterone (18-OH DOC), and tetrahydrocorticosterone isomers (ΣTHB) have been assayed by mass fragmentography.

The increase of THB isomers on day 4 in the female after birth could be similarly explained by the effect of the estrogen inhibition on the reductases. The finding that different isomers of THB are present in different amounts according to the age is not yet explained.

The fact that adrenal cells can be cultured more easily when the donor rat is 7 or 8 days old could be related to the relative lack of hormonal corticosteroids at this period.

ACKNOWLEDGMENTS

Financial help is gratefully acknowledged through research grants from: Délégation Générale à la Recherche Scientifique et Technique, Action Complémentaire Coordonnée, Développement Périnatal; Centre National de la Recherche Scientifique, ERA 267; Enseignement Supérieur, Vème et VIème Plan; Fondation pour la Recherche Médicale Française.

REFERENCES

Baillie, T. A., Brooks, C. J. W., and Middleditch, B. S. (1972): Comparison of corticosteroid derivatives by gas chromatography mass spectrometry. *Analytical Chemistry*, 44:30–37.

Bournot, P., Chessebeuf, M., Maume, G., Olsson, A., Maume, B. F., and Padieu P. (1974): Application of mass fragmentography to the study of sex-linked metabolism of testosterone and corticosterone in liver organ and in liver cells in culture. *This volume.*

Bournot, P., Maume, B. F., and Padieu, P. (1974): *Biomedical Mass Spectrometry (in press).*

Colby, M. D., and Kitay, J. I. (1972): Effects of gonadal hormones on adrenocortical secretion of 5α-reduced metabolites of corticosterone in the rat. *Endocrinology*, 91:1523–1527.

Fabre, L. F., Fenimore, D. C., Farmer, R. W., Davis, H. W., and Farell, G. (1969): Determination of aldosterone and tetrahydroaldosterone in blood by electron capture gas chromatography. In: *Advances in Chromatography*, edited by A. Zlatkis. Preston Technical Abstracts Company, Evanston, Illinois.

Hammar, C. G., and Hessling, R. (1971): Novel peak matching technique by means of a new and combined multiple ion detector-peak matcher device. Elemental analyses of compounds in submicrogram quantities without prior isolation. *Analytical Chemistry*, 43:298–306.

Hammar, C. G., Holmstedt, B., and Ryhage, R. (1968): Mass fragmentography identification of chlorpromazine and its metabolites in human blood by a new method. *Analytical Biochemistry*, 25:532–548.

Horning, E. C., and Maume, B. F. (1969): Derivatives of aldosterone for gas phase analysis. *Journal of Chromatographic Science*, 7:411–418.

Horning, E. C., VandenHeuvel, W. J. A., and Creech, B. G. (1963): Separation and determination of steroids by gas chromatography. In: *Methods of Biochemical Analysis*, Vol. 11, edited by D. Glick, pp. 69–147. Interscience, New York.

Kalavsky, S. M. (1971): Fetal rat adrenal steroidogenesis. *Biological Neonate*, 17:427–435.

Kittinger, G. W. (1964): Quantitative gas chromatography of 17-deoxycorticosteroids and other steroids produced by the rat adrenal gland. *Steroids*, 3:21–42.

Maume, B. F., Bournot, P., Lhuguenot, J. C., Baron, C., Barbier, F., Maume, G., Prost, M., and Padieu, P. (1973a): Mass fragmentographic analysis of steroids, catecholamines and amino acids in biological materials. *Analytical Chemistry*, 45:1073–1082.

Maume, B. F., Maume, G., Prost, M., Bournot, P., Lhuguenot, J. C., Durand, J., and Padieu, P. (1973b): New developments in steroid and catecholamine analysis in biological media by gas chromatography–mass spectrometry using a multiple ion detector. In: *Organisation des Laboratoires, Biologie Prospective*, IIᵉ Colloque de Pont à Mousson, pp. 637–654. Expansion Scientifique Française, Paris.

Maume, B. F., and Prost, M. (1974): *Comptes Rendus de la Societe de Biologie (in press).*

Padieu, P., Barbier, F., Chessebeuf, M., Cordier, D., Gerique, M., Lallemant, C., and Olsson, A. (1971a): Relations between cell differentiation and organogenesis in single cell culture from newborn rat organ. In: *Proceedings of the "1ère Conférence Internationale sur la différenciation cellulaire."* Nice, France.

Padieu, P., Lallemant, C., Barbier, F., and Chessebeuf, M. (1971b): Conservation of differentiated characteristics in eukaryotes in single cell culture. In: *Proceedings of the "7ème Congrès de la Fédération Européenne des Sociétés de Biochimie."* Varna, Bulgaria.

Padieu, P., Barbier, F., Bègue, R. J., Bournot, P., Desgrès, J., Durand, J., Maume, G. M., Lhuguenot, J. C., Prost, M., and Maume, B. F. (1973): Automation in clinical chemistry. An evaluation of the present and of the future development of automated separation methods. In: *Organisation des Laboratoires, Biologie Prospective*, IIᵉ Colloque de Pont à Mousson, pp. 741–754. Expansion Scientifique Française, Paris.

Palem, M., Lapière, C. L., Coninx, P., and Margoulies, M. (1970): A sensitive assay of 18-hydroxydesoxycorticosterone and aldosterone by gas chromatography with electron capture detection. *Revue Européenne d'Etudes Cliniques et Biologiques*, 15:851–856.

Pasqualini, J. R. (1964): Conversion of tritiated-18-hydroxycorticosterone to aldosterone by slices of human corticoadrenal gland and adrenal tumour. *Nature*, 201:501.

Prost, M., and Maume, B. F. (1973): Hormones stéroides de la surrénale de rat. Analyse des 18-hydroxycorticostéroides par chromatographie gaz-liquide couplée à la spectrométrie de masse et par fragmentographie de masse. *Journal of Steroid Biochemistry (accepted for publication).*

Sweeley, C. C., Elliott, W. H., Fries, I., and Ryhage, R. (1966): Mass spectrometric determination of unresolved components in gas chromatographic effluents. *Analytical Chemistry*, 38:1549–1553.

Thenot, J.-P., and Horning, E. C. (1972): MO–TMS-derivatives of human urinary steroids for GC and GC–MS studies. *Analytical Letters*, 5:21–33.

Mass Spectrometry in Biochemistry and Medicine,
edited by A. Frigerio and N. Castagnoli.
Raven Press, New York © 1974

Application of Mass Fragmentography to the Study of Sex-Linked Metabolism of Testosterone and Corticosterone in Liver Organ and in Liver Cells in Culture

P. Bournot, M. Chessebeuf, G. Maume, A. Olsson,
B. F. Maume, and P. Padieu

Centre de Biochimie de la Différenciation Cellulaire, Laboratoire de Culture Cellulaire et Laboratoire d'Application en Chromatographie Gazeuse et en Spectrométrie de Masse, Faculté de Médecine, 7 bd Jeanne d'Arc, 21033 Dijon, France

I. INTRODUCTION

The sexual dependence of steroid metabolism by the rat liver has been shown by several groups (Forchielli, Brown-Grant, and Dorfman, 1958; De Moor and Denef, 1968; Staib, Sonnenschein, and Staib, 1970; Eriksson and Gustafsson, 1971; Schriefers, Ghraf, Hoff, and Ockenfels, 1971; Eriksson, 1971). Denef and De Moor (1972) found that the sexual maturity of the liver occurs around the 30th day of life. The prepubescent steroid pattern arising from liver metabolism in the male is the same as that of a female.

Steroid metabolism studied *in vitro* with liver homogenates and liver slices is dependent on the physiological state (hormonal alterations, diet, and type of strain) of the animal at the time of tissue sampling.

Rat liver cells explanted from the organ and grown in culture under defined conditions as it is done in this laboratory (Padieu, Barbier, Chessebeuf, Cordier, Gerique, Lallemant, and Olsson, 1971*a*; Padieu, Lallemant, Barbier, and Chessebeuf, 1971*b*) provide a very interesting biological model for the study of this liver sexual dimorphism. But the small amount of tissue in a culture dish led us to combine tissue culture with gas phase analysis through gas-liquid chromatography (GLC) and mass spectrometry (MS). Both new methods of tissue culture and MS have been developed in close connection to do quantitative biology on eukaryote cells in a biological system working *in vitro*. For this reason, gas chromatography (GC) combined with mass fragmentography (GC–MF) has been found as a method of choice because it allows one to detect and to quantitate a few nanograms of steroids (Maume, Bournot, Durand, Lhuguenot, Maume, Prost, and Padieu, 1973*a;* Maume, Bournot, Lhuguenot, Baron, Barbier, Maume, Prost, and Padieu, 1973*b;* Padieu, Barbier, Bègue, Bournot, Desgrès, Durand, Maume,

Lhuguenot, Prost, and Maume 1973). We have studied the endocellular steroid metabolites of the liver in the first days of life and during adulthood for male and female Wistar rats. In this chapter we will report for the first time that liver cells in culture explanted from 8-day-old and 18-day-old rats retain their capacity to metabolize corticosterone and testosterone.

II. MATERIALS AND METHODS

A. Identification Procedure

The steroids were analyzed by GC–MS and by GC–MF using SE-30 as the stationary phase (Maume et al., 1973b). Fully silylated trimethyl ethers (TMS) and O-methyloxime (MO) derivatives were prepared.

B. Preparation of Steroids from Liver Organ

The animals were male and female rats of Wistar/US/Commentry inbred strain (Dr. J. Causeret, Station de Recherches sur l'Alimentation de l'Homme, Institut National de la Recherche Agronomique, Dijon). They were sacrificed by decapitation, and the livers were rapidly removed and homogenized in chloroform-methanol (1:1, v/v). After centrifugation, the extract was evaporated to dryness, and the residue was treated with methanol-water (70:30, v/v) at $-20°C$ for 20 hr to remove lipids. Then, the steroid conjugates were hydrolyzed by a combination of enzymatic and solvolytic methods. The free and the liberated steroids were extracted by dichloromethane and ethyl acetate. The steroid extract was analyzed by GC–MF on a 2/10 aliquot while the remaining material was purified by preparative thin-layer chromatography (TLC) and then run in GC–MS on the isolated TLC fractions. Because tetrahydrocortisone ($3\alpha,17\alpha,21$-trihydroxy-5β-pregnan-3,20-dione or THE) is not present in the normal rat, we used this steroid as an internal standard, adding it to the steroid extract in the amount of 1 μg for each 5 g of fresh liver.

C. Liver Cell Culture

The same strain of rat was used for the cultures which were prepared from 8- and 18-day-old postnatal rats. One or two litters were used to make a cell explantation. The overall procedure is the one generally in use in tissue culture methodology. The livers are removed rapidly and immediately minced with scissors in the same medium as used for trypsinization of the tissue. The mince was transferred to a sterile flask with a magnetic spinning bar. A solution of 0.1% trypsin (B grade from Calbiochem) (v/v) in Ham F10 medium without calcium and magnesium prepared extemporaneously is added to make three times the volume of minced liver. The enzymatic disintegration is carried out at 36 to 37°C for 10 min at 200 rpm. The supernatant is then transferred to a sterile tube containing the same volume of culture medium in order to stop the trypsin action. Cells are centrifuged in a clinical centrifuge at 600 rpm for 10 min at room temperature. They are suspended in 5 ml of culture medium and seeded in a 60-mm diameter plastic dish (Falcon). Trypsinizations are repeated 10 to 12 times. The culture medium is changed every 2 days. It is made from Ham F10 powder medium supplied by Gibco (Grand Island Biological Company, Rhode Island) to which 10% of fetal calf serum is added (Flow Laboratories, Irvine, Scotland) and 10% human pooled serum groups A and O from the local Blood Bank[a].

We found that the most important parameters for a successful explantation of isolated cells were the highest purity of water, the freshness of the powder and of the solution medium, and the protection of the cells from all the toxic vapors produced in the laboratory. The primary explantation was done in a sterile room under positive pressure. Trypsinization and culture media were routinely tested for toxicity on the postnatal beating heart cell cultures done on the same rats according to the overall procedure of Harary and Farley (1963).

When the best conditions were realized we found that from the 6th to the 8th cell pellets we obtained a high percentage of epithelial cells of two types. Most of the cells had a dark cytoplasm with many granulations, especially around the nucleus, and later they were more refringent and clearer with a sharp membrane and one or two dark nucleoli. Other cells had a clear cytoplasm with a less polygonal shape and a larger size. Some cells had two nuclei. Electromicographies have been done that show a highly organized structure (Périssel, Chessebeuf, Padieu, and Malet, 1973). When these two populations of cells are in a low-toxicity medium, they do not undergo much cytolysis and their doubling time is approximately 18 hr. The lag phase after seeding is short whereas fibroblasts start to divide actively after the 4th day. Therefore, the epithelial cells are able to outgrow the fibroblasts. Such a primary culture with a high yield of epithelial cells reaches confluency after 4 to 6 days of culture. The cells are then trypsinized the same way in the dish and transferred to another dish. The second subculture is split into two dishes and progressively the splitting ratio is increased to reach the ratio of 32 to 40 dishes from 1. We have observed that the remaining fibroblasts are eliminated by the epithelial cells at each passage. We have established three lines from unsexed postnatal rats, two from 8-day-old, and one from 18-day-old rats. Started in mid-September 1972, these lines are actually at their 22nd passage. Other lines are also being established either from adult or postnatal rats growing on normal or modified medium. In these cells the following hepatic functions have been found at the levels in normal organs

---

[a] We thank Dr. J. Devant, Director, for the continuous gift of typed sera.

at the same age: tyrosine aminotransferase activity induced by dexamethasone, biosynthesis of cholesterol from 1,2-^{14}C-acetate, and B aldolase activity (the ratio of aldolase A activity to aldolase B activity is between 1 and 2 as assayed by the NADH spectrophotometric method on six different cell lines). More on these observations will be reported later by Chessebeuf and Olsson *(to be published)*.

D. Incubation of 4-^{14}C-Corticosterone in Cultured Liver Cell

Cultured liver cells in 60-mm dishes at the 12th passage of the above-mentioned established lines were kept in normal culture medium. To each dish 6 μl of an ethanol solution containing 16,380 dpm 4-^{14}C-corticosterone (56.7 mC/mM) was added, corresponding to 0.14 nM and 2.9 nM of corticosterone. After 72 hr of incubation, the steroids of the medium were extracted by dichloromethane and ethyl acetate. The combined extracts were concentrated under vacuum and the residue subjected to TLC on silica gel in a chloroform/ethanol/water (172:26:2, v/v) system. Radioactive zones detected by radio-scanning were scraped off the glass plate and the radioactive steroids were eluted with a mixture of dichloromethane and methanol (1:1, v/v). Steroids as MO–TMS derivatives were characterized by GC–MF.

E. Incubation of 4-^{14}C-Testosterone in Liver Cell Culture

Incubations of 4-^{14}C-testosterone (specific activity 50 mC/mM) in physiological amounts (0.1 nM/ml of medium) for 72 hr in cultured liver cells in Ham F10 medium without fetal calf and human sera were performed. The cells were at their 17th passage. The metabolites were extracted from the medium with dichloromethane and purified by TLC (0.25 mm thick silica gel plates F254 from Merck). In order to eliminate the lipids a first run with isopropyl ether was carried out before the second migration with ethyl acetate-cyclohexane (v/v), which separated testosterone from its metabolites. Radio-scanning of the plate gave us an initial idea about the metabolites produced and allowed us to localize the radioactive areas. The radioactive areas were scraped off and eluted with ethyl acetate, and the usual MO–TMS derivatives formed. A mass spectrometer, LKB 9000 equipped with the new multiple-ion detection — peak matching unit (Hammar and Hessling, 1971) allowed the detection and the quantitation of both radioactive and nonradioactive metabolites. Internal standards playing the role of carrier were perdeuterated trimethylsilyl reference steroids: 17-MO-5α-androstane-3α-d9-TMS and 3-MO-5α-androstane-17β-d9-TMS.

III. RESULTS

A. Study of Age and Sex-Dependent Formation of Metabolites of Endogenous Corticosterone in Liver Organ

Multiple-ion monitoring was found suitable for the study of the following metabolites:

11β-21-dihydroxy-pregn-4-ene-3,20-dione (B)
11β,21-dihydroxy-5α-pregnan-3,20-dione (5α-DHB)
11β,21-dihydroxy-5β-pregnan-3,20-dione (5β-DHB)
3α,11β,21-trihydroxy-5α-pregnan-20-one (THB I)
3β,11β,21-trihydroxy-5α-pregnan-20-one (THB II)
3α,11β,21-trihydroxy-5β-pregnan-20-one (THB III)
3β,11β,21-trihydroxy-5β-pregnan-20-one (THB IV)
5α-pregnan-3α,11β,20α,21-tetrol (HHB I-20α)
5α-pregnan-3α,11β,20β,21-tetrol (HHB I-20β)
5α-pregnan-3β,11β,20β,21-tetrol (HHB II-20β)
pregnane-3,11,20,21-tetrol (HHB X, HHB Y)

The following ions are used: $m/e = 550$ (M) and/or $m/e = 519$ (M-31) for dihydrocompound DHB, $m/e = 595$ (M) and/or $m/e = 564$ (M-31) for tetrahydrocompound THB, $m/e = 537$ (M-103) and/or $m/e = 447$ (M-103–90) for hexahydrocompound HHB, $m/e = 683$ (M) and/or $m/e = 652$ (M-31) for hydroxylated THB (OH–THB).

The synthesis of corticosterone metabolites such as DHB, THB, HHB, and OH–THB isomers is compared in both male and female rats in the course of the postnatal development.

In agreement with previous results (Bournot, Maume, and Padieu, 1974; Eriksson and Gustafsson, 1971; Guillemant, Masson-Bobas, Barthelemy-Freneaux, and Desgrez, 1970; Lowy, Albepart, and Pasqualini, 1969; Schriefers, et al., 1971), the formation of reduced and hydroxylated metabolites of corticosterone shows distinct differences in liver of mature male and female rats whereas corticosterone is metabolized through the same pathway in female and male rats at the neonatal stage.

Development in the formation of corticosterone metabolites is observed with the age of the animal.

1. Female Rats

In newborns, two predominant hydrogenated metabolites were identified: THB I and THB II (Fig. 1). Mature female rats continued to metabolize corticosterone in THB I and THB II, but there was a marked increase of these metabolites. As shown in Fig. 2, THB I formation exceeded by three

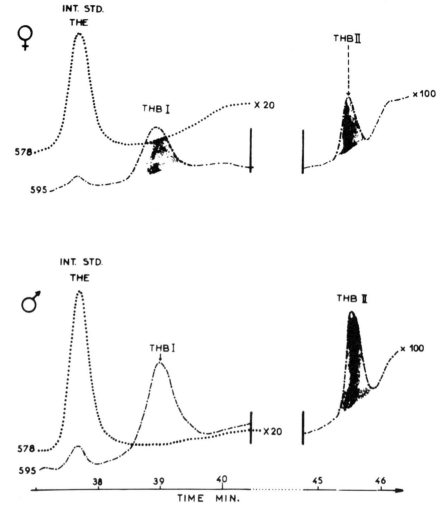

FIG. 1. Mass fragmentogram of steroid extracts from 200 mg of fresh liver tissue of new-born female (top) and male (bottom) rats. Isomeric THB are recorded by using the fragment ion at m/e = 595 (M), and the internal standard THE is recorded by using the fragment ion at m/e = 578 (M-31). Chromatographic conditions: 12-ft 1% OV-1 column, temperature programmed at 1°C/min from 200°C. Mass spectrometric conditions: electron-ionizing energy 28 eV.

times the formation of THB II. Due to the level of sensitivity and of chromatographic separation achieved in these experiments, it is not possible to prove the absence of THB III and THB IV. In our experiments, two hydroxylated THB's appear to be formed in significant amounts in mature female rats. They have been previously shown to be 3α(and 3β),11β,15α, 21-tetrahydroxy-5α-pregnan-20-one (15α-OH–THB I and 15α-OH–

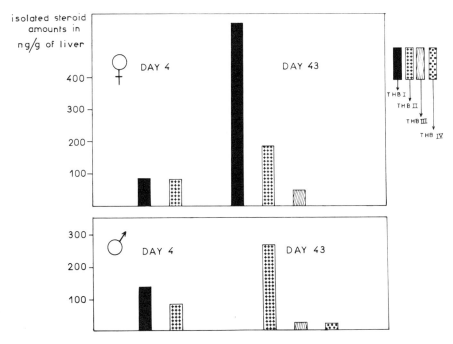

FIG. 2. Isomeric THB in livers of female and male rats at the ages of 4 and 43 days. The amounts of the isolated metabolites are expressed as nanograms per gram of fresh tissue.

THB II) (Gustafsson and Sjövall, 1968; Eriksson and Gustafsson, 1971).

Experiments are in progress to study the presence of DHB and HHB isomers in newborn rats.

2. Male Rats

In newborns, the production of corticosterone metabolites is the same as in newborn female rats: sex differences are not apparent (Schriefers et al., 1971). Although there is a low production of THB I and THB II in the liver of newborn rats, there is a significantly higher formation of THB II and no formation of THB I in the mature male animal.

Two THB isomers (THB III and THB IV) and five isomers of pregnan-tetrol (HHB I-20α, HHB I-20β, HHB II-20β, and two isomers of un-known structure) are produced in the mature male animals. Their presence in newborns is under investigation.

In conclusion, the formation of reduced and hydroxylated metabolites of corticosterone is age- and sex-dependent. The sexual difference of THB II and THB III is purely quantitative. But the formation of THB I, HHB isomers, and hydroxylated-THB is completely different from female to male. These metabolites show a sex specificity: THB I and hydroxylated THB are only produced in the livers of mature female rats, whereas HHB isomers are produced in the livers of male rats.

B. Metabolism of 4-¹⁴C-Corticosterone in Liver Cell Culture

After a 72-hr incubation of ^{14}C-corticosterone in liver cells, a major metabolite appeared. This metabolite exhibited chromatographic properties identical to 11-dehydrocorticosterone in the following TLC system: chloroform/ethanol/water (172:26:2, v/v). It was detected by mass fragmentography by using the molecular ion at m/e = 474 (M) and its isotopic peak at m/e = 475 (M+1). Ratios of the intensities of these two ions and retention times were compared to values of reference compounds. Our data are consistent with the presence of 11-dehydrocorticosterone in the incubation medium (Fig. 3).

When the radioactive corticosterone was incubated with the culture medium in the absence of cells, no formation of the radioactive metabolite

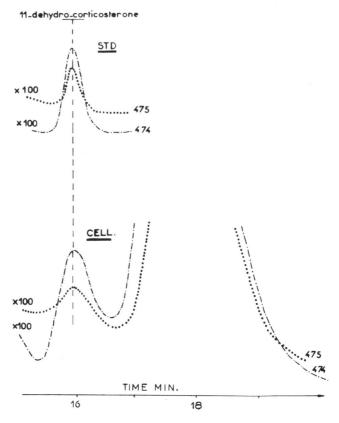

FIG. 3. Mass fragmentogram of a steroid extract from liver cells in culture after incubation with 4-¹⁴C-corticosterone. The metabolite 11-dehydrocorticosterone is detected by using the molecular ion at m/e = 474 and its isotopic peak at m/e = 475. Chromatographic conditions: 12-ft 1% OV-1; isothermal temperature, 250°C. Mass spectrometric conditions: electron-ionizing energy 28 eV.

was found. It can be concluded that the cells do metabolize radioactive corticosterone. This metabolism has been retained by cell lines established for 5 months and taken through the 17th passage.

C. Metabolism of 4-^{14}C-Testosterone in Liver Cell Culture

The incubated cellular extract was submitted to mass fragmentography after MO–TMS derivatization. Each compound was identified by its retention time, its response to at least two selected masses, and the ratio of two masses. The ratio of the intensities of two masses increases the specificity of the method and allows one to detect impurities through abnormal response which in this case would lead to a variable ratio.

In Table 1 the selected masses for the mass fragmentographic analyses are summarized.

TABLE 1. *Selected masses of testosterone and metabolites (labeled and unlabeled) for mass fragmentography*

Testosterone and metabolites	m/e values		$\dfrac{391}{360}$
3α-Hydroxy-5α-androstan-17-one (MO–TMS)	M$^+$ M-31	391 360	$\dfrac{1}{18}$
4-^{14}C-3α-Hydroxy-5α-androstan-17-one(MO–TMS)	M$^+$ M-31	393 362	
Testosterone (MO–TMS)	M$^+$	389	
4-^{14}C-Testosterone (MO–TMS)	M$^+$ M-31	391 360	4
17β-Hydroxy-3α-androstan-3-one (MO–TMS)	M$^+$ M-31	391 360	4
4-^{14}C-17β-Hydroxy-3α-androstan-3-one (MO–TMS)	M$^+$ M-31	393 362	
3α-Hydroxy-5α-androstan-17-one (MO–d9–TMS)	M$^+$	400	
17β-Hydroxy-5α-androstan-3-one (MO–d9–TMS)	M$^+$	400	

The radio-scan in Fig. 4 shows the radioactive peak of testosterone (A) and two other compounds (B and C). The TLC mobility of the compounds in peak B is the same as for 3α-hydroxy-5α-androstan-17-one and 17β-hydroxy-5α-androstan-3-one, and in peak C as for 5α/5β-androstan-3,17-dione.

Figure 5 shows the fragmentogram of TLC peaks A + B as MO–TMS derivatives. For the first time we found that liver cells in culture metabolize 4-^{14}C-testosterone mainly to 17β-hydroxy-5α-androstan-3-one (5α-DHT) and to a lesser extent to 3α-hydroxy-5α-androstan-17-one (An).

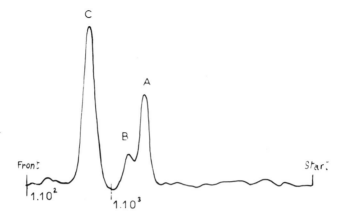

FIG. 4. Radio-scan (radio-scanner Berthold, Germany) of a TLC (silica gel Merck F254 0.250 mm thick) separation of the total steroid extract (800,000 dpm) from liver cells in culture after incubation with 4-¹⁴C-testosterone. Eluents: first run, isopropyl ether; second run, ethyl acetate-cyclohexane, 1:1, v/v. Liver cells came from 8-day-old rats taken through the 17th generation.

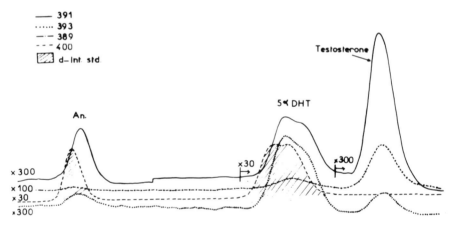

FIG. 5. Mass fragmentogram of the A+B TLC purified fraction from liver cells. Testosterone and the metabolites 3α-hydroxy-5α-androstan-17-one (An) and 17β-hydroxy-5α-androstan-3-one (5α-DHT), all as MO–TMS derivatives, are detected by the molecular ions m/e = 389, 391, 393 (see Table 1). The internal standards: 3α-hydroxy-5α-androstan-17-one and 17β-hydroxy-5α-androstan-3-one as MO–d9–TMS (200 ng each) are recorded by the molecular ion m/e = 400 (stippled area). Chromatographic conditions: 12-ft 1% OV-1 column; isothermal temperature 220°C. Mass spectrometric condition: electron-ionizing energy 28 eV.

This appears therefore to be a genuine hepatic pattern for the metabolism of testosterone (Schriefers, Ghraf, and Lax, 1972). These postnatal liver cells in culture retain their initial undifferentiated state in an established 5-month-old cell line. At the present time, the metabolic pathway of testosterone in liver cell culture can be schematized according to Fig. 6. It must

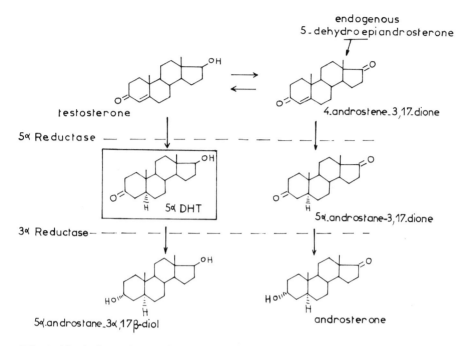

FIG. 6. Metabolic pathway of testosterone in hepatic cell cultures. 17β-Hydroxy-5α-androstan-3-one (5α-DHT) is the major metabolite.

be pointed out that without the help of mass fragmentography it would have been impossible to investigate these small amounts of different steroids with such high specificity.

IV. CONCLUSION

This work proves that liver cells can now be put in culture with a high percentage of epithelial cells and can be kept in established lines by splitting culture dishes at confluency.

Besides specific metabolic patterns such as tyrosine aminotransferase induction, cholesterol biosynthesis, and aldolase B activity, these cells retain several enzymes which regulate steroid metabolism through sex hormone dependence.

From these observations, it can be concluded that these liver cell lines can be used to study molecular processes which are involved in the regulation of specific metabolism depending on the genome expression. These *in vitro* biological systems can now be produced in large amounts. Advanced gas phase analytical methods will solve the problems of identification and quantitative estimation of picomoles of metabolites in order to achieve quantitative molecular biology on eukaryote-isolated cells.

ACKNOWLEDGMENTS

Financial help is gratefully acknowledged through research grants from: Institut National de la Santé et de la Recherche Médicale, Action Thématique Programmée no. 2; Centre National de la Recherche Scientifique, ERA 267; Enseignement Supérieur, Vème et VIème Plan; Fondation pour la Recherche Médicale Française.

REFERENCES

Bournot, P., Maume, B., and Padieu, P. (1974): Sex-linked specificity of the hepatic metabolism of steroids in the rat. I. Mass fragmentography as a method for the assay of hydrogenated metabolites of corticosterone in the liver. *Biomedical Mass Spectrometry (in press)*.

De Moor, P., and Denef, C. (1968): Puberty of the rat liver. Feminine pattern of cortisol metabolism in male rats castrated at birth. *Endocrinology*, 82:480–492.

Denef, C., and De Moor, P. (1972): Sexual differentiation of steroid metabolizing enzymes in the rat liver. Further studies on predetermination by testosterone at birth. *Endocrinology*, 91:374–384.

Eriksson, H. (1971): Steroids in germfree and conventional rats. Metabolites of 4-^{14}C-pregnenolone and 4-^{14}C-corticosterone in urine and faeces from male rats. *European Journal of Biochemistry*, 18:86–93.

Eriksson, H., and Gustafsson, J. A. (1971): Metabolism of corticosterone in the isolated perfused rat liver. *European Journal of Biochemistry*, 20:231–236.

Forchielli, E., Brown-Grant, K., and Dorfman, R. I. (1958): Steroid Δ4-hydrogenases of rat liver. *Proceedings of the Society for Experimental Biology and Medicine*, 99:594–596.

Gustafsson, J. A., and Sjövall, J. (1968): Steroids in germfree and conventional rats. Identification of 15α- and 21-hydroxylated C21 steroids in faeces from germfree rats. *European Journal of Biochemistry*, 6:236–247.

Guillemant, S., Masson-Bobas, F., Barthelemy-Freneaux, C., and Desgrez, P. (1970): Métabolisme de la corticostérone *in vivo* chez le rat. II. Etude des métabolites excrétés dans la bile. *Bulletin de la Société de Chimie Biologique*, 52:35–49.

Hammar, C. G., and Hessling, R. (1971): Novel peak matching technique by means of a new and combined multiple ion detector-peak matcher device. *Analytical Chemistry*, 43:298–306.

Harary, I., and Farley, B. (1963): *In vitro* studies on single beating rat heart cells. I. Growth and organization. *Experimental Cell Research*, 29:451–465.

Lowy, J., Albepart, T., and Pasqualini, J. R. (1969): ^{3}H-Corticosterone metabolism in the rat. *Acta Endocrinologica*, 61:483–493.

Maume, B., Bournot, P., Durand, J., Lhuguenot, J. C., Maume, G., Prost, M., and Padieu, P. (1973a): New developments in steroid and catecholamine analysis in biological media by gas chromatography-mass spectrometry using a multiple ion detector. In: *Organisation des Laboratoires, Biologie Prospective*, IIe Colloque de Pont à Mousson, pp. 637–654. Expansion Scientifique Française, Paris.

Maume, B., Bournot, P., Lhuguenot, J. C., Baron, C., Barbier, F., Maume, G., Prost, M., and Padieu, P. (1973b): Mass fragmentographic analysis of steroids, catecholamines and amino acids in biological materials. *Analytical Chemistry*, 45:1073–1082.

Padieu, P., Barbier, F., Bègue, R. J., Bournot, P., Desgrès, J., Durand, J., Maume, G., Lhuguenot, J. C., Prost, M., and Maume, B. (1973): Automation in clinical chemistry. An evaluation of the present and of the future development of automated separation methods. In: *Organisation des Laboratoires, Biologie Prospective*, IIe Colloque de Pont à Mousson, pp. 741–754. Expansion Scientifique Française, Paris.

Padieu, P., Barbier, F., Chessebeuf, M., Cordier, D., Gerique, M., Lallemant, C., and Olsson, A. (1971a): Relations between cell differentiation and organogenesis in single cell culture from newborn rat organ. *Proceedings of the First International Conference on Cell Differentiation*. Nice, France.

Padieu, P., Lallemant, C., Barbier, F., and Chessebeuf, M. (1971b): Conservation of differentiated characteristics in eukaryotes in single cell culture. *Proceedings of the 7th Congress of the Federation of European Biochemical Societies.* Varna, Bulgaria.

Périssel, B., Chessebeuf, M., Padieu, P., and Malet, P. (1973): Quelques caractères ultra-structuraux des cultures primaires de foie de rat nouveau-né. *Colloque de la Société Française de Microscopie Electronique.* Dijon.

Schriefers, H., Ghraf, R., Hoff, H. G., and Ockenfels, H. (1971): Einfluss von Alter and Geschlecht auf die Entwicklung und Differenzierung der Aktivitätmuster von Enzymen des Steroid-hormon-Stoffwechsels in der Leber von Ratten zweier verschiedener Tierstämme. *Hoppe-Seyler's Zeitschrift für Physiologische Chemie,* 352:1363–1371.

Schriefers, G., Ghraf, R., and Lax, E. R. (1972): Sex-specific aglucone patterns of testosterone metabolism in rat liver and their alteration following interference with sexual differentiation. *Hoppe-Seyler's Zeitschrift für Physiologische Chemie,* 353:371–377.

Staib, R., Sonnenschein, R., and Staib, W. (1970): Der Testosteronstoffwechsel in der isoliert perfundierten Rattenleber. *European Journal of Biochemistry,* 13:142–148.

Mass Spectrometry in Biochemistry and Medicine,
edited by A. Frigerio and N. Castagnoli.
Raven Press, New York © 1974

Analysis of Natural and Synthetic Hormonal Steroids in Biological Fluids by Mass Fragmentography

H. Adlercreutz

Department of Clinical Chemistry, University of Helsinki, Meilahti Hospital, SF-00290 Helsinki 29, Finland

I. INTRODUCTION

The great potential of the mass spectrometer as a sensitive detector in gas chromatography (GC) was already realized before the first combined gas chromatograph–mass spectrometer (GC–MS) became commercially available (Henneberg, 1961). In the field of steroid analysis the technique was first used by Sweeley, Elliott, Fries, and Ryhage (1966) for the quantitation of unresolved peaks in GC effluents. The monitoring of single ions in the quantitative determination of drug metabolites was introduced by Hammar, Holmstedt, and Ryhage (1968). They referred to the technique as mass fragmentography (MF). Few studies on the use of MF in the quantitative analyses of steroids have been made (Adlercreutz, 1969; Breuer, Nocke, and Siekmann, 1970; Siekmann, Hoppen, and Breuer, 1970; Brooks and Middleditch, 1971; Kelly, 1971; Adlercreutz and Ervast, 1973; Adlercreutz and Hunneman, 1973). If the instrument is connected to a computer, computerized quantitation of the compounds can be made in various ways (Reimendal and Sjövall, 1972; Baczynskyj, Duchamp, Zieserl, and Axen, 1973). However, practically no studies exist in which MF has been applied in an effort to solve medical or biological problems related to the steroid hormones (Brooks, Thawley, Rocher, Middleditch, Anthony, and Stillwell, 1971; Adlercreutz and Ervast, 1973). This is probably due not only to the expensive instrumentation required but also to the many technical problems which are encountered with this technique (Brooks and Middleditch, 1971; Sjöquist and Änggård, 1972; Adlercreutz, 1973).

This chapter deals with several MF methods for the analysis of natural and synthetic steroids and also with some preliminary results obtained in the analysis of various biological materials.

II. INSTRUMENTATION

In most of the experiments described a Varian Mat CH7 GC–MS instrument was used. It was equipped with a peak-matching device capable of simultaneously monitoring two ions whose intensity could be recorded and amplified independently on a two-pen recorder. In some experiments the data system Spectrosystem 100 MS was used. The stability of the magnetic field was so good that calibration was only necessary once a day and in fact sometimes only every second day. Because of the simplicity of operation, the instrument allows continuous uninterrupted use over a considerable period of time. The separate amplification of the two ions is of great advantage when working with unknown samples to which an internal standard has been added. The amount of the internal standard can always be kept relatively small because the peak can be amplified to a suitable level independent of the concentration of the unknown, provided both measurements are carried out in the linear part of the calibration curve.

In some experiments we have also used the LKB 9000 GC–MS equipped with an accelerating voltage alternator (AVA) which allows simultaneous measurement of three separate ions. When this system is used the ions cannot be amplified separately. A single-pen potentiometric recorder was used instead of the U.V.-recorder. Also, some modifications in this instrument have been made in order to decrease the noise level.

The frequency of the AVA has been altered by increasing both capacitances of the multivibrator (C_3–C_6; integrated circuit N 2) to about 60 μF, resulting in a change in the voltage of the AVA every 4th sec. The signal from the preamplifier of the multiplier to the potentiometric recorder is taken through a passive filter with a time constant of about 3 to 3.5 sec.

The modified LKB 9000 system is slightly more sensitive than the Varian Mat CH7, but the instability of the instrument makes frequent calibration necessary (see also Sjöquist and Änggård, 1972). This causes considerable loss of effective working time. Other problems encountered include obstruction of the jets and differences in the degree of opening of the separator valve which result in differences in the amount of substance entering the mass spectrometer in repetitive injections. Internal standardization is an absolute necessity when using this instrument. In this connection it may be mentioned that we do not recommend manual focusing during the analysis, because this results in very low precision despite internal standardization.

III. DETERMINATION OF UNCONJUGATED ESTRIOL IN PREGNANCY PLASMA

Several techniques for the measurement of estriol in pregnancy plasma have been tested. It is possible to determine estriol by direct silylation of the estriol in 1 to 2 μl of plasma. However, in this way only high values may be measured and care must be taken not to hydrolyze the sulfates during derivatization. If 100 μl of plasma is extracted, the extract evaporated to dryness, and the dry residue silylated, measurements can be made; this procedure provides results within 30 min. However, when the molecular ion at m/e 504 is monitored there are so many other compounds giving this ion later in the chromatogram that before the next analysis is carried out the operator must wait for about 20 min. In order to eliminate this rate-limiting step a simple chromatography of the estriol extract on partially deactivated alumina is carried out. When this purification step is included, injections may be made every 6 to 8 min. The method is as follows (Adlercreutz, Nylander, and Hunneman, *unpublished*): 100 μl of plasma is diluted with 200 μl of 1.5M sodium acetate buffer, pH 4.1, and extracted with 2×2 ml of diethyl ether. The ether extract is evaporated almost to dryness and the extract dissolved in 5 ml of benzene and the remaining ether blown off. The benzene extract is transferred to a 2-g alumina column (acid alumina, activity grade I, Merck, Darmstadt, Germany). The alumina is prewashed with distilled ethyl acetate, reactivated at 100°C overnight (Siegel, Adlercreutz, and Luukkainen, 1969), and then deactivated with 7% distilled water. When the extract has percolated through the column, 10 ml of 1% ethanol in benzene (v/v) is added and the fraction obtained discarded. The estriol is eluted with 30 ml of 20% ethanol in benzene (v/v) to which 1% distilled water (v/v) has been added. The fraction is evaporated to dryness in a stream of nitrogen. The extract is then silylated with 20 μl of hexamethyldisilazane and 2 μl trimethylchlorosilane in 200 μl of pyridine overnight (silylation can also be performed in a few minutes, if necessary). The solvents are evaporated and the silylated steroid fraction extracted with n-hexane. A known amount of standard estriol is silylated in the same way with hexamethyl-d_{18}-disilazane and trimethyl-d_9-chlorosilane (Supelco, Inc.) and the n-hexane extract of the silylated product is combined with the n-hexane extract of the sample in a graduated microtube. MF is carried out on a 1% QF-1 column (250°C). The ions m/e 504 and 531 (deuterized internal standard derivative) are monitored.

The recovery of estriol added to plasma was 86% (eight determinations). The precision, expressed as the coefficient of variation obtained in 11 analyses of a plasma pool, was 8%. The practical sensitivity limit of the method is 10-pg estriol/injection; however the standard curve is usually made for 30- to 200-pg estriol. None of the other 11 estrogens known to occur in the unconjugated form in pregnancy plasma (see below) interferes with the determination.

Some preliminary results are shown in Table 1. In various late normal and pathological pregnancies, values have been obtained in the range 4.9 to 16.3 μg/l (correction was not made for methodological losses). It is not yet possible to establish the normal range for unconjugated estriol in late pregnancy plasma using this method but the values obtained hitherto suggest that the mean concentration is somewhat lower than has been obtained by other investigators using radioimmunoassay. This difference may be due to the presence of other unconjugated estrogens in plasma which cross-react with the antibody (see below).

TABLE 1. *Unconjugated estriol in pregnancy plasma*

Patient	Pregnancy week	Diagnosis	Estriol (μg/l)	Birth weight (g)
LL	40	Normal (day of delivery)	9.3	3,360
EK	40	Normal (day of delivery)	4.9	3,520
MU	40	Normal (day of delivery)	7.7	3,110
Pool I (n > 10)	36–40	Normal	7.1	
HK	38	Toxemia, Cesarean section	6.3	3,270
VL	38	Hypertonia, Diabetes mellitus latens	10.8	4,150
''	39	Same	16.3	''
SV	38	Pruritus gravidarum	11.1	3,420
HK	33	Diabetes mellitus	7.9	3,740
LR	35	Proteinuria	5.6	3,310
PS	39	Pruritus gravidarum	12.5	4,220
''	40	Same	6.6	''
SM	40	Diabetes mellitus	8.3	4,680
''	''	Same	7.7	''

Until cheaper instruments are made available it is unlikely that a method for estriol determination based on MF can be applied in routine work. However, the method's potential as a research tool is great because its specificity is greater than that any other method previously used. If deuterized estriol were available, the method could be further improved. By adding known amounts of this compound to plasma it could be used both, for correcting losses incurred during the procedure and as an internal standard in MF (Siekmann et al., 1970).

IV. DETERMINATION OF 11 UNCONJUGATED AND CONJUGATED ESTROGENS IN PREGNANCY PLASMA

The GC method of Adlercreutz and Luukkainen (1968) with slight modifications (Adlercreutz and Luukkainen, 1967; Adlercreutz, 1974) was employed but instead of using GC with a hydrogen flame ionization detector the final determination was made by MF using the same technique as described for estriol (see above). The results were corrected for methodological losses by simultaneously carrying out a duplicate recovery experiment for nine estrogens added to plasma and correcting the results according to the mean value obtained for each estrogen.

The results obtained in a plasma pool from more than 10 normal women in late pregnancy are shown in Table 2. For this analysis a 20-ml plasma sample was used and all 11 estrogens could be quantitated both in the

TABLE 2. *Estrogens in late pregnancy plasma determined by MF in a 20-ml pooled specimen and by GC in five 100- to 200-ml pooled samples (mean values)*

	Unconjugated (μg/l)		Conjugated (μg/l)	
Estrogen	MF	GC[a]	MF	GC[a]
Estriol	7.1	6.1	106	124
Estrone	7.0	9.9	45.5	80
2-Methoxyestrone	0.14	1.5	0.41	1.3
Estradiol-17β	9.6	15.0	4.8	4.5
Estradiol-17α	0.16	$-$[b]	0.27	(0.5)[c]
16-Epiestriol	0.58	1.1	3.19	5.4
17-Epiestriol	0.06	$-$	0.30	$-$
16α-Hydroxyestrone	1.02	2.1	49.3	40
16β-Hydroxyestrone	0.93	$-$	7.3	4.3
16-Oxoestradiol	0.59	2.3	26.3	20
15α-Hydroxyestrone	0.79	$+$[d]	3.4	2.1

[a] According to Adlercreutz and Luukkainen (1971).
[b] A $-$ sign means that no GC peak could be detected.
[c] Includes another estradiol of unknown structure.
[d] A $+$ sign means that the estrogen was present as judged by GC–MS, but could not be quantitated by GC.

unconjugated and conjugated estrogen fraction. In Table 2 comparison is also made with the earlier results obtained in this laboratory using GC (Adlercreutz and Luukkainen 1971). In general there is very good agreement between the two sets of data. However, it seems that some of the estrogens occurring in very small amounts have been either overestimated or not detected at all by the GC method, the sensitivity limit of which is about 1 μg/l. It is likely that because of its extreme sensitivity the MF method will prove to be much more precise in low-level determinations. The lowest sensitivity in the assay was found in the case of 2-methoxy-estrone-TMS. The base peak in its mass spectrum is the molecular ion (m/e 372), but, in order to allow 2-methoxyestrone-TMS measurement simultaneously with estrone-TMS (molecular ion m/e 342) without re-calibration of the instrument, the M-30 fragment was used. The mass frag-mentograms obtained for this fraction are shown in Fig. 1. Because of the large amounts of estrone compared to 2-methoxyestrone in the samples, simultaneous determination of both estrogens in the same GC run could not be carried out; the analysis had to be repeated. If the molecular ion of 2-methoxyestrone-TMS (m/e 372) is monitored, the sensitivity is three times greater.

As can be seen from Table 2 the estrogens studied are conjugated to different degrees. The ring D-ketolic estrogens all occur in unconjugated form in approximately equal concentrations but great differences are seen in the concentrations of their conjugates. Approximately one-third of the estradiol-17α occurs unconjugated but two-thirds of the estradiol-17β is present in this fraction. Only some 6% of the estriol occurs in the un-

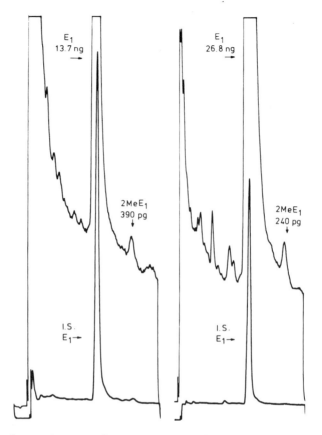

FIG. 1. Mass fragmentograms of two estrone–2-methoxyestrone fractions, which were obtained by analyzing 20 ml of pooled late pregnancy plasma. Mass fragmentogram of un-conjugated estrone-TMS (E_1) and 2-methoxyestrone-TMS (MeE_1) (left), and of the con-jugated estrone and 2-methoxyestrone (right). I.S.E_1 = deuterized trimethylsilyl ether derivative of estrone internal standard. 1% QF-1 was used as stationary phase, the tem-perature of the GC column was 225°C, and the ions m/e 342 and 351 were monitored using the Varian Mat CH7 instrument.

conjugated form. It is highly likely that these differences have biological significance but too little is known about the metabolism of the estrogens to be able to relate differences to specific physiological situations.

It may be concluded that the MF studies of this plasma pool confirm to a large degree previous results obtained with the GC method of Adlercreutz and Luukkainen (1971) and extend our knowledge regarding the presence and concentration of several metabolites occurring in very small amounts.

V. DETERMINATION OF 11 CONJUGATED ESTROGENS IN BILE AND URINE

Samples of bile (100 to 150 ml) from a 53-year-old man and from three postmenopausal women obtained after cholecystectomy for gallstones were processed in the same way as the pregnancy plasma (see above). The same procedure excluding the first step (precipitation of fatty material in cold methanol) was used for the three 50-ml urine samples obtained from a normal woman on the 14th, 17th, and 20th day of the menstrual cycle. All 11 estrogens could be determined without difficulty in each of the samples, and it was found that the sample size could have been reduced to about 50 ml in the case of bile and 20 ml with urine. Figure 2 shows two mass fragmentograms of the estrone–2-methoxyestrone fraction obtained from bile and urine. Figure 3 shows the mass fragmentograms of the ring D-ketolic estrogen fractions obtained from one bile and one urine sample, and Fig. 4 shows mass fragmentograms of the 16-epiestriol–17-epiestriol fraction obtained from these two biological fluids.

The values obtained for the male bile samples are shown in Fig. 5, and the mean values of the three analyses carried out in the bile of postmenopausal

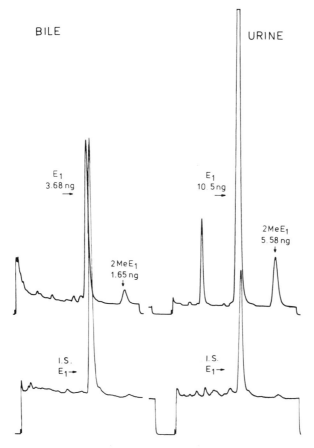

FIG. 2. Mass fragmentograms of two estrone–2-methoxyestrone fractions, which were obtained by analyzing a bile (left) and a urine (right) sample. The abbreviations and the experimental conditions are the same as in Fig. 1.

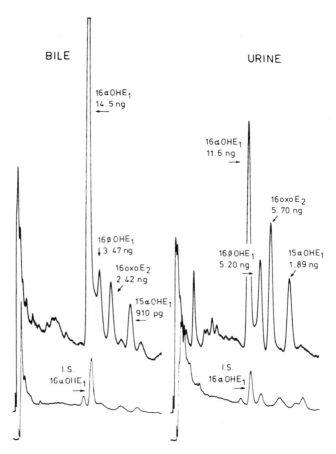

FIG. 3. Mass fragmentograms of two fractions containing the ring D-ketolic estrogens obtained from bile (left) and urine (right). 16αOHE₁ = 16α-hydroxyestrone-TMS; 16βOHE₁ = 16β-hydroxyestrone-TMS; 16oxoE₂ = 16-oxoestradiol-17β-TMS; 15αOHE₁ = 15α-hydroxyestrone-TMS; I.S. = Internal Standard (deuterized TMS derivative). 1% QF-1 was used as stationary phase, the temperature of the GC column was 225°C and the ions m/e 430 and 448 (= molecular ions) were monitored using the Varian Mat CH7 instrument.

women are shown in Fig. 6. In Fig. 7 the mean values for the three determinations in urine of a normal female subject are presented.

The most striking finding with regard to bile, was the very high concentration of 16α-hydroxyestrone as compared to the other estrogens determined. It represented about 50% of the total estrogens. This finding emphasizes the key position of this compound in the hepatic metabolism and enterohepatic circulation of estrogens. When the biliary estrogen pattern is compared to the urinary pattern, it can be seen that 16α-hydroxyestrone represents only 10% of total urinary estrogens determined which suggests that some of the 16α-hydroxyestrone participating in the enterohepatic circulation is later converted to estriol as previously proposed by

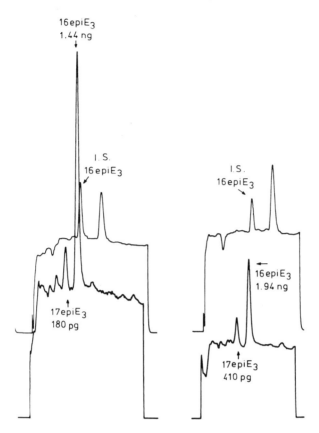

FIG. 4. Mass fragmentograms of two 16-epiestriol–17-epiestriol fractions obtained from bile (left) and urine (right). 16epiE$_3$ = 16-epiestriol (acetonide-TMS derivative); 17epiE$_3$ = 17-epiestriol (acetonide-TMS derivative); I.S. = Internal Standard (deuterized acetonide-TMS derivative). The epimeric estriols were first converted to their 16,17-acetonides and thereafter silylated. 1% SE-30 was used as stationary phase, the temperature of the GC column was 250°C, and the ions m/e 400 and 409 (= molecular ions) were monitored using the Varian Mat CH7 instrument.

Adlercreutz and Luukkainen (1967) and Adlercreutz and Tikkanen (1973). The results also indicate conversion of biliary 16β-hydroxyestrone to 16-epiestriol during later phases of its metabolism.

It may be concluded that the new method utilizing MF has enabled the detection and quantitative determination of a number of endogenous estrogen metabolites in bile and urine, which have never been assayed before on account of their extremely low concentrations.

VI. DETERMINATION OF MEGESTROL ACETATE IN PLASMA

In order to facilitate investigation of the absorption, plasma levels, and metabolism of orally administered megestrol acetate (MA), a rapid method

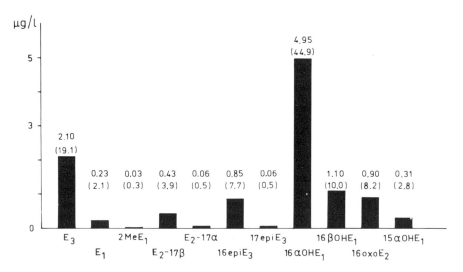

FIG. 5. Results of estrogen determination in a sample of male bile. The concentration and the relative proportion of each estrogen (in percent of total estrogens, in parentheses) are shown on the top of the bars. The abbreviations are the same as in previous figures.

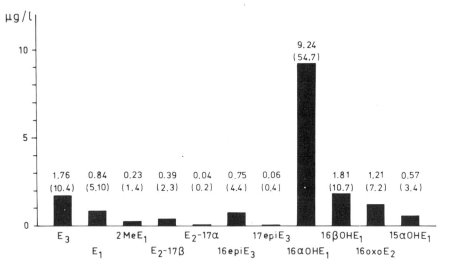

FIG. 6. Results of estrogen determinations in three bile samples obtained from post-menopausal women (mean values). See legend of Fig. 5.

for its assay in plasma was developed. The method (Adlercreutz and Ervast, 1973) consists of the following steps:

1. Extraction of plasma (1 to 5 ml) with diethyl ether:chloroform (3:1 v/v) (3 × 10 ml). The extract is washed with distilled water (0.5 ml) and evaporated to dryness (recovery = 96%).

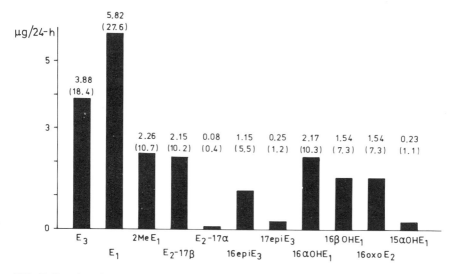

FIG. 7. Results of estrogen determinations in three samples of urine obtained from a normal woman at different stages of the menstrual cycle (mean values). See legend of Fig. 5.

2.   Column chromatography on a 2-g silica gel column. The column is washed three times with 5 ml of ethyl acetate:benzene (5:95 v/v) and the sample is then applied to the column in 3 × 1.5 ml portions of the same solvent. The column is first eluted with 30 ml of this solvent to remove cholesterol. The MA fraction is eluted with 20 ml of ethanol:benzene (5:95 v/v). The recovery of MA, added to a plasma extract, for this step is 90%.

3. A known amount of medroxyprogesterone acetate (MPA) is added to the fraction containing MA as an internal standard, and the solvents evaporated. The monomethoxime derivatives of the steroids are formed by adding 1 ml of saturated methoxyaminehydrochloride in pyridine to the dry sample. After standing overnight, the pyridine is evaporated, 2 ml of distilled water added, and the steroid derivatives extracted with 1 × 5 ml and 2 × 2.5 ml portions of toluene. The toluene is concentrated and transferred to a graduated microtube.

4. Mass fragmentography: Both 1% SE-30 and 3% OV-210 have been used as stationary phases and the column temperature was 230°C. The ions m/e 310 (MA) and 312 (MPA) were monitored.

At present the practical sensitivity limit of the method is 30 pg/injection. The specificity of the method was proved by isolating the unconjugated MA from plasma following oral administration of the compound and identifying it by mass spectrometry. There are no interfering compounds in plasma with the same retention times as the monomethoxime derivatives of MA and MPA which give fragments at m/e 310 and 312, even if larger amounts of plasma are analyzed. No metabolites of MA, which interfere in its estimation, occur in the plasma extract containing the unconjugated steroids.

Analyses of plasma following oral administration of MA revealed great individual variation in plasma concentrations. For example, after administration of 50 mg of MA twice daily for therapeutic purposes to women with gynecological cancer, the plasma concentration varied from 10 to 600 nmole/l, the most common concentration being 150 to 400 nmole/l (eight subjects). These great individual variations in absorption, metabolism, or both may be of significance when MA is used for therapeutic and contraceptive purposes.

VII. GC–MS WITH CONTINUOUS SCANNING AND COMPUTER ANALYSIS OF ESTROGEN FRACTIONS OBTAINED FROM PREGNANCY URINE

The GC method, which was used in the studies described above but in combination with MF, has been used in a great number of previous investigations where it has proved its specificity. However, in the five fractions (estrone–2-methoxyestrone, ring D-ketolic estrogens, estradiol, estriol, and epimeric estriols), which are obtained using this method, several unknown compounds occur, usually in low concentrations that are not very well separated from each other on the liquid phases used. In a collaborative study with the Department of Chemistry, Stanford University, California (Smith, Buchanan, Engelmore, Adlercreutz, and Djerassi, 1973) it was found by computerized interpretation of a variety of mass spectral data obtained from all five fractions (pregnancy urine) that they obviously contain unknown estrogens in addition to some neutral steroids. In order to get further information as to the content and structure of the unknown steroids, these fractions were analyzed with the Varian Mat CH7 GC–MS combined with Spectrosystem 100 MS. The fractions were scanned continuously and all spectra stored in the computer and analyzed in various ways. The most useful presentation of the results for interpretation could be made by computer plotting of the fragment ion currents of some selected ion typical of the steroids investigated.

In this connection only a few examples of the use of this technique will be given. Figure 8 shows the computer-plotted total ion current of the estrone–2-methoxyestrone fraction obtained from pregnancy urine. Peak no. 7 is estrone and peak no. 10 is 2-methoxyestrone, the other peaks contain unknown steroids or other compounds. Figures 9–11 show the computer-plotted ion currents for some selected ions.

In Fig. 9 the ion currents for m/e 312, 372, and 430 are plotted. The ion m/e 312 is typical of estrogen trimethylsilyl ether derivatives (TMS) and two peaks corresponding to estrone-TMS and 2-methoxyestrone-TMS are

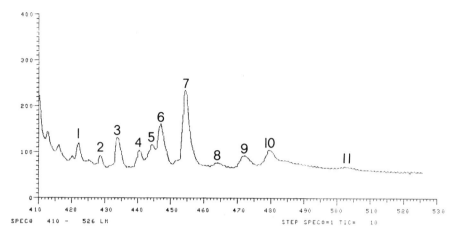

FIG. 8. Computer plot of total ion current in GC–MS analysis (continuous scanning) of the estrone–2-methoxyestrone fraction of pregnancy urine. Peak no. 7 corresponds to estrone-TMS and peak no. 10 to 2-methoxyestrone-TMS. The spectrum numbers are seen below. The stationary phase was 1% QF-1, the temperature 225°C. The Varian Mat CH7 coupled to Spectrosystem 100 MS was used.

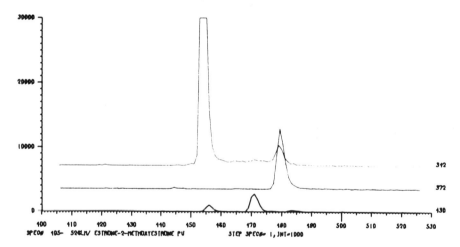

FIG. 9. Computer plots of the currents of three selected ions from the analysis shown in Fig. 8. The m/e values are shown at the right end of the chromatograms. The spectrum numbers correspond to those in Fig. 8.

seen. The ion m/e 372 is the molecular ion for 2-methoxyestrone-TMS and the ion m/e 430 is the molecular ion of estrogens with one oxo-group and two silylated hydroxyl groups. Three small peaks with maxima in spectra no. 456, 471, and 483 can be seen in Fig. 9. Spectrum no. 471 shows at least 10 ions characteristic of an estrogen and similar to that of 16-oxoestradiol but the retention time is shorter and this compound should not occur in this

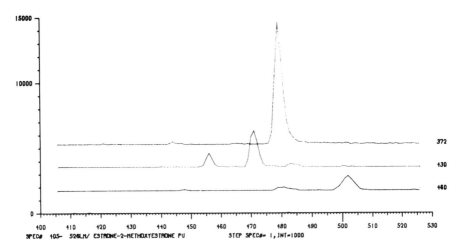

FIG. 10. Computer plots of the currents of three selected ions from the analysis shown in FIG. 8. The m/e values are shown at the right end of the chromatogram. The spectrum numbers correspond to those in Fig. 8.

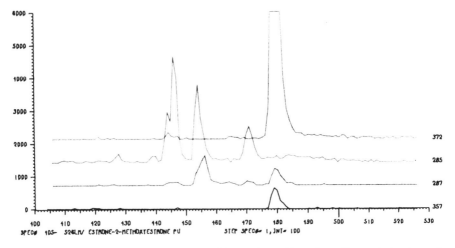

FIG. 11. Computer plots of the currents of four selected ions from the analysis shown in Fig. 8. The m/e values are shown at the right end of the chromatogram. The spectrum numbers correspond to those in Fig. 8.

fraction. Because the tail of the strong estrone peak occurs in spectrum no. 456, it is impossible to obtain any idea of the spectrum of the compound with the ion, m/e 430. Because the last peak is very small, it seems that only the compound with a peak maximum of m/e 430 in spectrum no. 471 should be investigated more carefully with regard to its structure.

In Fig. 10 both the current of the molecular ion of 2-methoxyestrone-TMS and of the ion m/e 430 are plotted. In addition the current from the ion m/e

460 corresponding to a methoxylated dihydroxy-mono-oxo-estrogen-TMS is plotted. A peak in spectrum no. 502 suggests the presence of such a compound. The compound corresponds to peak no. 11 in Fig. 8. The spectrum shows several ions characteristic of an estrogen and it is probably a previously undetected compound. The spectrum is similar to that of other 2-methoxylated estrogens which means that the abundance of the ions is small with the exception of the molecular ion which is the base peak.

In Fig. 11 the computer has plotted the currents for ions m/e 372, 285, 287, and 357. All except m/e 285 can be found in 2-methoxyestrone-TMS and m/e 285 and 287 can be found in estrone-TMS. The ion m/e 285 occurs in spectrum no. 471, which was discussed above. In spectrum no. 444 and 446 the ion m/e 285 occurs but further examination of the spectra revealed that these fragments do not belong to estrogens.

The results revealed that the estrone–2-methoxyestrone fraction contains at least two estrogens previously undetected. However, further studies are necessary before it will be possible to obtain a definite structure for these compounds. However, at least definite data such as retention time, molecular weight, and some typical fragments of these compounds are now known. The results also reveal that no unknown estrogen interferes to any significant extent with the measurement of estrone and 2-methoxyestrone in this fraction.

VIII. SUMMARY

MF methods for the determination of unconjugated estriol in pregnancy plasma, 11 unconjugated and conjugated estrogens in pregnancy plasma, 11 conjugated estrogens in bile and urine, and megestrol acetate in plasma are described. New information as to the concentration of many estrogens in various biological fluids were obtained using these methods. The utilization of continuous scanning of GC effluent with a mass spectrometer, connected to a computer, for the detection of new steroids in biological fluids is demonstrated.

It may be concluded that MF because of its extreme sensitivity will have a great impact on the development of research in the steroid field in the near future. Measurements of many steroid metabolites which some few years ago seemed impossible may now be carried out without any great difficulty and without working at the sensitivity limit of the technique.

ACKNOWLEDGMENTS

For the careful preparation of most of the samples I am very indebted to Mrs. Anja Manner and for excellent work with the mass spectrometers I wish to thank Mrs. Sirkka Tiainen. This work was supported by the Ford Foundation, New York.

REFERENCES

Adlercreutz, H. (1969): Combined gas chromatography–mass spectrometry in steroid research. In: *Abhandlungen der Deutschen Akademie der Wissenschaften zu Berlin, Klasse für Medizin*, edited by K. Schubert. Akademie Verlag, Berlin.

Adlercreutz, H. (1973): Application of combined gas chromatography and mass spectrometry to steroid analysis in clinical chemistry. *Chemische Rundschau*, 26:12–13.

Adlercreutz, H. (1974): Determination of some new oestrogens in various biological fluids by gas chromatography. In: *Methods of Hormone Analysis*, edited by H. Breuer, D. Hamel, and H. L. Krüskemper. Georg Thieme Verlag, Stuttgart *(in press)*.

Adlercreutz, H., and Ervast, H.-S. (1973): Mass fragmentographic determination of megestrol acetate in plasma. *Acta Endocrinologica (København)*, Suppl. 177:32.

Adlercreutz, H., and Hunneman, D. H. (1973): Quantitation of up to 12 estrogens in 1–50 μl of pregnancy urine. *Journal of Steroid Biochemistry*, 4:233–237.

Adlercreutz, H., and Luukkainen, T. (1967): Biochemical and clinical aspects of the enterohepatic circulation of oestrogens. *Acta Endocrinologica*, Suppl. 124:101–140.

Adlercreutz, H., and Luukkainen, T. (1968): New methods for estimation of oestrogens in urine of non-pregnant individuals. In: *Gas Chromatography of Hormonal Steroids*, edited by R. Scholler and M. F. Jayle. Dunod, Paris, and Gordon and Breach, New York.

Adlercreutz, H., and Luukkainen, T. (1971): Methodik der Gaschromatographie und Massenspektrometrie der "neueren Östrogene." *Zeitschrift für Klinische Chemie und Klinische Biochemie*, 9:421–426.

Adlercreutz, H., and Tikkanen, M. J. (1973): Defects in hepatic uptake and transport and biliary excretion of estrogens. Some new concepts of liver metabolism. *Médecine & Chirurgie Digestives*, 2:59–65.

Baczynskyj, L., Duchamp, D. J., Zieserl, J. F., Jr., and Axen, U. (1973): Computerized quantitation of drugs by gas chromatography–mass spectrometry. *Analytical Chemistry*, 45:479–482.

Breuer, H., Nocke, L., and Siekmann, L. (1970): Neue Methoden zur Bestimmung von Steroidhormonen. *Zeitschrift für Klinische Chemie und Klinische Biochemie*, 8:329–338.

Brooks, C. J. W., and Middleditch, B. S. (1971): The mass spectrometer as a gas chromatographic detector. *Clinica Chimica Acta*, 34:145–157.

Brooks, C. J. W., Thawley, A. R., Rocher, P., Middleditch, B. S., Anthony, G. M., and Stillwell, W. G. (1971): Characterization of steroidal drug metabolites by combined gas chromatography–mass spectrometry. *Journal of Chromatographic Science*, 9:35–43.

Hammar, C.-G., Holmstedt, B., and Ryhage, R. (1968): Mass fragmentography. Identification of chlorpromazine and its metabolites in human blood by a new method. *Analytical Biochemistry*, 25:532–548.

Henneberg, D. (1961): Eine Kombination von Gaschromatograph und Massenspektrometer zur Analyse organischer Stoffgemische. *Zeitschrift für Analytische Chemie*, 183:12–23.

Kelly, R. W. (1971): The measurement by gas chromatography–mass spectrometry of oestra-1,3,5,-triene-3,15α,16α,17β-tetrol (oestetrol) in pregnancy urine. *Journal of Chromatography*, 54:345–355.

Reimendal, R., and Sjövall, J. (1972): Analysis of steroids by off-line computerized gas chromatography–mass spectrometry. *Analytical Chemistry*, 44:21–29.

Siegel, A. L., Adlercreutz, H., and Luukkainen, T. (1969): Gas chromatographic and mass spectrometric identification of neutral and phenolic steroids in amniotic fluid. *Annales Medicinae Experimentalis et Biologiae Fenniae*, 47:22–32.

Siekmann, L., Hoppen, H.-O., and Breuer, H. (1970): Zur gas-chromatographische-massenspektrometrischen Bestimmung von Steroidhormonen in Körperflüssigkeiten unter Verwendung eines Multiple Ion Detectors (Fragmentographie). *Zeitschrift für Analytische Chemie*, 252:294–298.

Smith, D. H., Buchanan, B. G., Engelmore, R. S., Adlercreutz, H., and Djerassi, C. (1973): Applications of artificial intelligence for chemical inference. IX. Analysis of mixtures without prior separation as illustrated for estrogens. *Journal of the American Chemical Society*, 95:6078–6084.

Sjöquist, B., and Änggård, E. (1972): Gas chromatographic determination of homovanillic acid in human cerebrospinal fluid by electron capture detection and by mass fragmentography with a deuterated internal standard. *Analytical Chemistry*, 44:2297–2301.

Sweeley, C. C., Elliott, W. H., Fries, I., and Ryhage, R. (1966): Mass spectrometric determination of unresolved components in gas chromatographic effluents. *Analytical Chemistry*, 38:1549–1553.

Mass Spectrometry in Biochemistry and Medicine,
edited by A. Frigerio and N. Castagnoli.
Raven Press, New York © 1974

Analysis of Some Quaternary Ammonium Compounds by Gas Chromatography–Mass Spectrometry

W. Bianchi, L. Boniforti, and A. di Domenico

Istituto Superiore di Sanità, Laboratori di Chimica, Viale Regina Elena 299, 00161 Roma, Italy

Low-resolution mass spectrometry (MS) was used to identify some of the major components obtained by thermal decomposition of four quaternary ammonium compounds with germicide and antiseptic activity. Small amounts of the ammonium salts were quickly heated in a Curie-point pyrolyzer set up on a gas chromatographic (GC) inlet where pyrolysis was alternatively carried out. The mixtures of decomposition gases were then analyzed by GC–MS. Mass spectra showed that both forms of pyrolysis led to similar decompositions. The following compounds were studied: cetylpyridinium chloride, Desogen, cetyltrimethylammonium bromide, benzethonium chloride.

I. INTRODUCTION

In recent years some quaternary ammonium compounds have found extensive medical use for their germicide and disinfectant action, and this has stimulated researchers to develop appropriate analytical techniques for their identification and purity control. In our laboratory, gas-liquid chromatography (GLC) was successfully applied to the determination of four ammonium compounds (Tucci, Cardini, Cavazzutti, and Quercia, 1968) which were injected into the GC inlet without any previous chemical treatment. Each compound was characterized by the presence in the gas chromatogram of one main peak which could be used for quantitative and qualitative analyses. On the whole, GLC was suggested by its simplicity and reproducibility even though, because of the extremely low vapor pressure of the ammonium salts, it seemed obvious to assume that the ammonium compounds would undergo pyrolytic decomposition before entering the GC column. Thus, we thought it of interest to undertake a study in order to interpret cracking processes as well as to ascribe the proper structural formula to the major GC peaks.

Several authors (Kourovtzeff, 1966; Laycock and Mulley, 1966; Jennings and Mitchner, 1967; Mitchner and Jennings, 1967; Kojima and Oka, 1968; Hummel, 1962) have already dealt with the analysis of long-chain quaternary ammonium salts. In most cases the compound is first subjected to a degrading chemical process that leads to simpler structures, and then the reaction mixture is resolved into components by GLC. Although this procedure may be useful to identify the alkyl substituents in the quaternary ammonium derivatives and provides for the determination of the ammonium salt content and impurities, it is somewhat too complex to apply to the analysis of commercial pharmaceutical products. So, wherever a reproducible GC pattern could be obtained, the direct use of GLC would be recommended.

We had previously observed that injection port temperatures seemed to affect the chromatographic response. Therefore, pyrolysis was carried out at various temperatures – 375, 450, 510, 610, and 770°C – by a Pye Unicam Curie-point pyrolyzer, the outlet of which was set up in the GC inlet and kept isothermal with the column. Alternatively, the cracking process was performed by injecting samples into the injection port at about 400°C. The mixtures of decomposition gases were then analyzed by GC–MS.

II. EXPERIMENTAL

A Perkin-Elmer 900 GC equipped with a linear temperature programmer and glass columns was employed in tandem with a Perkin-Elmer 270B MS. Watson-Biemann molecular separator, gas inlet, and ionization chamber temperatures were kept at 250°, 150°, and 100°C respectively. The filament emission current was set at 150 μA with a trap current of about 80 μA and ionizing voltage of 70 volts. Spectra reported in Figs. 2, 4, 6, and 8 were recorded with a Honeywell visicorder, at a scanning rate of 10 s/decade, when each GC peak was at its maximum. Attention was paid to eliminate the background whenever possible. However, it is to be expected that the resultant bar graphs are not completely free from interference. GC's were obtained by connecting an L&N Speedomax W recorder to the total ion monitor system of the MS.

By comparing spectra taken before, at, and after the maximum peak heights, it could be ascertained that each peak was due to a single component.

For the study of cetylpyridinium chloride (CPC), Desogen, and cetyltrimethylammonium bromide (CMAB), the PE 900 GC was provided with a 300 cm × 3 mm i.d. glass column containing 3% OV-17 supported on Chromosorb G HP of 80/100 mesh size. Helium flow rate was 12 ml/min with temperature programs of 70° to 195°C and 120° to 160°C at a rate of 4°C/min for CPC and CMAB respectively, and Desogen decomposition was analyzed with the oven isothermal at 200°C. To avoid long and impractical retention times, a 120 cm × 3 mm i.d. glass column containing 5%

OV-61 supported on Gas-Chrom Q of 80/100 mesh size was employed for benzethonium chloride (BC). Here, temperature was programmed from 200° to 250°C at 4°C/min.

The following commercial products were supplied by the manufacturer (shown in parentheses) and used with no further purification: (1) cetyl-pyridinium chloride, 100% (Richardson Merrell S.p.A., Naples, Italy); (2) Desogen, 94% (Geigy S.A., Milan, Italy), mixture of Desogen with isopropyl alcohol; (3) cetyltrimethylammonium bromide, 95% (Hopkin & Williams Ltd., Chadwell Heath, Essex, England); (4) benzethonium chloride (Hyamine 1622), 98% (BDH, Milan, Italy); (5) Chromosorb G HP, 80/100 mesh size (Johns-Manville Products Corp., distributed by Analabs, North Haven, Connecticut, USA); (6) Gas-Chrom Q, 80/100 mesh size (Carlo Erba, Milan, Italy); (7) OV-17 and OV-61 stationary phases were both sup-plied by Carlo Erba, Milan, Italy.

III. RESULTS AND DISCUSSION

A. Cetylpyridinium Chloride

The cracking process leads to a very simple chromatogram (Fig. 1) where only one major component is visible. A smaller peak with higher retention

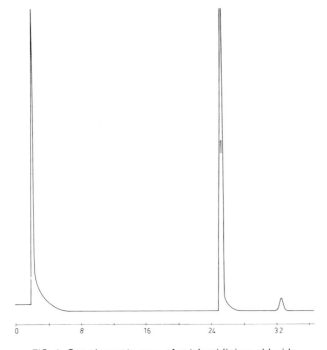

FIG. 1. Gas chromatogram of cetylpyridinium chloride.

time is probably due to an impurity because its surface is strictly proportional to the first peak area and does not depend upon pyrolysis conditions.
For this reason and because the second component has never been employed in analytical determinations, we disregarded it in this study.

The mass spectrum reported in Fig. 2 refers to the largest peak in the GC
of CPC and shows the fragmentation pattern of n-hexadecyl chloride which
is already well known in the literature (McLafferty, 1962). With the exception of fragments 91 and 105, respectively $C_4H_8^{35}Cl^+$ and $C_5H_{10}^{35}Cl^+$ cyclic
ions, the homologous halide compounds with $m/e = 105 + n(14)$, where $n =$
1, 2, . . . , have very small intensities, as expected. The most abundant
masses in the spectrum are clearly due to hydrocarbon fragments and belong
to the $m/e = 29 + n(14)$ series, the base peak being $C_4H_9^+$. These ions may
lose one or more hydrogen atoms as well as giving rise to the quite intense
$(C_nH_{2n-1})^+$ ions.

On the whole, we may conclude that cetylpyridinium chloride breaks
down in the GC injection port according to the following cracking process:

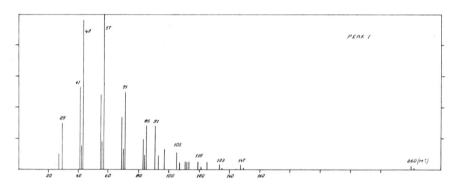

Cetylpyridinium   Pyridine   Cetyl chloride
chloride   mw = 79   mw=260 (^{35}Cl)

It has finally to be noted that no proof has been collected as to whether
pyridine is released as such or undergoes structural changes during the
breakdown process.

FIG. 2. Mass spectrum of the largest component in the gas chromatogram of cetylpyridinium chloride: n-hexadecyl chloride.

B. Desogen

Two peaks, (1) and (2), are displayed in the GC of Desogen (Fig. 3). As
the area ratio (1) to (2) is unaffected by changes in the pyrolysis temperature,

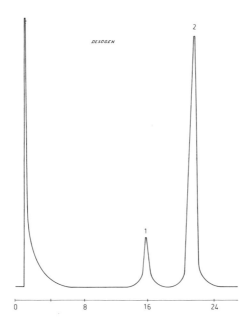

FIG. 3. Gas chromatogram of Desogen.

we think that, despite the quite large amount of (1) with respect to (2), the smaller component is an impurity. It may also be noted that mass spectra of compounds (1) and (2), reported in Fig. 4, are extremely similar and both exhibit intense fragments at m/e = 258, which were the heaviest detectable masses. The conclusion might be drawn that (1) and (2) are due to isomeric structures; however, further evidence should be collected in order to confirm it.

The cracking process that produces component (2) is not yet clear but from its mass spectrum and by analogy with the other ammonium salts, the following pyrolysis pathway might be proposed:

$$H_3C \overset{}{-\!\!\!\bigcirc\!\!\!-} CH - CH_2(CH_2)_9CH_3$$
$$\underset{N^+(CH_3)_3 \ \ ^-SO_3OCH_3}{|}$$

Desogen mw = 429

|
Heat
↓

$$N(CH_3)_3 \ + \ HSO_3OCH_3 \ + \ H_3C \overset{}{-\!\!\!\bigcirc\!\!\!-} CH{=}CH(CH_2)_9CH_3$$

mw = 59   mw = 112   1-(*p*-tolyl)dodecene mw = 258

This process would involve the rupture of an N—C bond, which is not surprising since it occurs in all three instances of this study, and would

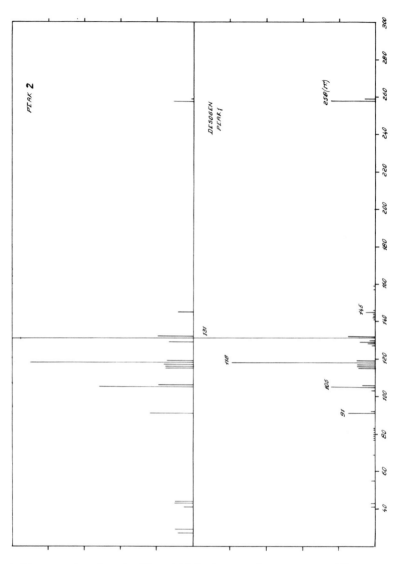

FIG. 4. Mass spectra of peaks (1) and (2) in the gas chromatogram of Desogen: L-(p-tolyl)dodecene (top).

also require a hydrogen atom to migrate to restore the charge balance on the right side of the equation and yield neutral substances. If the hydrogen atom stems from C-2 of the alkyl chain, then a tolyl derivative of 1-dodecene should be expected. The mass spectrum exhibits only a few main fragments, the formation of which is briefly discussed hereafter according to the proposed structure.

The tropylium ion, $m/e = 91$, can be explained on the basis of the presence in the molecule of $CH_3C_6H_4$— which easily undergoes rearrangement, upon electron impact, to yield the cyclic $C_7H_7^+$ species. However, a preferential process takes place and results in the base peak, $m/e = 131$, probably owing to the cleavage of the allylically activated C—C bond. Thus, $(CH_3C_6H_4CH=CHCH_2)^+$ is the most intense fragment in the spectrum, whereas the heavier mass 145 is much less abundant. The intervention of a McLafferty rearrangement could account for the rather high intensity of mass 132. Fragments 105 and 118 cannot be justified in such a simple way because their production entails hydrogen rearrangements with shifting of the double bond. These processes in alkenes are, however, well known and do not clash with the earlier assumptions. Hence, even if more study should be carried out, it seems reasonable to conclude that the pyrolytic decay of Desogen occurs according to one main reaction channel with yield of trimethylamine, methyl sulfate, and 1-(p-tolyl)dodecene. No search for the first two compounds has been undertaken.

C. Cetyltrimethylammonium Bromide

A more complicated decomposition pattern is presented in Fig. 5. Upon heating, cetyltrimethylammonium bromide undergoes pyrolytic degradation and yields at least four major components which can be fully resolved by GLC. From the analysis of the mass spectral data it can be assumed that two different cracking pathways determine the pyrolysis products of CMAB visible in Fig. 5:

$(H_3C)_3 \overset{+}{N} - (CH_2)_{15}CH_3 \xrightarrow{\text{Heat (I)}} (H_3C)_3 N + Br CH_2(CH_2)_{14}CH_3$

Br$^-$

Cetyltrimethylammonium bromide mw = 363 (^{79}Br)

Cetyl bromide

mw = 304 (^{79}Br)

Heat (II)
$- CH_3 Br$

$(H_3C)_2 N - (CH_2)_{15}CH_3$ mw = 269

$-CH_2=CH_2$

$(H_3C)_2 N - (CH_2)_{13}CH_3$ mw = 241

$-CH_2=CH_2$

$(H_3C)_2 N - (CH_2)_{11}CH_3$ mw = 213

It will be seen later that the four main compounds with molecular weights between 213 and 304 are all represented in the gas chromatogram of Fig. 5. As usual, the lighter decomposition masses, such as $(CH_3)_3N$, CH_3Br, and $CH_2=CH_2$, have been neglected in this study. Furthermore, it must be pointed out that no major peak fails to arise from CMAB, as can be ascer-

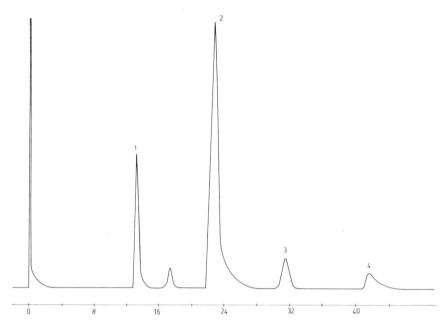

FIG. 5. Gas chromatogram of cetyltrimethylammonium bromide.

tained by a quick calculation of their percentages in the decomposition mixture; and the small amount of impurity might be responsible for the one peak which has not been taken into consideration.

The mass spectrum of peak (3) is reported in Fig. 6. From a comparison with the literature (McLafferty, 1962), the peak can be recognized as due to n-hexadecyl bromide, the fragmentation pattern of which offers no cause for particular attention. We would only like to emphasize the great intensity of masses 135 and $149 - C_4H_8{}^{79}Br^+$ and $C_5H_{10}{}^{79}Br^+$ cyclic ions respectively – and the absence, in our spectrum, of the molecular ion which could not be resolved from the background noise. For further comment, we refer to the cetylpyridinium chloride section.

The mass spectrum of peak (4) is shown in Fig. 6 also, and confirms the hypothesis that a second cracking channel occurs yielding cetyldimethyl-amine. In fact, it is known that in aliphatic amines the simple cleavage of the C—C bond adjacent to the nitrogen atom provides an alkyl radical and the most abundant ion in the spectrum. In our case, fragment 58, $(CH_3)_2\overset{+}{N}{=}CH_2$, arises through the loss of a pentadecyl radical. Three more nitrogen-containing ions can be identified at higher masses on the basis of known fragmentation processes: fragment 114 corresponds to the $\overline{CH_2CH_2CH_2CH_2CH_2\overset{+}{N}(CH_3)_2}$ ion with a six-membered cyclic structure; fragment 84 may be due to $\overline{CH_2CH_2CH_2CH{=}\overset{+}{N}CH_3}$, the production of

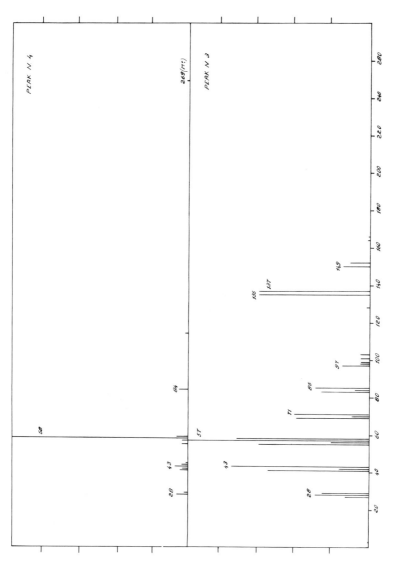

FIG. 6. Mass spectra of peaks (3) and (4) in the gas chromatogram of cetyltrimethyl-ammonium bromide: n-hexadecyl bromide (bottom) and cetyldimethylamine (top).

which would proceed through the formation of a five-membered ring (mw = 100) with subsequent loss of a methyl group and a hydrogen atom, and, finally, the molecular ion at m/e = 269. No other ion has any analytical importance in the spectrum. Its features, on the other hand, well support the hypothesis that peak (4) in the gas chromatogram is determined by the afore-said n-alkyl tertiary amine.

For the sake of brevity, we have not reported mass spectra of peaks (1) and (2) in Fig. 6, the latter being the largest and most important GC component. These spectra are very similar to that of compound (4), except for the molecular ion which exhibits masses 213 and 241 for peaks (1) and (2) respectively. As there is a constant difference of 28 amu between the molecular weights of (1), (2), and (4), we believe that the cracking process does not stop when cetyldimethylamine appears but progresses continually with gradual expulsion of one, two, and, perhaps, although this has not been observed, more neutral molecules of ethylene from the alkyl chain. This is shown in the sequence at the beginning of this section. It has to be noted that the first expulsion, leading to mass 241, occurs to a great extent because component (2) is the most abundant—even more abundant than (4) which descends from CMAB by the simple loss of CH_3Br and would be expected to be in a stable form.

If our assumptions are correct, the release of ethylene is quite a surprising phenomenon and would be worthy of further attention to understand how, and in which point of the molecule, the breakaway process takes place. However, it must be pointed out that the relative amounts of components (1), (2), (3), and (4) do not seem to depend greatly on the pyrolysis temperature: in other words, once the cracking process has taken place, the amount of ethylene released is unexpectedly not temperature-dependent.

D. Benzethonium Chloride

This compound undergoes pyrolysis yielding only two main components, (1) and (2). After GC resolution (Fig. 7), both peaks were analyzed by MS (Fig. 8), and the following cracking channels were suggested:

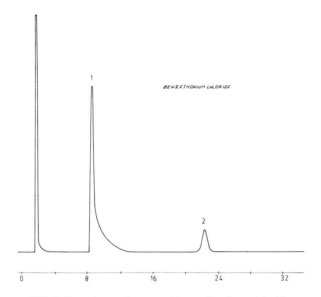

FIG. 7. Gas chromatogram of benzethonium chloride.

Fragmentation of tertiary amines has already been commented on; thus, one would expect the base peak of compound (1) to be fragment 58, and it is; whereas mass 72 is probably due to $(CH_3)_2\overset{+}{N}{=}CHCH_3$ although a cyclic ether with formula $(\overline{CH_2CH_2OCH_2CH_2})^{\ddot{+}}$ could also have a bearing on this fragment. The second major fragment in the mass spectrum of compound (1) has m/e = 116 and can be reasonably attributed to a six-membered cyclic ion with two hetero atoms in the ring and structural formula $\overline{CH_2CH_2OCH_2CH_2\overset{+}{N}(CH_3)_2}$. Three more low-intensity ions are visible at m/e = 134, 250, and 321. Although it is not clear which species is responsible for the first mass,

$$\left[(CH_3)_2C{=}\!\!\left\langle\rule{0pt}{8pt}\right\rangle\!\!{=}O\right]^{\ddot{+}}$$

could be proposed. No doubt exists that loss of a neopentyl radical $(CH_3)_3CCH_2$ yields the fragment at m/e = 250, and that the last mass is the molecular ion of compound (1).

As to component (2), the mass spectrum is modified because of the presence of the benzyl group in place of CH_3—. Hence, the base peak turns out to be

$$\underset{\overset{|}{CH_3}}{H_2C{=}\overset{+}{N}{-}CH_2}\!\!\left\langle\rule{0pt}{8pt}\right\rangle,\ m/e = 134,$$

whereas mass 148 is due to the addition of one methylene unit,

$$H_3CCH{=}\overset{+}{\underset{\underset{CH_3}{|}}{N}}{-}CH_2{-}\langle\rangle;$$

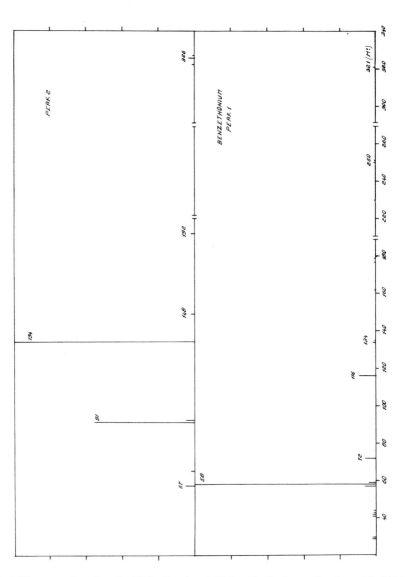

FIG. 8. Mass spectra of peaks (1) (bottom) and (2) (top) in the benzethonium chloride gas chromatogram.

fragments 192 and 326 are respectively due to a cyclic compound formation of structure

$$\overset{\lceil}{C}H_2CH_2OCH_2CH_2\overset{\rceil^+}{N}{-}CH_2{-}\bigcirc$$
$$\underset{CH_3}{\mid}$$

and to loss from the molecular ion of a neopentyl radical. Finally, on the left side of the base peak, two fragments can be easily recognized: $C_4H_9^+$, t-butyl ion, m/e = 57, and $C_7H_7^+$, tropylium ion, m/e = 91. The former is also present in the mass spectrum of component (1) and is due to the presence of a t-butyl radical in the molecule. The latter clearly originates from the benzyl group attached to the nitrogen atom.

IV. CONCLUSION

In this investigation the main products, obtained by pyrolytic decomposition of four quaternary ammonium compounds, have been studied and identified on the basis of MS results.

It is known that quaternary ammonium bases decompose on heating into tertiary amines, water, and an unsaturated hydrocarbon (Hofmann's degradation), and only in tetramethylammonium hydroxide does the fission take another course: methanol and trimethylamine are formed. Our results show that when the hydroxyl of the ammonium base is replaced by halide ions, pyrolysis does not yield an alkene but the corresponding alkyl halide with a mechanism which would be worthy of further investigation.

Finally, it has to be pointed out that, for cetyltrimethylammonium bromide and benzethonium chloride, peaks due to tertiary amines are used for GC analyses, and that cetylpyridinium chloride and Desogen are identified by their hydrocarbon derivatives.

REFERENCES

Hummel, D. (1962): *Identification and Analysis of Surface-Active Agents.* Interscience Publishers, New York.

Jennings, E. C., Jr., and Mitchner, H. (1967): Modified Hofmann degradation for the analysis of n-alkylbenzylammonium chlorides by gas chromatography: I. C_{14}- to C_{18}-alkyl compounds. *Journal of Pharmaceutical Sciences,* 56:1590-1594.

Kojima, T., and Oka, H. (1968): Gas-liquid chromatographic analysis of long chain quaternary ammonium salts. *Kogyo Kagaku Zasshi,* 71:1844-1847.

Kourovtzeff, K. (1966): Utilisation de la chromatographie dans l'analyse des agents de surfaces cationiques: Amines, polyamines, sels d'ammonium quaternaire: I. Chromatographie en phase gazeuse. *Revue Française des Corps Gras,* 13:271-276.

Laycock, H. H., and Mulley, B. A. (1966): Determination of the homologue composition of some alkyltrimethyl quaternary ammonium antibacterial agents by gas chromatography. *Journal of Pharmacy and Pharmacology,* 18:9s-11s.

McLafferty, F. W. (1962): Mass spectrometric analysis—Aliphatic halogenated compounds. *Analytical Chemistry*, 34:2–15.

Mitchner, H., and Jennings, E. C., Jr. (1967): Modified Hofmann degradation for the analysis of n-alkylbenzyldimethylammonium chlorides by gas chromatography: II. Benzalkonium chloride. *Journal of Pharmaceutical Sciences*, 56:1595–1598.

Tucci, B., Cardini, C., Cavazzutti, G., and Quercia, V. (1968): Applicazioni della gas cromatografia in analisi farmaceutica: VII. Determinazione di sali ammonici quaternari in preparazioni farmaceutiche. *Bollettino Chimico Farmaceutico*, 107:629–635.

Mass Spectrometry in Biochemistry and Medicine,
edited by A. Frigerio and N. Castagnoli.
Raven Press, New York © 1974

Structure Elucidation of Pheromones Produced by the Pharaoh's Ant, Monomorium pharaonis L.

E. Talman, F. J. Ritter, and P. E. J. Verwiel

Centraal Laboratorium TNO, Delft, The Netherlands

I. INTRODUCTION

Pharaoh's ant, *Monomorium pharaonis* L., originally is a tropical insect. Nevertheless it has established itself as a persistent nuisance in countries with a moderate, nontropical climate. Pharaoh's ant is now becoming a threat to the inhabitants of large cities in Western Europe and the United States. The pest is not easy to control by methods used against other kinds of ants. As is easily understood because Pharaoh's ant is a tropical insect, it is mainly found in or near well-heated rooms or installations. It constitutes a serious problem in hospitals, because it is a distributor of various micro-organisms pathogenic to man. It is known that the ants feed on, for example, wound exudates and baby slobber, and that they are even able to penetrate bandages. They are often found in toilets, in foodstores and other hospital stocks. Their tiny size (2 to 3 mm) allows them to find their way through small holes, and so they can be found inside sterile packs. It is extremely difficult to trace their nests. The queens rarely leave the nests; they are catered to, like the *larvae* and *pupae,* by worker ants, which themselves are not fertile. If the animals are sprayed with insecticides outside the nests, the plague will vanish only temporarily. In the nests the ants repro-duce rapidly (the cycle is only 6 weeks). The only effective control method would be to kill the queens and the breed (eggs, *pupae* and *larvae*), but it is nearly impossible to localize them. In view of these problems a search for an effective and safe control method is urgent. A means for controlling the insects in a specific manner might be found in the use of their pheromones (Ritter, Rotgans, Talman, Verwiel, and Stein, 1973). Pheromones are chemical messengers (semiochemicals) produced by exocrine glands and used for communication within a species.

Experience in our laboratory with trail-following substances for termites and their applications has taught us that these pheromones can be useful in controlling social insects. The behavior of Pharaoh's ants suggests that they produce a trail-following substance (Fig. 1). Blum (1966) already observed

FIG. 1. Trail-following Pharaoh's ants on a wall near an oven in a bakery.

that the trail odor of Pharaoh's ants is rather persistent and that ants followed it even after 24 hr.

The first ant trail-following substance whose structure was elucidated was that of the leaf-cutting ant, *Atta texana* (Tumlinson, Silverstein, Moser, Brownlee, and Ruth, 1971). It is a nitrogen-containing substance: methyl 4-methylpyrrole-2-carboxylate.

II. REARING OF THE ANTS, BIOLOGICAL ASSAY OF THE PHEROMONE, AND ITS EXTRACTION FROM THE ANTS

The biological assay was developed as a choice test (Fig. 2). The apparatus consists of a glass reservoir provided with two glass tubes. In each of these a paper strip is placed, one containing the material to be tested, and the other serving as a blank. Figure 3 shows the result for an active extract. Although the compound to be discussed here may elicit trail-following activity, its biological activity is more clearly demonstrated by the choice test.

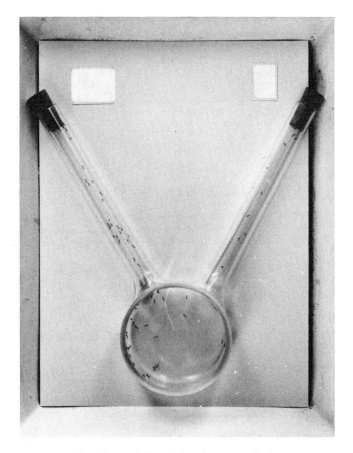

FIG. 2. General view of the choice test device.

Active material was collected from ants by extracting a homogenate of thousands of ants with methylene chloride. Gas chromatography of this extract on OV-17 resulted in three biologically active zones in the chromatogram (Fig. 4). OV-101 gave a better resolution (Fig. 5). This figure shows that MP 3 consists of two compounds, one of which, MP 3–1, appears to be the active component. The structures of MP 1, MP 2–1, and MP 3–1 have been elucidated. This chapter is mainly concerned with the structure of MP 1.

III. SPECTROMETRIC INVESTIGATION OF MP 1

This compound was isolated by trapping it in a capillary tube connected to the outlet of the gas chromatograph and cooled with liquid nitrogen. The mass spectrum of this compound is represented in Fig. 6. The bar graph is a representation of the LRP spectrum obtained with an LKB 9000 GC–MS

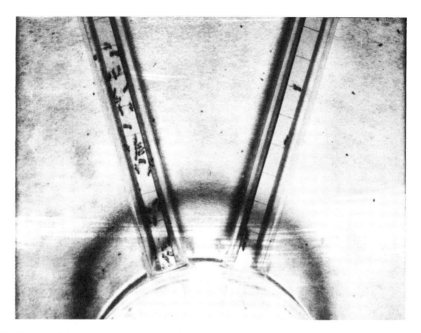

FIG. 3. Part of the apparatus of Fig. 2. In the left tube, paper impregnated with pheromone extract, in the right one blank paper.

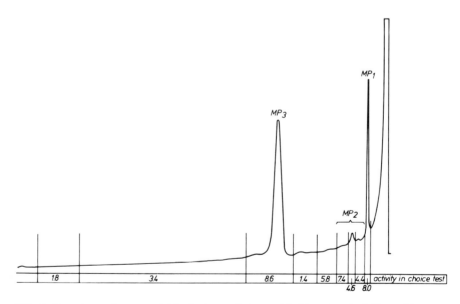

FIG. 4. Gas chromatogram of 10 µl of a methylene chloride extract of Pharaoh's ants. Column: 2 m 3% OV-17; column temp.: 120°C. The activity is the difference between the average number of ants on the sample and on the blank.

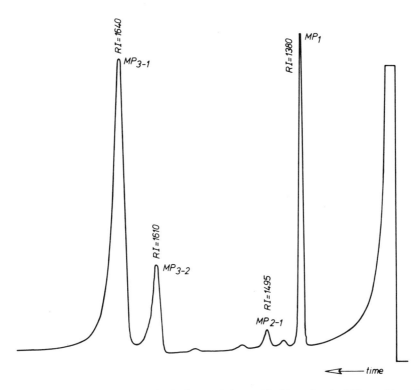

FIG. 5 Gas chromatogram of 100 μl of a methylene chloride extract of Pharaoh's ants. Column: 6 m 5% OV-101; column temp.: 130°C.

combination; the list of elementary compositions of the fragment ions has been derived from HRP spectra. These four metastable transitions could be determined unambiguously:

$$195 \rightarrow 138 \qquad M \; -C_4H_9$$
$$138 \rightarrow 136 \qquad 138 - H_2$$
$$138 \rightarrow 95 \qquad 138 - C_2H_5N$$
$$138 \rightarrow 68 \qquad 138 - C_5H_{10}.$$

The following conclusions were drawn from the mass spectrometric data of the purified compound:

(1) Elementary formula $C_{13}H_{25}N$: two elements of unsaturation (double bonds or rings)

(2) Very prominent peak at m/e 138: α-splitting of an amine

$$\diagdown N-C-\!\!\!\!\big\{C_4H_9 \rightarrow \diagdown N^{+}\!\!=\!\!C$$

(3) Prominent peak at m/e 180: α-splitting of an amine

$$\diagdown N-C-\!\!\!\!\big\{CH_3 \rightarrow \diagdown N^{+}\!\!=\!\!C$$

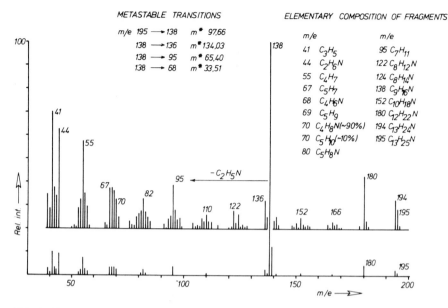

FIG. 6. Mass spectrum of the first active compound MP 1 from the gas chromatogram represented in Fig. 5.

(4) (M-1) peak higher than M-peak: this possibly indicates a cyclic amine.

These conclusions suggest the partial structure

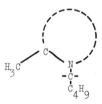

The mass spectrum of a compound such as N-pentylpiperidine is expected to be very similar to that of MP 1.

The spectrum of the former is shown in Fig. 7, which indeed features M-1 (m/e 154), M-57, C_4H_9 (m/e 98), and 98–43, C_2H_5N (m/e 55). These fragmentations have been proved by metastable transitions. The removal of C_2H_5N from the ion m/e 98 can be understood as follows:

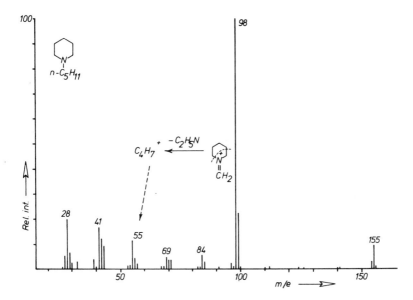

FIG. 7. Mass spectrum of N-pentyl piperidine.

If the removal of C_2H_5N from the ion m/e 138 of MP 1 is an analogous process, two structures for MP 1 remain to be considered:

OR

The mass spectra thus suggest, that in MP 1, the nitrogen atom is bonded to a CH_2 group. However, the NMR spectrum (Fig. 8) of the compound precludes this structure. We clearly see the signals of three different N—CH protons, and so we arrive at the structures for MP 1:

(I)   or   (II)

The loss of a C_2H_5N fragment from m/e 138 may then be depicted by

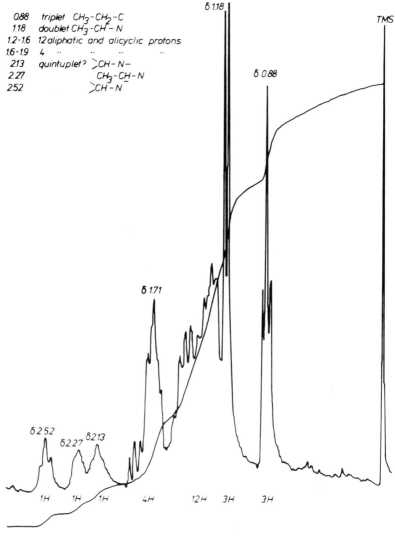

The infrared spectrum of MP 1 is consistent with structures of the kind outlined in the foregoing: CH_3— and CH_2 groups, no double bonds, no N—H bonds, possibly a secondary or tertiary amine. The Sadtler catalogue of IR spectra contains a number of compounds whose band frequencies in the fingerprint region correspond well with the spectrum of MP 1 (Table 1).

FIG. 8. NMR spectrum (220 MHz) of the active compound MP 1.

TABLE 1. *IR spectra of compounds related to MP 1 in Sadtler catalogue*

Nr.	formula	name	structure	bandfrequencies in fingerprint region *)
5102	$C_{11}H_{21}N$	1-cyclohexyl-piperidine		1240 (1220) 1220 (1200) 1180 (1190) 1160 (1160) 1120 (1120) 1100 (1100) 970 (960)
15285	$C_7H_{15}N$	1 ethyl-piperidine	C_2H_5-N	1310 (1310) 1100 (1120) 1080 (1100) 1040 (1030) 1020 (1000) 960 (960) 930 (930)
5189 K	$C_8H_{15}N$	octahydro indolizine		1150 (1160) 1120 (1120) 1090 (1100) 1060 (1050) 1050 (1030)
15042	$C_7H_{13}N$	1 aza-bicyclo (4,2,0) octane (conidine)		1360 (1370) 1310 (1310) 1290 (1295) 1290 (1295) 1240 (1220) 1210 (1200) 1150 (1160) 970 (960)
16581	$C_5H_{11}N$	1-methyl-pyrrolidine	$N-CH_3$	1220 (1220) 1200 (1200) 1150 (1120) 1110 (1100) 1040 (1030) 960 (960)
17073	$C_{10}H_{17}N$	1-H quinolizine, 4 methyl,2,6,7,8, 9,9A hexahydro	CH_3	

*) Wave numbers in parenthesis are maxima in the spectrum of MP 1.

IV. SYNTHESIS OF 3-METHYL-5-BUTYL-OCTAHYDROINDOLIZINE

The spectra do not allow a distinction to be made between I and II for the structure of MP 1.

The synthesis of II was expected to be easier than that of I. This synthesis was carried out by condensing 4-aminopentanal diethylacetal, pentanal, and diethyl-3-oxo-glutarate to yield 3-methyl-5-butyl-6,8-dicarbethoxy-7-oxo-octahydroindolizine. Subsequent saponification, decarboxylation, and reduction yielded 3-methyl-5-butyl-octahydroindolizine. A schematic representation is given in Fig. 9. The details of this synthesis will be published elsewhere. The reaction product showed biological activity toward Pharaoh's ants.

It was shown, however, to be a mixture of many compounds, and so the question arose: which is the active component? A gas chromatogram of the mixture is given in Fig. 10. Only component B was biologically active, probably being identical with the natural product MP 1. The fractions have been analyzed by GC–MS. Figure 11 shows the mass spectrum of component A. This compound apparently is 2-methyl-6-butyl piperidine, a by-product of the synthesis; it may have been formed by reaction of only two of the three reactants:

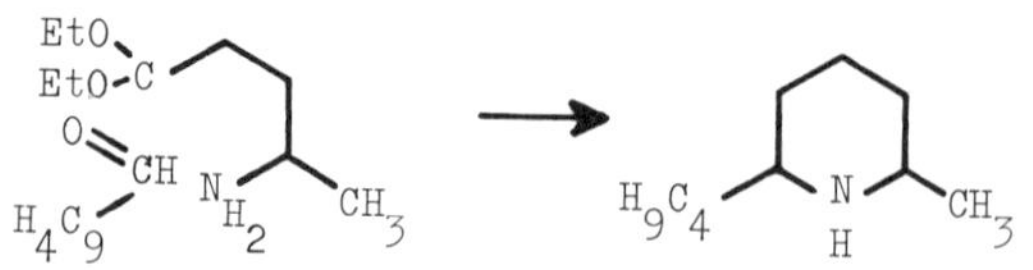

The mass spectrum of component B is identical to that of MP 1 (Fig. 12). The small peaks at m/e 182 and m/e 140 are due to a minor impurity of molecular weight 197, probably having the structure

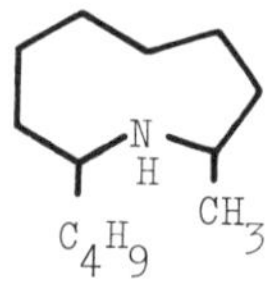

Components C and D showed nearly identical mass spectra (Fig. 13) which differed slightly, but significantly, from the mass spectrum of B or MP 1. B shows rather prominent peaks at m/e 68 and m/e 95, which are more or less lacking in the spectra of C and D, whereas in these spectra m/e 110 is relatively prominent.

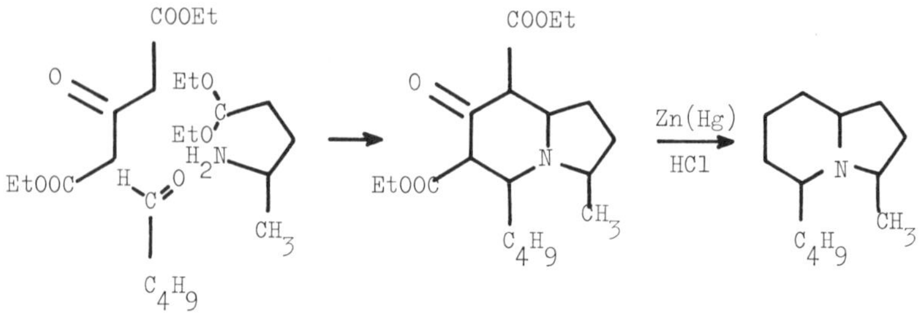

FIG. 9. Synthesis of 3-methyl-5-butyl-octahydroindolizine.

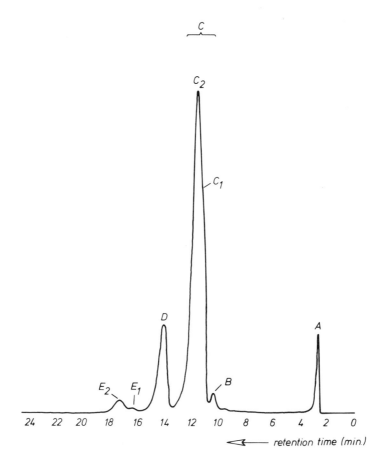

FIG. 10. Gas chromatogram of ca. 5 μg of the synthetic product.

The products found in the GC–MS analysis of the reaction mixture and their molecular weights are listed in Table 2 (see also Fig. 10). Although the synthesis was aimed at obtaining II, we suspect that the active component B has structure I. We have two arguments in favor of this:

(1) By GC–MS analysis of one of the other active fractions (fraction MP 2 in Fig. 4) of the ant extract, we came across a compound (MP 2–1 in Fig. 5) which itself had little or no biological activity, and which had the mass spectrum represented in Fig. 14, molecular weight 197, and two very large fragment ion peaks at m/e 140 (M-57, C_4H_9) and m/e 126 (M-72, C_5H_{11}). The proposed structure of this compound is the following:

TABLE 2. GC–MS analysis of the synthetic reaction
mixture

Component	Molecular weight		
A	155		
B		195	197
C$_1$		195	
C$_2$		195	
D		195	
E$_1$ } E$_2$ }	193	195	197

A is a by-product of the synthesis.
B is a minor component with M-195, contains a
small amount of a component of M-197.
C$_1$, C$_2$, D are three components of M-195.
E$_1$ and E$_2$ are two peaks, consisting of compo-
nents of M-193, M-195, and M-197.

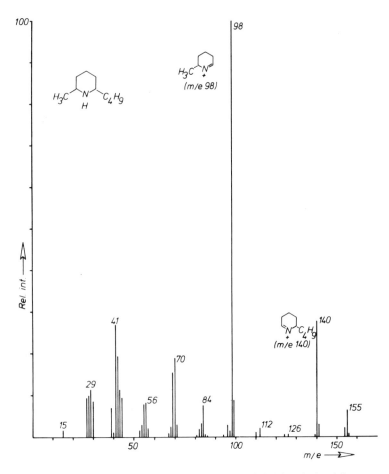

FIG. 11. Mass spectrum of compound A, 2-methyl-6-butyl-piperidine.

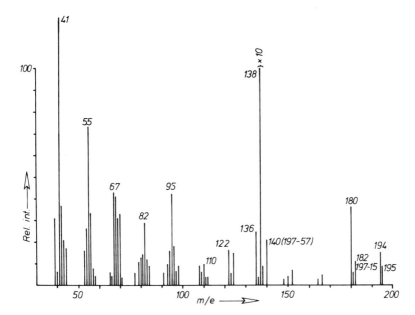

FIG. 12. Mass spectrum of compound B.

The structure of this molecule can be represented in such a way that a biogenetic relation with MP 1 becomes apparent:

If this relation does exist, MP 1 must have the structure with the methyl group in the six-membered ring.

(2) A second argument is that the yield of the active compound is small, and that the synthesis produces four more compounds of molecular weight 195 (see Table 2), whereas only four DL-pairs of stereoisomers were expected.

V. DEHYDROGENATION EXPERIMENTS

The structure of MP 1 could be proved with the aid of (de)-hydrogenation experiments. Beroza (1962) published a method for determining the carbon skeleton of a compound by exhaustive hydrogenation. The heteroatoms are eliminated and a saturated hydrocarbon remains. The compounds under investigation, when hydrogenated according to the Beroza method, should give the straight-chain aliphatic hydrocarbon tridecane, $C_{13}H_{28}$. This hydrocarbon gives no clues to the location of the substituents in the octa-

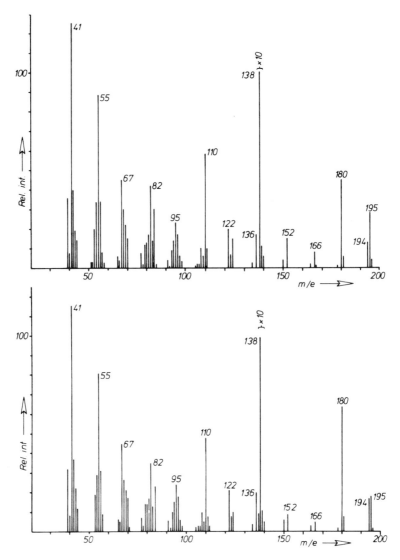

FIG. 13. Mass spectra of compounds C and D.

hydroindolizine skeleton. Fortunately, however, some cyclic compounds on dehydrogenation give rise to aromatic products already found by Beroza and Sarmiento (1964; see also Kepner and Maarse, 1972). We found that treatment of the active natural compound and of the synthetic components according to the Beroza technique also gave unsaturated products, having molecular weights of 4, 6, and 8 mass units less than the parent compound.

Figure 15 shows the gas chromatograms (on OV-101) of the dehydrogenation products of MP 1 and of the synthetic product D. These chromato-

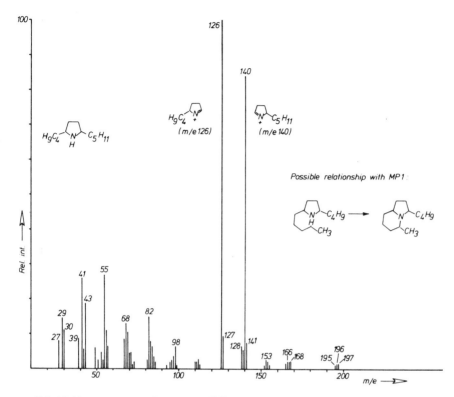

FIG. 14. Mass spectrum of compound MP 2-1 in Fig. 6, 2-butyl-5-pentyl-pyrrolidine.

grams are a composite picture of the gas chromatograms from the singly injected components of each mixture on the GC–MS. The products with the shortest retention times and a molecular weight 191 showed a striking difference in their mass spectra (Fig. 16). The product of MP 1 showed a large peak at m/e 107, whereas that of D showed one at m/e 149. A satisfactory explanation can only be given if in MP 1 the methyl group and in D the butyl group is attached to the six-membered ring:

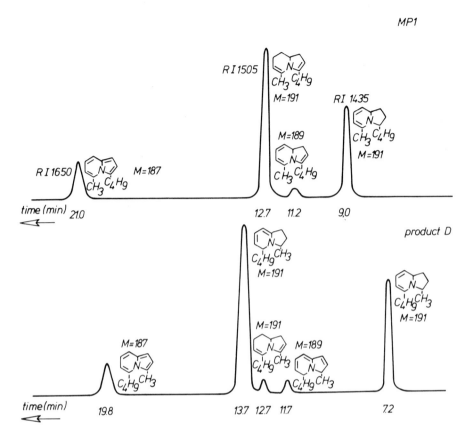

FIG. 15. Reconstructed gas chromatograms of dehydrogenation products of the natural compound MP 1 and of the synthetic compound D. Stationary phase OV-101.

The other products formed and their mass spectra are shown in the Figs. 17–20. A detailed study of the mass spectra enabled us to determine the probable location of the double bonds.

The position of the substituents at the octahydroindolizine skeleton having been cleared, it can be understood how 3-butyl-5-methyloctahydroindolizine might be formed as a by-product in the synthesis of 3-methyl-5-butyloctahydroindolizine:

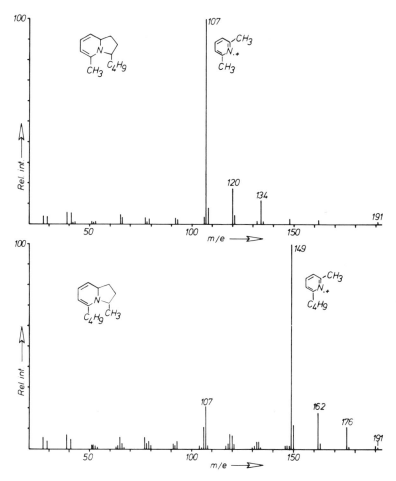

FIG. 16. Mass spectra of dehydrogenation products of MP 1 and compound D, molecular weight 191.

VI. DISCUSSION OF THE STEREOCHEMICAL CONFIGURATION

The final problem is that of the stereochemistry. Infrared and NMR provide clues to the possible stereochemical configuration of the molecule.

In the infrared spectra of *trans*-quinolizidine ring systems, Bohlmann (1958) observed a characteristic group of absorption bands at 2,700 to 2,800 cm⁻¹ if the compound, in a stable conformation, has at least two C—H bonds *trans* relative to the lone electron pair of the nitrogen atom. If this observation also holds for an octahydroindolizine ring system, we may conclude, from the very weak intensity of the "Bohlmann bands" in the infrared spectrum of MP 1, that in the most stable conformation not

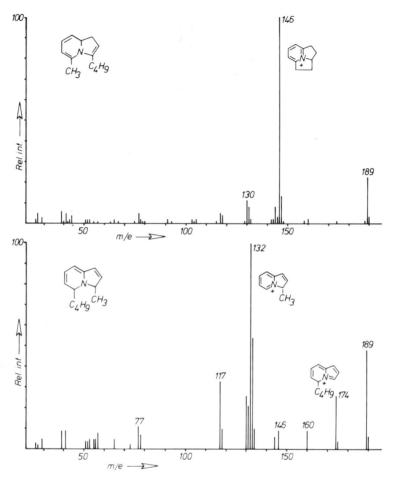

FIG. 17. Mass spectra of dehydrogenation products of MP 1 and compound D, molecular weight 189.

more than one α-H atom is *trans* relative to the electron pair. From the chemical shifts in the NMR spectrum, it follows that all α-protons are probably axial or (in the five-membered ring) pseudo-axial. The only possible configuration then is:

or its mirror image.

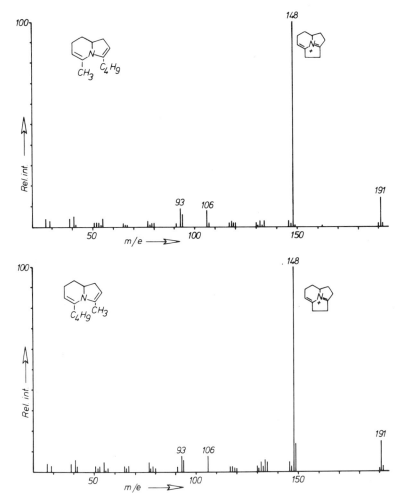

FIG. 18. Mass spectra of dehydrogenation products of MP 1 and compound D, molecular weight 191.

ACKNOWLEDGMENT

This work was supported by the Netherlands Ministry of Health and Environmental Hygiene. Pharaoh's ants were supplied by Dr. A. Busch-inger, Bonn, Germany, and Mr. W. R. C. M. van der Loo, Rotterdam. High resolution mass spectra were run by Dr. H. A. H. Craenen, Chemisch Laboratorium RVO–TNO, Rijswijk, on a JEOL 01SG-2, and by Dr. W. Heerma, Analytisch Chemisch Laboratorium, University, Utrecht, using an AEI MS 902.

The team working on this subject includes: Miss I. E. M. Rotgans (biochemistry, gas chromatography), S. J. Spijk (gas chromatography–

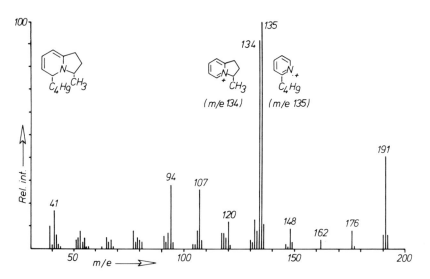

FIG. 19. Mass spectrum of dehydrogenation product of compound D, molecular weight 191.

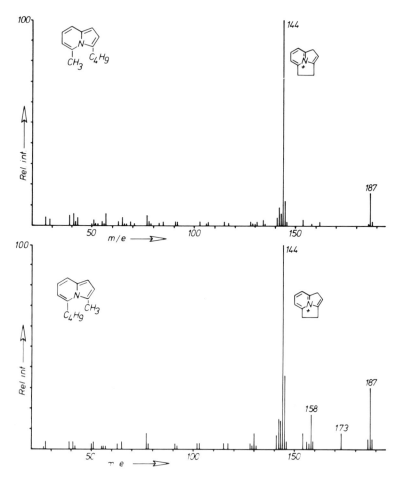

FIG. 20. Mass spectra of dehydrogenation products of MP 1 and compound D, molecular weight 187.

mass spectrometry), Dr. R. Deen and P. J. F. Nooijen (infrared spectrometry), and F. Stein and J. M. Timmner (synthesis).

REFERENCES

Beroza, M. (1962): Determination of the chemical structure of microgram amounts of organic compounds by gas chromatography. *Analytical Chemistry*, 34:1801–1811.

Beroza, M., and Sarmiento, R. (1964): Carbon skeleton chromatography using hot-wire thermal-conductivity detection. *Analytical Chemistry*, 36:1744–1750.

Blum, M. S. (1966): The source and specificity of trail pheromones in *Termitopone, Monomorium, and Huberia,* and their relation to those of some other ants. *Proceedings of the Royal Entomological Society of London,* (A) 41:155–160.

Bohlmann, F. (1958): Zur Konfigurationsbestimmung von Chinolizidin-Derivaten. *Chemische Berichte,* 91:2157–2167.

Kepner, R. E., and Maarse, H. (1972): Hydrogenolysis of terpenes in the injection port of a gas chromatograph. I. Monoterpenes. *Journal of Chromatography,* 66:229–237.

Ritter, F. J., Rotgans, I. E. M., Talman, E., Verwiel, P. E. J., and Stein, F. (1973): 3-Butyl-5-methyl-octahydroindolizine, a novel type of pheromone attractive to Pharaoh's ants, *Monomorium pharaonis* (L). *Experientia,* 29:530–531.

Tumlinson, J. H., Silverstein, R. M., Moser, J. C., Brownlee, R. G., and Ruth, J. M. (1971): Identification of the trail pheromone of a leaf-cutting ant, *Atta texana. Nature,* 234:348–349.

Mass Spectrometry in Biochemistry and Medicine,
edited by A. Frigerio and N. Castagnoli.
Raven Press, New York © 1974

Structure Elucidation of Pyrolytic Products of Cannabidiol by Mass Spectrometry

W. Heerma, J. K. Terlouw, A. Laven, G. Dijkstra, F. J. E. M. Küppers, R. J. J. C. Lousberg, and C. A. Salemink

Laboratories for Analytical Chemistry and Organic Chemistry, University of Utrecht, The Netherlands

I. INTRODUCTION

The increasing consumption of hashish, the well-known cannabis resin, has led to a growing interest in the chemical and medicinal properties of the material. Progress has been rapid, especially since 1965. The active compound, which is responsible for the effect when the drug is taken, is Δ-1(2) tetrahydrocannabinol (THC). However, one of the riddles in the properties of hashish is the psychotomimetic effect produced when it is smoked, independently of the presence or near absence of this active compound. Because the pyrolysate of the total resin forms a necessarily extremely complicated mixture, we decided to investigate the pyrolysis products of cannabidiol (CBD), an inactive but main cannabis constituent which may produce new active compounds.

II. RESULTS AND DISCUSSION

The pyrolysis was performed with pure isolated CBD in a quartz tube at a temperature range of 400 to 700°C. The gas chromatogram (GC) of the collected material indicated the formation of a mixture of several components next to the unchanged CBD (Fig. 1).

We isolated each peak by preparative GC followed by consecutive mass spectra that were recorded during the slow evaporation of each sample at a relatively low ion source temperature. Most of the starting material remained unchanged (about 90%); peaks I and II with retention times smaller than CBD yielded mass spectra of compounds with lower molecular weights than that of CBD and must be considered as cracking products. We, therefore, primarily considered peak III (5% of the starting material) of which an average spectrum is presented in Fig. 2. The abnormal broadness of the peak and the continuously changing fragmentation pattern, combined with the unusual mass differences in the spectra (330, 314, 295), clearly indicate that the GC peak consists of a mixture of compounds.

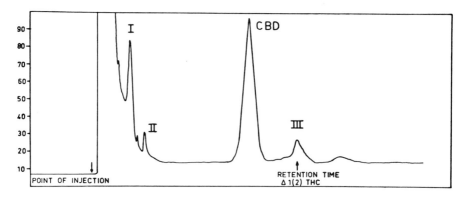

FIG. 1. Gas chromatogram of the pyrolysate of CBD.

Indeed, thin-layer chromatography of peak III revealed about 10 spots which after subsequent extraction were submitted to mass spectral analysis. One of the spots corresponds to a component with a molecular weight of 330. The other spots orginate from four compounds with a molecular weight of 314, one with a molecular weight of 312, and another with a molecular weight of 310. The amount of the sample of the other spots was too small to permit the recording of well-defined mass spectra.

From the mass spectra of a number of cannabinoids reported earlier (Budzikiewicz, Alpin, Lichtner, Djerassi, Mechoulam, and Gaoni, 1965; Claussen and Korte, 1965), it is clear that the choice of recording conditions is very critical (Fig. 3). To avoid thermal rearrangements and/or degradations it is imperative to keep the ion source and probe temperature at the lowest level which still permits a proper evaporation of the sample and thus allows the recording of the most reliable and characteristic mass spectra. The mass spectrum of the main pyrolytic component (molecular weight 330) is presented in Fig. 4. Its composition ($C_{21}H_{30}O_3$) has been established by exact mass measurement, and its purity is guaranteed by

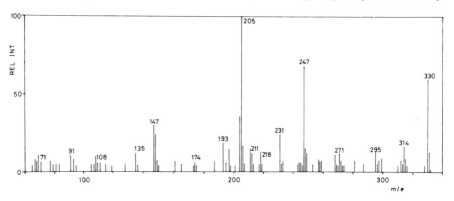

FIG. 2. 70 eV spectrum of GLC peak III.

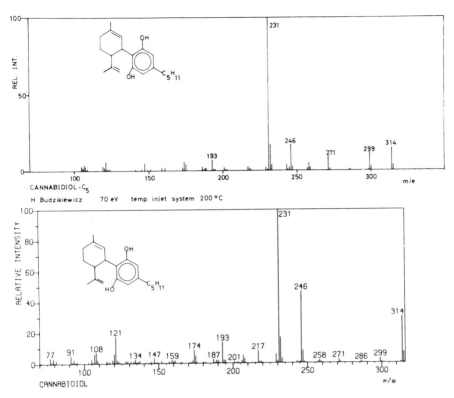

CANNABIDIOL-C₅

H. Budzikiewicz   70 eV   temp. inlet system 200 °C

CANNABIDIOL

FIG. 3. 70 eV temperature inlet system 200°C (Budzikiewicz et al., 1965) (top) and 70 eV temperature probe and ion source 65°C (bottom).

the constancy of several peak intensity ratios. The absence of this compound in the pyrolysate when oxygen was excluded during the pyrolysis of CBD, indicated that oxygen from the air had been introduced into the molecule. The composition of all fragment ions has been determined, and main fragmentations have been traced by means of metastable measurements,

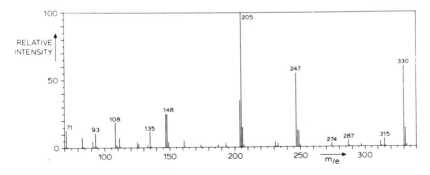

FIG. 4. Compound-330.

using the defocusing technique. Silylation of the compound under mild conditions yielded a monosilylated product, whereas more rigorous silylation resulted in a small amount of a disilylated product. From the mass spectra of both derivatives, as compared with that of the original compound, evidence could be obtained that only one phenolic hydroxyl group was present. Since an $(M-H_2O)^{+\cdot}$ peak is completely absent in the spectrum of CBD, the presence of the $(M-H_2O)^{+\cdot}$ peak at m/e 312 in the spectrum of the "330-compound" indicated that the second hydroxyl group was most likely situated in the terpene moiety of the molecule. This substantial amount of structural information together with the knowledge of the mass spectral behavior of other cannabinoids suggested a structure closely related to the decarboxylated product of cannabielsoic acid, a naturally occurring cannabinoid earlier isolated by Shani and Mechoulam (1970). At elevated temperatures this cannabielsoic acid, of which two stereoisomers are presented in Fig. 5, was easily converted into cannabielsoin by elimination of carbon dioxide. The principal fragment ions of cannabielsoin, together with its measured fragmentation pathways (Fig. 6), fit its chemical structure very well.

The majority of the fragmentation reactions are governed by the hydroxyl group in the nonaromatic part of the molecule. The fragmentation is initiated by ring opening, hydrogen shifts, and elimination of different neutrals. In most fragment ions the aromatic part of the molecule is still present, because fragmentation of this moiety is not an energetically favorable process. The abundance of these fragment ions can be explained by a stabilization due to the presence of the aromatic part in the ion, which also can expand to yield a tropylium structure. The ion at m/e 71 ($C_4H_7O^+$), although of relatively small abundance, yielded important structural information about the position of the hydroxyl group, because it originates directly from the molecular ion (Fig. 6). In order to decide which isomer has been obtained through pyrolysis of CBD, cannabielsoic acid from natural sources has been isolated and decarboxylated in sufficient quantities to allow NMR experiments. Based upon these NMR data, Rf-values, and

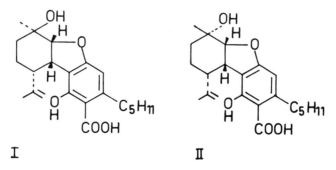

FIG. 5. C1-stereoisomers of cannabielsoic acid A.

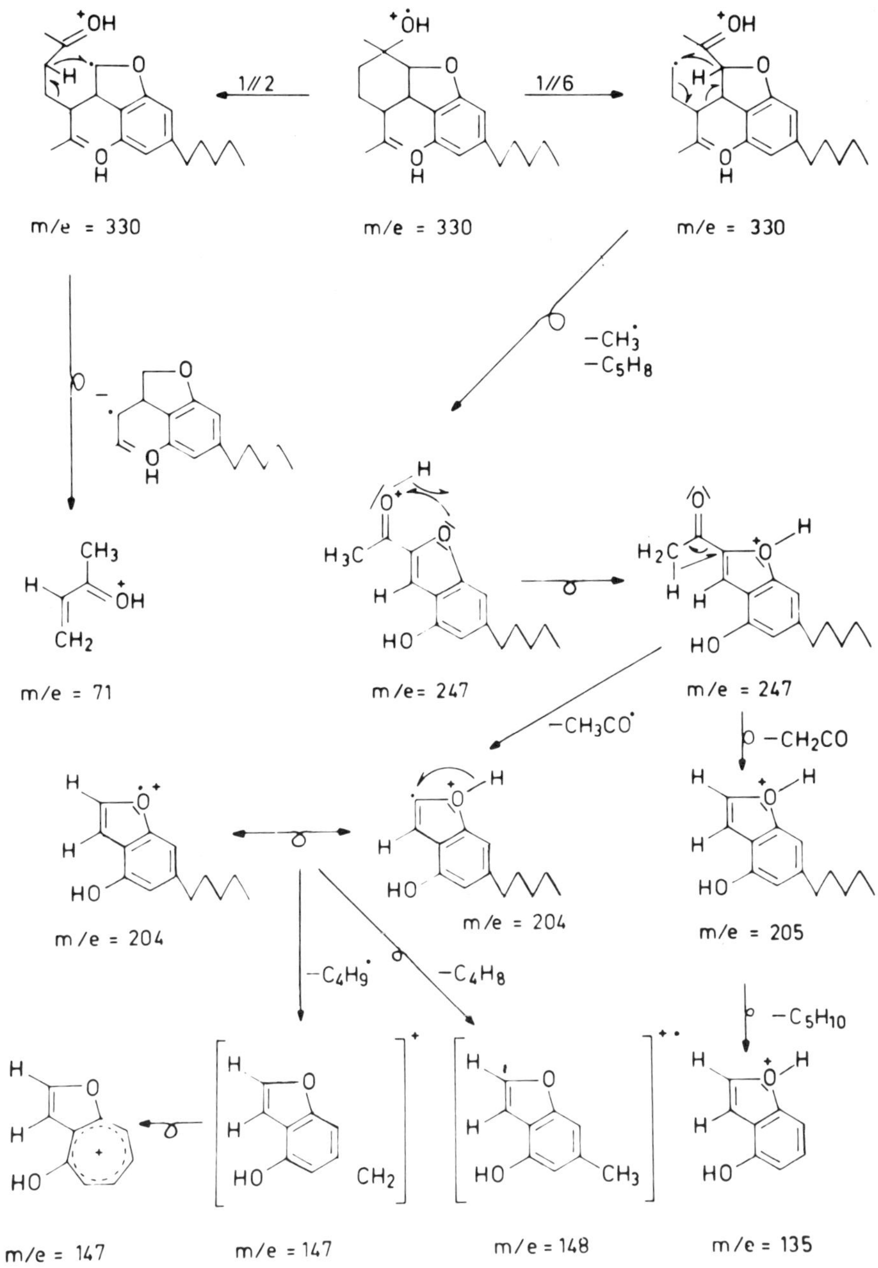

FIG. 6. Proposed fragmentation mechanism of cannabielsoin.

comparison of mass spectra (Fig. 7), the hydroxyl group at C_1 was determined to be in an equatorial position. Hydrogen atoms at C_2 and C_3 were found to be in a *cis* orientation, in agreement with Mechoulam's proposal.

The four pyrolysis compounds with molecular weight 314 in peak III (Fig. 1) were found to be CBD due to the large tailing of the unconverted starting material, Δ-1(2) THC in small amounts, and two so far unknown compounds of which the mass spectra and tentative structures (merely based upon NMR data) are presented in Fig. 8.

Finally it should be noted that there is still much obscurity in the fragmentation behavior of well-known cannabinoids. For instance, Claussen and Korte (1965) found that the position of the double bond in the terpene moiety is responsible for a typical fragmentation pattern. Although we agree with their view that the position of the double bond is important, we still cannot understand the magnitude of the differences in the mass spectra (Claussen, Fehlhaber, and Korte, 1966). This can be illustrated in the spectra of Δ-1(2) THC and Δ-1(6) THC, where the abundances of the $(M-CH_3)^+$ ions differ by a factor of 10 (Fig. 9).

Starting from a chemical structure of a cannabinoid, the presence of distinct fragment ions cannot be predicted, let alone their abundances. Therefore, it is almost impossible to elucidate the structure of an unknown cannabinoid from its mass spectrum alone. All spectrometric information

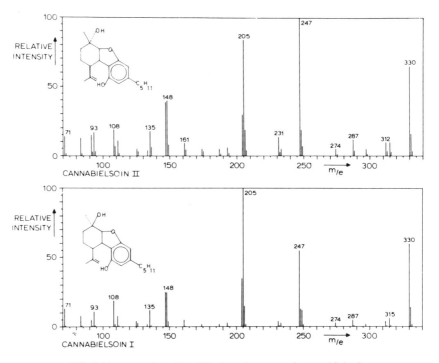

FIG. 7. Mass spectra of the C1-stereoisomers of cannabielsoin.

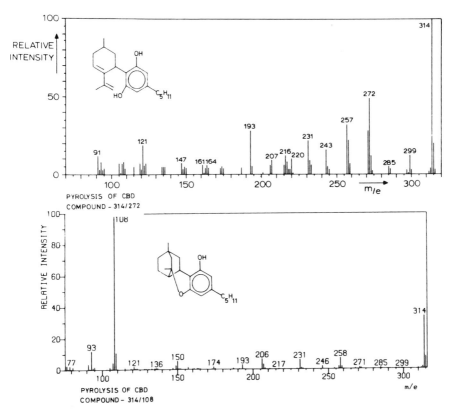

FIG. 8. Mass spectra and tentative structures of two unknown products.

should be compiled to obtain a good basis for structural proposals. Better use of mass spectrometric information in this field can be made when more insight has been gained into the fragmentation behavior of cannabinoids and the factors influencing this behavior. This information must be obtained by thoroughly studying the mass spectra of cannabinoids labeled at distinct positions in the molecule and utilizing energetic data such as ionization and appearance potentials and information on metastables.

III. EXPERIMENTAL

The detailed pyrolysis of CBD and the chromatographic experiments have been described elsewhere (Küppers, Lousberg, Bercht, Salemink, Terlouw, Heerma, and Laven, 1973). Mass spectra were recorded using an AEI MS 902 instrument at an ion source temperature of about 50 to 70°C and an electron energy of 70 eV. HRP measurements were made using a Ferranti Argus 500 computer on line with the mass spectrometer. LRP plots were made by averaging at least 10 spectra.

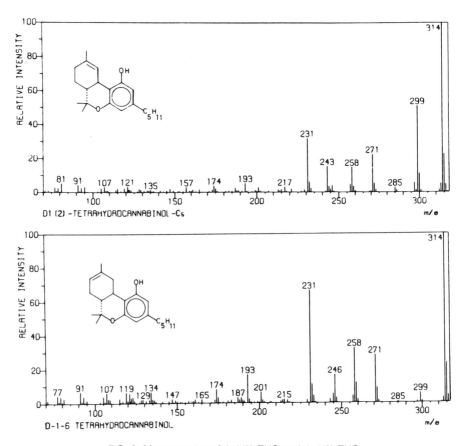

FIG. 9. Mass spectra of Δ-1(2)-THC and Δ-1(6)-THC.

ACKNOWLEDGMENT

We are indebted to Miss N. C. Schut, Mrs. J. S. Vaarkamp, and Mr. C. Versluis for valuable assistance.

This work was supported by the Netherlands Ministry of Public Health and Environmental Hygiene.

REFERENCES

Budzikiewicz, H., Alpin, R. T., Lichtner, D. A., Djerassi, C., Mechoulam, R., and Gaoni, Y., (1965): Massenspektroskopie und ihre Anwendung auf strukturelle und stereochemische Probleme—LXVIII Massenspektroskopische Untersuchung der Inhaltsstoffe von Haschisch. *Tetrahedron,* 21:1881–1888.

Claussen, U., Fehlhaber, H. W., and Korte, F. (1966): Haschisch—XI Massenspektrometrische Bestimmung von Haschisch-Inhaltsstoffe II. *Tetrahedron,* 22:3535–3543.

Claussen, U., and Korte, F. (1965): Zur chemischen Klassifizierung von Pflanzen—XXX.

Haschisch – IX Massenspektrometrische Bestimmung von Haschisch-Inhaltsstoffe. *Tetrahedron*, Suppl. 7:89–96.

Küppers, F. J. E. M., Lousberg, R. J. J. C., Bercht, C. A. L., Salemink, C. A., Terlouw, J. K., Heerma, W., and Laven, A. (1973): Cannabis-VIII pyrolysis of cannabidiol. Structure elucidation of the main pyrolytic product. *Tetrahedron*, 29:2797–2802.

Shani, A., and Mechoulam, R. (1970): A new type of cannabinoid: Synthesis of cannabielsoic acid A by a novel photo-oxidative cyclisation. *Chemical Communications*, 273–274.

Mass Spectrometry in Biochemistry and Medicine,
edited by A. Frigerio and N. Castagnoli.
Raven Press, New York © 1974

Mass Spectral Studies in Phenoxazine and Azaphenoxazine Series

J. Cassan, M. Rouillard, and M. Azzaro

Laboratoire de Chimie Physique Organique, Université de NICE, 06034 NICE Cédex

The therapeutic import of phenothiazine derivatives is well established. In order to establish a correlation between structure and pharmacodynamic activity, several studies of this molecule and homologues such as the phenoxazines and azaphenoxazines have provided a better knowledge of structures and physical properties.

Our current research program focuses on the use of mass spectrometry to investigate a series of substituted 10-alkyl phenothiazines (Audier, Azzaro, Cambon, and Guedz, 1968), a series of phenothiazines substituted on the aromatic rings (Mital, Jairn, Azzaro, Cambon, and Rosset, 1970), and, very recently, a series of similarly substituted phenoxazines.

In this chapter we compare and contrast the mass spectral behavior of these oxygen and sulfur heterocyclic systems. The mass spectra of mono-, di-, tri-, and tetra-substituted phenoxazines and azaphenoxazines are presented and discussed.

I. PHENOXAZINE

In order to rationalize fragmentation modes, we have discussed the degradation of phenoxazines whose spectra show similarities with phenothiazines (Gilbert and Millard, 1969; Heiss and Keller, 1969).

The phenoxazinium ion stability is confirmed by the predominance of the parent peak as for phenothiazines. Phenoxazine looses an amino hydrogen atom more rapidly than phenothiazine; this elimination is induced by electron loss from the oxygen atom.

The localization of charge on the heterocyclic oxygen is confirmed by the absence of any oxygen loss by the molecular ion, whereas phenothiazine loses sulfur atom easily. This difference in the behavior of this compound is explained by the electronegativity of sulfur and oxygen.

The next significant peak at $m/e = 154$ corresponds to the ion formed by loss of [CO and H·] and confirms the bond strength for the bridging oxygen

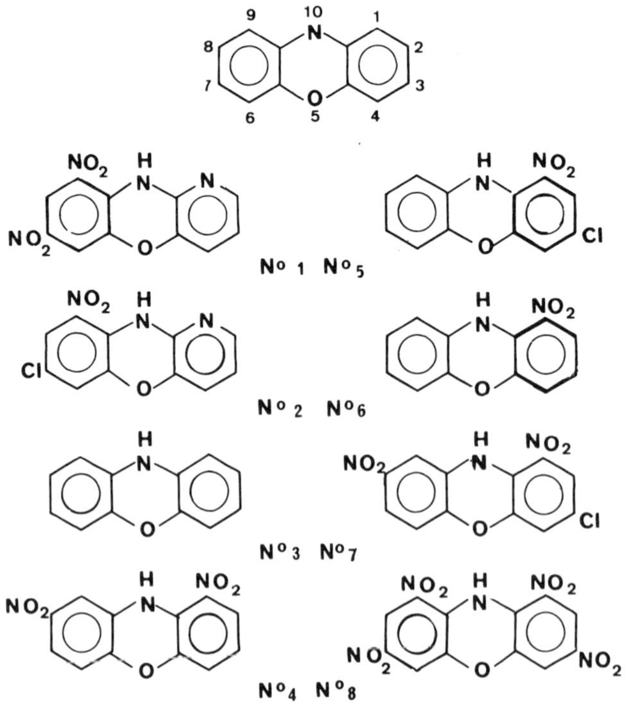

FIG. 1. Mono-, di-, tri-, and tetra-substituted phenoxazines and azaphenoxazines and the substitution site numbering.

atom in the central six-membered ring. Both pathways are possible for m/e = 154:

$$m/e\ 154 \begin{array}{c} \nearrow m/e\ 127 \searrow \\ \searrow m/e\ 128 \nearrow \end{array} m/e\ 101$$

Peaks at m/e 127 and 128 are also found in the phenothiazine mass spectra. Fragments arising from aromatic moieties are noticeable. Mechanisms reported by Morita (1966) for more conjugated and more coplanar compounds than phenoxazine and phenothiazine explain the formation of some weak intensity peaks.

II. SUBSTITUENT EFFECTS IN MASS SPECTROMETRIC FRAGMENTATION PROCESSES OF PHENOXAZINE

A. Loss of Amino Hydrogen

Unlike phenoxazine, the investigated series gives minor peaks (0.5%) corresponding to $[M^{+\cdot} - H\cdot]$. The hydrogen bond between NO_2 in the 1 or 9 position and amino hydrogen corroborates this degradation. If we compare hydrogen mobility in the phenoxazine series, we could give the following order: phenoxazine > phenothiazine > phenothiazine substituted > phenoxazine substituted.

B. Nitrogen and N–H Elimination

The direct loss of N or NH from the molecular ion is not observed. A neutral molecule of hydrogen cyanide is the major mode of nitrogen elimination. In our investigated series, this pathway is followed only after the loss of HCO (Simov and Taulov, 1971) like phenoxazine. Concerning lower intensity peaks, we observe that direct loss of N and NH is found only after substituents breakdown.

C. Bridged Oxygen Loss

The relative abundance of ions $[M^{+\cdot} - 16]$ is around 0.1%. Oxygen expulsion is similar in both the substituted and unsubstituted phenoxazine series. In every case the major pathway is $[CO + H\cdot]$. A nitro group (in position 3) extends conjugation (Giulieri, Rouillard, Mital, and Azzaro, 1970) so that it induces electron π delocalization and decreases π character of the CO group, making oxygen loss easier.

In the phenothiazine series the same result is obtained; effectively, fragment $[M^{+\cdot} - 32]$ is very abundant (42%). If the substituent is a nitro group, the intensity of $[M^{+\cdot} - 32]$ is lower (1%) and if there is a chlorine or bromine atom, the relative abundance of $[M^{+\cdot} - 32]$ is of the order of 6%.

The transconjugation is confirmed by ESCA. We use "transconjugation" to describe π or p interactions between substituent groups through the aromatic system. The loss of sulfur for compounds 9 and 10 is in contradiction with this hypothesis.

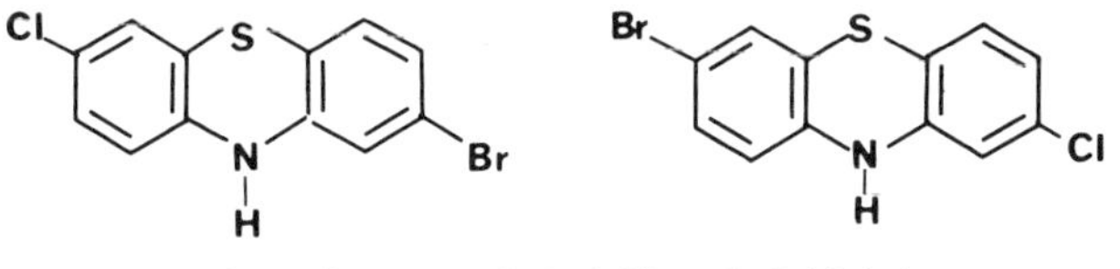

FIG. 2. Compounds 9 (left) and 10 (right).

Bromine (compound 9) and chlorine (compound 10) stabilize the sulfur compounds and prevent almost completely the loss of sulfur. ESCA confirms that simultaneous transconjugation through sulfur and nitrogen is in opposition. Transconjugation through sulfur favors the folded structure of the molecule whereas through nitrogen the effect is opposite. The molecule tends toward coplanarity. That is why sulfur loss is easier for phenothiazine than oxygen loss for phenoxazine.

III. DEGRADATION PATHWAYS INFLUENCED BY CHLORINE AND NO$_2$ GROUPS

A peak at $[M^{+\cdot} - 17]$ is present in all the series and comes from elimination of $-OH$ (confirmed by high resolution). The origin of this peak is the loss of an oxygen from the nitro group in position 1 with a hydrogen from amino group. Intensity of this peak is variable according to substitutions and becomes very small for the 1,3- or 7,9-dinitro substitution. Transconjugation of the nitro group in the 3 position destabilizes the bond between NO$_2$ in the 1 position and the aromatic rings, and so orientates loss of NO$_2$H, promoting the formation of peak $[M^{+\cdot} - 47]$ instead of $[M^{+\cdot} - 17]$ (0.3%).

An elimination mechanism of OH from NO$_2$ in the 3 or 7 position and a hydrogen in (2 or 4) or in (6 or 8) is not probable. The peak at $[M^{+\cdot} - 17]$ is noticeable in the phenothiazine series for compounds with a nitro in 1. This NO$_2$ group *is* necessary for a six-membered ring to promote a rearrangement similar to Mc LAFFERTY's (Meyerson, Puskas, and Fields, 1966).

This rearrangement is only possible if the interatomic distance between the hydrogen and oxygen acceptor is less than 1.8 Å. For an easier delocalization, the six atoms must be coplanar. These stereochemical conditions when operative, like for phenothiazine, correct the orientation of functional hydrogen. Amino hydrogen in 10 can be either in an intra or extra position. The presence of the peak $[M^+ - 17]$ and also the hydrogen bond in the IR show an intra position for 1-nitro phenoxazine.

IV. MECHANISMS CONCERNING IONS $[M^{+\cdot} - 30]$, $[M^{+\cdot} - 46]$, AND $[M^{+\cdot} - 47]$

These peaks are preponderant in all the spectra investigated here, and agree with loss of nitroso, nitro, and nitrous acid fragments. Such mechanisms are argued in other similar aromatic and heterocyclic series.

Loss of the nitroso fragment is realized after stabilization of a hydroxy group by hydrogen bonding but is also explained by nitroso isomerization. For every position of the nitro group, this mechanism is appropriate. The $[M^{+\cdot} - 46]/[M^{+\cdot} - 47]$ ratio is a function of hydrogen mobility (for nitro substitution in 1, the hydrogen eliminated is the amino hydrogen.

The presence of chlorine instead of NO$_2$ (compounds No. 5 and No. 6) gives little difference in the spectra. Like *m*-chloronitrobenzene, chlorine prevents loss of NO but supports loss of NO$_2$ and NO$_2$H. Chlorine is only eliminated after the NO$_2$ group. If there are two NO$_2$ groups on a molecule, the loss of chlorine decreases after the first NO$_2$ elimination and increases after the second NO$_2$.

V. 1,3,7,9-TETRANITRO PHENOXAZINE

The degradation of compound No. 8 is a study of great interest because it summarizes earlier remarks. The elimination of the four NO_2 groups prevails and it is possible to determine the sequence of breakdowns. N.M.R. confirms that the hydrogen bond can be established only with one NO_2. The unchelated NO_2 in 1 is the first to be eliminated. That explains the dominant loss of a nitro group instead of a nitrous acid fragment. After the loss of NO_2H (NO_2 in 9) follows the elimination of the other NO_2 groups situated in 3 and 5. These substituents being eliminated, the loss of HCO according to classical mechanisms gives a preponderant peak according to the following fragmentation.

$$m/e = 178 \ [M^{+\cdot} - NO_2H - 3(NO_2)] \xrightarrow[-29]{-HCO} m/e = 149$$

VI. 1-NITROPHENOXAZINE

The compound No. 6 gives a particular spectrum. There are peaks at $[M^{+\cdot} - 47]$ and at $[M^{+\cdot} + 47]$ of which relative intensities are analogous to the one of molecular peak $[M^{+\cdot}]$. It is evident that an insert temperature which is too high causes duplication phenomena. As a matter of fact, this spectrum of No. 6 is the superposition of three degradation pathways.

(1) $[M^{+\cdot}]$: corresponding to the compound No. 6
(2) $[M^{+\cdot} - 47]$: like phenoxazine
(3) $[M^{+\cdot} + 47]$: like a dinitrophenoxazine

VII. DOUBLY CHARGED IONS

In phenothiazines substituted on the aromatic rings the presence of dicharged ions is only important for dihalogenated compounds (No. 9 and 10). Chlorine and bromine ionization potentials are very close. The impossibility of transconjugation increases electronic density on the chlorine and bromine. Simultaneous losses give dipositive ions. Bromine and chlorine are ionized but not sulfur and nitrogen.

There are doubly charged ions in all investigated spectra. The transconjugation effect observed for dihalogenated phenothiazines is not found in the phenoxazine series. The abundance of dicharged ions increases with the presence of substituents favoring the planarity of the molecule. The correlation between molecular folding and dicharged ions is confirmed by investigations on plane conjugated systems such as quinoxaline and phenazine.

VIII. AZAPHENOXAZINE DEGRADATION

In the case of azaphenoxazine, we observe an increase of the peak $[M^{+\cdot} - 46]$ which is probably related to the basicity of the pyridine nitrogen. We still observed an abundance of dicharged peaks. This fact is in good agreement with the coplanarity of azaphenoxazine molecules.

IX. SUMMARY

The stability of the phenoxazine nucleus leads to complex mechanisms of fragmentation. Substituents favor some pathways and simplify the spectra. Transconjugation effects and the presence of dicharged ions provide fundamental differences between the phenoxazine and phenothiazine series.

The pharmacological properties of phenothiazine appear to be correlated with the folded structure of the nucleus and (side chain) decreasing hydrophobic character of the molecule.

The phenoxazine series is more coplanar and their pharmacological properties are of less importance.

REFERENCES

Audier, L., Azzaro, M., Cambon, A., and Guedj, R. (1968): Phénothiazines à chaines aliphatiques sur l'azote en 10. *Bulletin de la Société Chimique de France,* 3:1013–1020.

Gilbert, J. N. T., and Millard, B. J. (1969): High resolution mass spectra of phenothiazines. *Organic Mass Spectrometry,* 2:17–32.

Giulieri, N., Rouillard, M., Mital, R. L., and Azzaro, M. (1970): Etude structurale d'une série homologue d'azaphénoxazines et de phénoxazines. *Bulletin de la Société Chimique de France,* 11:4194–4197.

Heiss, J., and Keller, K. P. (1969): Mass spectrometry investigation of heterocyclic compounds. *Organic Mass Spectrometry,* 2:819–833.

Meyerson, S., Puskas, I., and Fields, E. (1966): Mass spectra of nitroarenes. *Journal of American Chemical Society,* 88:4974.

Mital, R. L., Jairn, S. K., Azzaro, M., Cambon, A., and Rosset, J. P. (1970): Structure du noyau phénothiazine—Mise en évidence de la configuration dans le réarrangement de Mac LAFFERTY. *Bulletin de la Société Chimique de France,* 6:2195–2198.

Morita, Y. (1966): Studies on phenazines. *Chimical and Pharmaceutical Bulletin,* 14:1:426–432.

Simov, D., and Taulov, I. G. (1971): Ten alkylphenothiazines. *Organic Mass Spectrometry,* 5:1133–1144.

Mass Spectrometry in Biochemistry and Medicine,
edited by A. Frigerio and N. Castagnoli.
Raven Press, New York © 1974

Mass Spectrum of the New Antibiotic Purpuromycin

L. F. Zerilli, M. Landi, G. G. Gallo, and M. R. Bardone

Laboratori Ricerche Lepetit S.p.A., 20158 Milano, Italia

Purpuromycin is a metabolite produced by the microorganism *Actinoplanes ianthinogenes* and has *in vitro* activity against Gram-positive and Gram-negative bacteria and fungi (Coronelli, Pagani, Bardone, and Lancini, 1973). On the basis of the chemical and physical data (functional group analysis, absorption spectra, polarographic behavior, etc.) a striking similarity appeared with the antibiotic γ-rubromycin (Brockmann and Zeeck, 1970).

γ-rubromycin I, R=H
purpuromycin II, R=OH

However, some significant differences indicated the novelty of purpuromycin. In particular, the elemental analysis corresponded to an additional oxygen atom and the mass spectrum did not show the expected molecular ion at m/e 538. The purpose of this chapter is the description of the work carried out with mass spectrometry on purpuromycin and on its degradation products during the structural study (Bardone, Martinelli, Zerilli, and Coronelli, 1973). The mass spectrum of γ-rubromycin, formerly studied by Greul (1971), was reinvestigated and more fully interpreted.

The mass spectrum of γ-rubromycin is shown in Fig. 1. The molecular ion is very intense, particularly if one considers the relatively high molecular weight of the compound. The other few intense peaks of the spectrum are characteristic of the naphthazarin moiety and are interpreted as shown in Scheme I. The base peak at m/e 274 is explained as deriving via a retro Diels Alder type decomposition from the molecular ion with a complete charge localization on the naphthazarin moiety. The fragment of mass 273 can also come directly from the molecular ion. For the other peaks we suggest the following interpretations. The ion of mass 256 derives from the

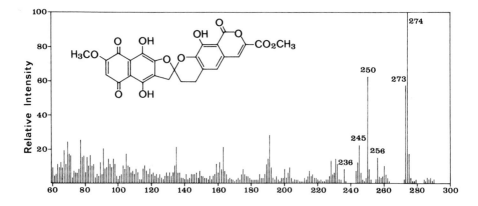

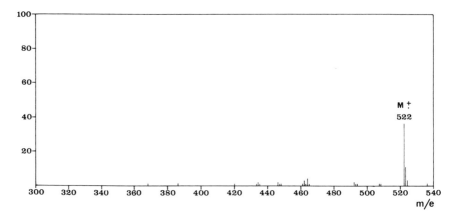

FIG. 1. Mass spectrum of γ-rubromycin (I).

naphthazarin ion of mass 274 by loss of water, a decomposition that will
be described in detail later for a model compound. The peaks at m/e 245,
244, and 243 correspond to the losses of $\cdot CHO$, CH_2O, $CH_3O \cdot$, respectively,
from the fragment of mass 274. The composition of the ions of mass 250 and
236 was demonstrated by exact mass measurements and we suggest that
they originate from the molecular ion. It is interesting to point out that all
the fragments described correspond to the naphthazarin moiety and that the
isocoumarin moiety is not significantly indicated by any peak in the spectrum.

The mass spectrum of purpuromycin, to which structure II was assigned
(Bardone at al., 1973), is shown in Fig. 2. This spectrum was carried out at
250°C (direct inlet system). The first feature we want to discuss is the
absence of the $M^{\pm}$ at the expected value of 538, as indicated by the elemental analysis. The highest mass fragment at m/e 520 is interpreted as
corresponding to the loss of water from purpuromycin itself. This interpreta-

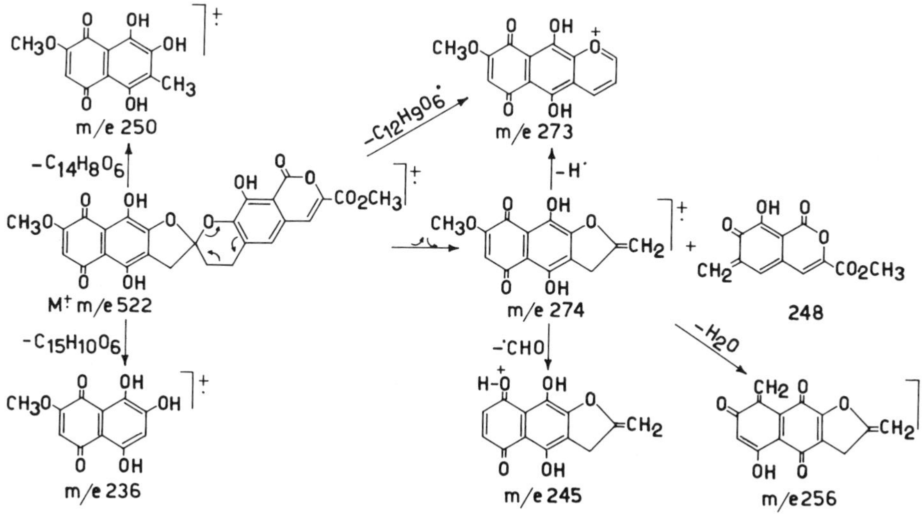

Scheme I

tion is unequivocally confirmed by the mass spectrum of the peracetyl derivative in which the M^+ at m/e 706 accounts for the introduction of four acetyl groups in a molecule having a molecular weight of 538. The origin of the ion of mass 520 has been explained by studying the spectra at different ionization voltages and at different temperatures of the inlet system. Thus, the intensity of this peak does not substantially change with the voltage (from 70 to 20 eV) and it definitely increases in the spectrum obtained at 300°C. This indicates that the m/e 520 peak could be the molecular ion of a product obtained by thermal decomposition of purpuromycin in the ion source. This hypothesis was confirmed by the mass spectrum of the product isolated after thermal decomposition of purpuromycin at 240°C. In fact, as shown by differential thermal analysis, an exothermic decomposition takes place at that temperature with loss of water. The elemental analysis of the product (obtained only in micro amounts) indicated the formula $C_{26}H_{16}O_{12}$, and the structure 3,4-dehydro-γ-rubromycin was safely assigned to the compound. The mass spectrum of this compound is substantially equal to that of purpuromycin at 300°C. It is worthwhile to mention that (Scheme II) the peak at m/e 264 that is present in the spectrum at 250°C is absent at 300°C and that this peak has been taken as characteristic of the decomposition of the molecular ion of purpuromycin. In conclusion, the fragmentation of purpuromycin is interpreted as starting from two initial processes: 1) the formation of the unstable molecular ion, unnoticeable at m/e 538; 2) the thermal loss of water to form the stable molecular ion at m/e 520.

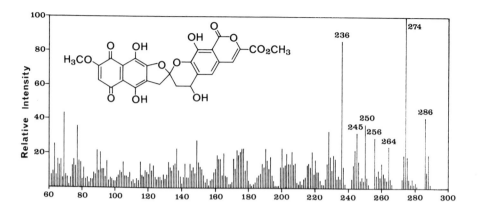

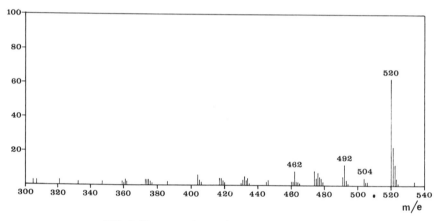

FIG. 2. Mass spectrum of purpuromycin (II).

The fragments of mass 274, 273, 256, 250, 245, and 236 are characteristic of the naphthazarin moiety and are formulated as for γ-rubromycin. The exact mass measurements confirmed the elemental composition of every fragment represented in Scheme II. The peaks 286 and 264, which are not present in the spectrum of γ-rubromycin, together with the peak 236, are discussed in detail. The ion of mass 286 originates from that of mass 520 and accounts for the presence of the double bond in the 3,4-position, which triggers the cleavage at the isocoumarin ring differently from γ-rubromycin, for which the common benzylic cleavage is observed. The ion of mass 264 is characteristic of the isocoumarin moiety and its origin is understood as a retro aldol type cleavage.

In purpuromycin the presence of the additional oxygen atom favors the charge localization on the isocoumarin moiety too. The ion of mass 236 originates from the ion of mass 264, as demonstrated by a metastable peak

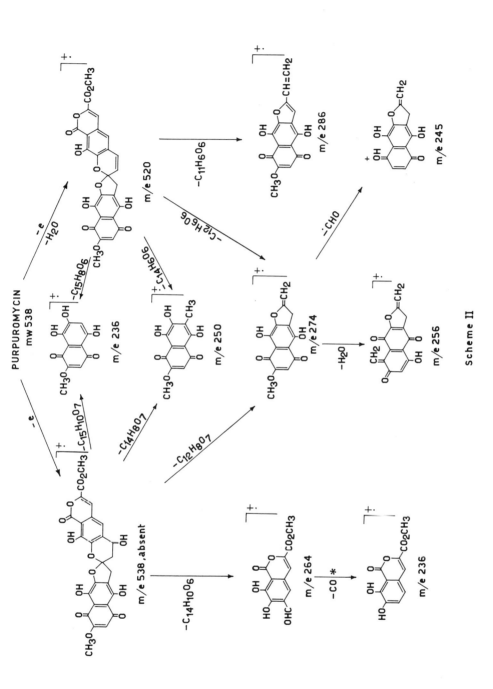

Scheme II

(indicated in Scheme II with an asterisk), and this explains the higher abundance of this ion in the spectrum of purpuromycin in respect to that of γ-rubromycin, where the ion at m/e 236 is interpreted as deriving only from the molecular ion.

During the chemical degradation studies of purpuromycin, three products were isolated and their mass spectra were important in this context. In fact, III and IV derive from the naphthazarin moiety, and V comes from the isocoumarin moiety.

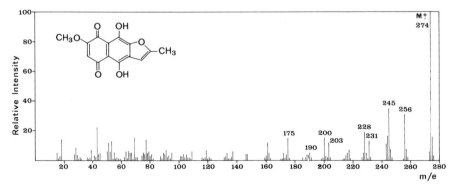

The mass spectrum of III is shown in Fig. 3. The molecular ion at m/e 274 is the base peak, as expected for this type of molecule. The main fragmentation processes are interpreted as shown in Scheme III. The loss of water from the molecular ion is typical for methoxynaphthazarins and involves the quinone oxygen, one hydrogen of the O-methoxy group and the hydrogen of the perihydroxy group. This fragmentation was demonstrated by Becher, Djerassi, Moore, Singh, and Scheuer (1966) for 2-methoxynaphthazarin using the deuterated compound. This ion of mass 256 then decomposes by consecutive losses of CO, giving rise to the ions of mass 228 and 200, as expected for the quinone system. The peaks at m/e 245, 244, and 243 correspond to the losses of CHO, CH_2O, and CH_3O, respectively, from the molecular ion 274, involving the methoxy group. The peak at m/e 231 likely originates from the molecular ion by loss of

FIG. 3. Mass spectrum of III.

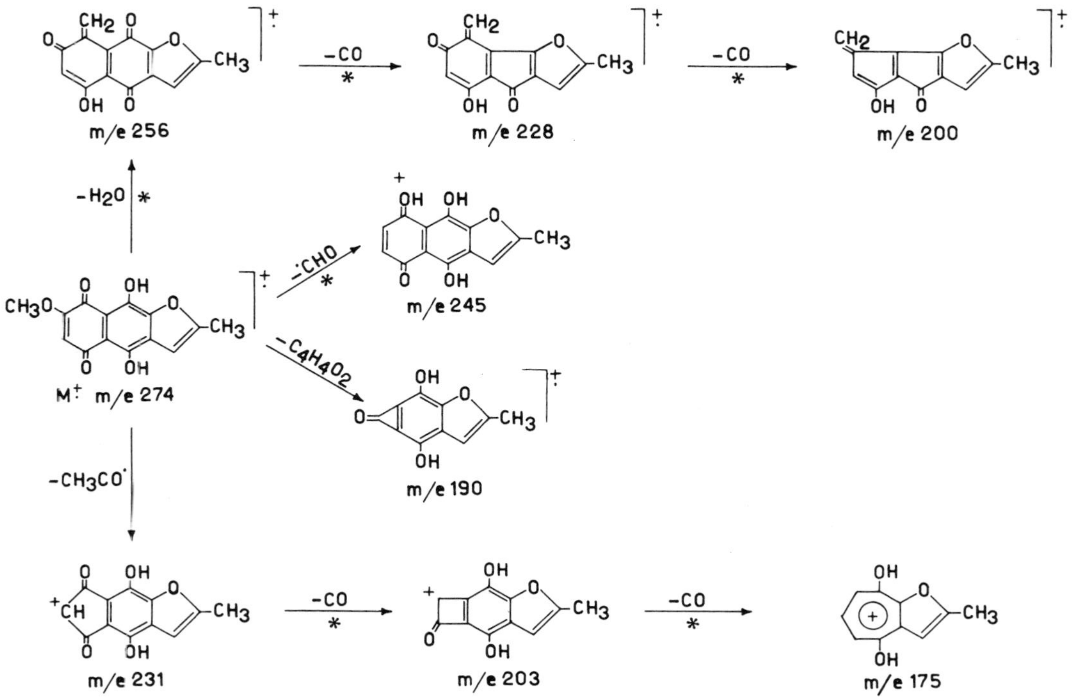

Scheme III

$CH_3CO·$ and this process is supported by its absence in the fragmentation of demethoxy-hydroxy derivative IV. This ion then decomposes by successive losses of CO giving rise to the ions of mass 203 and 175. The ion of mass 190 corresponds to the familiar breakdown of 1,4-quinones.

The mass spectrum of IV is shown in Fig. 4. The molecular ion is the base peak as observed in III. The main fragmentation processes are interpreted as shown in Scheme IV and elucidate the behavior of a quinone compound. In fact, the two peaks at 232 and 190 correspond to the successive losses of CO and C_2H_2O which are characteristic for a quinone compound.

The mass spectrum of V is shown in Fig. 5. The molecular ion at m/e 444 is the base peak, as expected for this type of derivative. The main fragments are interpreted as shown in Scheme V. The peak at m/e 261 corresponds to the loss of dinitroaniline and is characteristic of the isocoumarin moiety. In fact, the loss of the carbomethoxy radical originates the peak at m/e 202 which, by consecutive losses of two CO molecules, gives rise to the peaks at m/e 174 and 146 (Porter and Baldas, 1971).

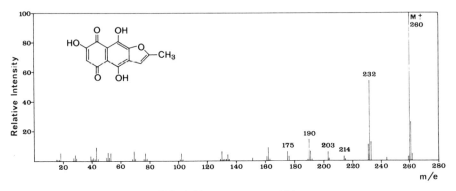

FIG. 4. Mass spectrum of IV.

Scheme IV

EXPERIMENTAL

The mass spectra of the compounds I through IV were taken with a Hitachi-Perkin Elmer RMU-6L instrument, at 70 eV, with the direct inlet system heated at 160 to 300°C. The mass spectrum of V was obtained with a Perkin Elmer 270 instrument, at 70 eV, with the direct inlet system heated to 300°C.

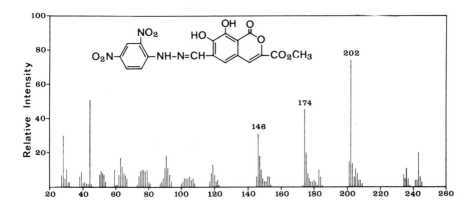

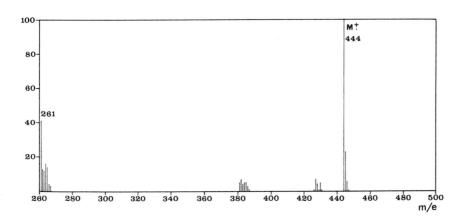

FIG. 5. Mass spectrum of V.

Scheme V

The high-resolution measurements were performed with a double-focusing Hitachi-Perkin Elmer RMU-6D instrument by the peak-matching technique.

REFERENCES

Bardone, M. R., Martinelli, E., Zerilli, L. F., and Coronelli, C. (1973): Structure determination of purpuromycin, a new antibiotic. *Tetrahedron (submitted for publication)*.

Becher, D., Djerassi, C., Moore, R. E., Singh, H., and Scheuer, P. J. (1966): The mass spectrometric fragmentation of substituted naphthoquinones and its application to structural elucidation of echinoderm pigments. *Journal of Organic Chemistry*, 31:3650–3660.

Brockmann, H., and Zeeck, A. (1970): Die konstitution von α-rubromycin, β-rubromycin, γ-rubromycin und γ-iso-rubromycin. *Chemische Berichte*, 103:1709–1726.

Coronelli, C., Pagani, H., Bardone, M. R., and Lancini, G. C. (1973): Purpuromycin, a new antibiotic produced by *Actinoplanes ianthinogenes* n. sp. *Journal of Antibiotics (submitted for publication)*.

Greul, V. (1971): Isolierung und konstitutionsaufklärung neuer rubromycine sowie synthese von rubromycin-partialstrukturen. *Dissertation zur Erlangung des Doktorgrades*. Göttingen.

Porter, Q. N., and Baldas, J. (1971): *Mass Spectrometry of Heterocyclic Compounds*. Wiley-Interscience, New York.

Mass Spectrometry in Biochemistry and Medicine,
edited by A. Frigerio and N. Castagnoli.
Raven Press, New York © 1974

Investigation of Poly-β-Alanines by Pyrolysis and Electron Impact*

I. Lüderwald, M. Przybylski, and H. Ringsdorf

Organisch-Chemisches Institut der Universität Mainz, 65 Mainz, Johann-Joachim-Becher-Weg 18–20, West Germany

I. INTRODUCTION

The use of polymers in pharmacology and medicine is a new field of increasing interest. Studies of synthetic polymers as enzyme carriers (Orth and Brümmer, 1972) or even highly effective enzyme models (Kiefer, Longdon, Scarpa, and Klotz, 1972), of polyvinyl-pyridine-N-oxides against silicoses (Schlipköter and Brockhaus, 1968), and intense investigations of polymers as long-lasting radiation prophylactics (Ringsdorf, 1967; Ringsdorf, Heisler, Müller, Graul, and Rüther, 1971) have been reported in the literature.

One major problem in· this area is the detectibility of polymers incorporated in cell systems. The high sensitivity of mass spectrometric techniques might provide a means of detection as was shown by preliminary investigations of polyvinylpyrolidone-containing liver cells. In this connection the degradation reactions of various substituted poly-β-alanines as possible carriers were studied *in vitro* (Table 1). The structures of these polymers have been established by independent methods and therefore might be appropriate as a model system.

TABLE 1. *Degradation reactions of substituted poly-β-alanines*

$$\text{-}[NH\text{---}CR_1R_2\text{---}CHR_3\text{---}CO]\text{-}_n$$

	I	II	III	IV	V	VI	VII
R_1	CH_3	C_2H_5	CH_3	CH_3	H	C_2H_3	D
R_2	H	H	CH_3	H	H	H	H
R_3	H	H	H	CH_3	H	H	D

---

* Luderwald and Ringsdorf (1973).

II. EXPERIMENTAL

The poly-β-alanines were investigated by direct pyrolysis in the ion source of a VARIAN CH 4 mass spectrometer. The investigation of polymers by thermal degradation and electron impact requires an exact differentiation of the thermal degradation mechanisms and the fragmentation reactions of the ionized monomers and oligomers. To avoid superposition of the pyrolytic degradation products and the fragments induced by electron impact, the pyrolysis was carried out under mild conditions, by increasing the temperature until the total ion current began to increase. The electron energy was reduced in order to establish that the monomers and oligomers were formed thermally and detected directly as the "molecule ions" and to get information about the polymer structure. Since the ionization energies of organic amides are about 10 eV, we worked with 13 eV. It could be shown that under these conditions only a few selective fragmentation steps take place.

The poly-β-alanines were prepared by anionic ring-opening polymerization of the corresponding β-lactames in order to obtain polymers of high purity. The pyrolyses were carried out by direct insertion at a constant heating rate of 50°C/min.

III. RESULTS AND DISCUSSION

Ions corresponding to multiples of the monomer units were found in the mass spectra of the poly-β-alanines. To study the fragmentation reactions of these pyrolytically formed oligomeric acid amides, dimeric, trimeric, and tetrameric β-lactames,

$$\text{R}\!-\!\!\left[\text{NH}\!-\!\text{CHR}'\!-\!\text{CHR}''\!-\!\text{CO}\right]_{1,2,3}\!\!\overset{\displaystyle \overset{\text{CHR}'\ \text{CHR}''}{|\qquad\ \ |}}{\text{N}\!-\!\!-\!\text{CO}}$$

$$\text{R} = -\text{CO}-\text{CH}_3, (\text{XIII,IX,X})-\text{CO}-\text{CF}_3$$
$$\text{R}^1 = -\text{H},-\text{CH}_3, \qquad \text{R}'' = -\text{H},-\text{CH}_3$$

which contain monomer units of β-alanine and which could be volatilized without thermal degradation, have been initially investigated.

The fragmentation reactions of these oligomeric β-alanines are generally induced by preferential ionization of the hetero atoms. Under our experimental conditions, mainly four typical fragmentation pathways take place ∢(1)–(4)].

$$\alpha_0\text{-cleavage (Spiteller, 1968):}$$

(1) $\sim\!\sim\!\text{NH}-\overset{\text{R}}{\underset{|}{\text{CH}}}-\text{CH}_2-\overset{\overset{\displaystyle \oplus\cdot}{\displaystyle \text{O}}}{\underset{\displaystyle \|}{\text{C}}}-\text{NH}\!\sim\!\sim \longrightarrow \sim\!\sim\!\text{NH}-\overset{\text{R}}{\underset{|}{\text{CH}}}-\text{CH}_2-\text{C}\!\equiv\!\overset{\displaystyle \oplus}{\text{O}} + \cdot\text{NH}\!\sim\!\sim$$

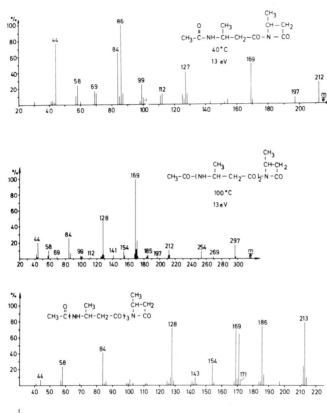

FIG. 1. Mass spectra of the dimeric, trimeric, and tetrameric N-acetylated 4-methyl-propiolactam.

β_0-cleavage (McLafferty rearrangement, 1959):

α_N-cleavage (McFadden, Seifert, and Wasserman, 1965; Bengelmans, Williams, Budzkiewicz, and Ojerassi, 1964):

$$(3) \quad \sim\!\!\sim CH\!-\!CH_2\!-\!CO\!-\!\overset{\oplus}{N}H\!-\!CH\!-\!CH_2\!\sim\!\!\sim \longrightarrow \sim\!\!\sim CH\!-\!CH_2\!-\!CO\!-\!\overset{\oplus}{N}H\!=\!CH + \cdot CH_2\!\sim\!\!\sim$$

(with R substituents on the CH carbons)

"Ketene-cleavage":

$$(4) \quad \sim\!\!\sim CH\!-\!CH \cdots NH\!-\!C\!-\!CH_2\!\sim\!\!\sim \longrightarrow \sim\!\!\sim CH\!-\!CH\!=\!C\!=\!O + H_2N\!=\!\overset{\oplus}{C}\!-\!CH_2 \sim\!\!\sim$$

The "ketene-cleavage" pathway was found in all oligomeric and polymeric β-alanines investigated containing at least one hydrogen atom at the β-carbon; however we cannot say if this fragmentation is a single- or two-stage reaction.

Nearly all peaks in the mass spectra of the oligomeric N-acetylated and N-trifluoroacetylated 4-methyl-propiolactams can be definitely interpreted by fragments according to the reactions (1)–(4). The main fragments are listed in Table 2. On the basis of the clear-cut fragmentation pattern of these model compounds both pyrolytic and fragmentation pathways of the poly-β-alanines could be interpreted.

TABLE 2. *Fragment table of the oligomeric 4-methyl-propiolactams VIII, IX, X*

m/e	Fragment+	VIII	IX	X	Reaction
382	M	—	—	57	—
339	$NH_2\!=\!C\!-\!CH_2\!-\!CO(NH\!-\!CH\!-\!CH_2\!-\!CO)_2N\!-\!CO$ (with CH_3, CH_3, $CH\!-\!CH_2$ / CH_3 substituents)	—	—	16	(4)
297	M	—	18	—	—
297	$CH_3\!-\!CO(NH\!-\!CH\!-\!CH_2\!-\!CO)_2N\!-\!CO$ (with CH_3, $CH\!-\!CH_2$ / CH_3)	—	—	11	(2)
256	$CH_3\!-\!CO(NH\!-\!CH\!-\!CH_2\!-\!CO)_2NH\!=\!CH$ (with CH_3, CH_3)	—	—	50	(3)
254	$NH_2\!=\!C\!-\!CH_2\!-\!CO\!-\!NH\!-\!CH\!-\!CH_2\!-\!CO\!-\!N\!-\!CO$ (with CH_3, CH_3, $CH\!-\!CH_2$ / CH_3)	—	10	100	(4)

TABLE 2. (Continued)

m/e	Fragment⁺	VIII	IX	X	Reaction
213	CH₃—CO—NH—CH(CH₃)—CH₂—CO—NH—CH(CH₃)—CH₂—C=O	–	5	81	(1)
212	CH₃—CO—NH—CH(CH₃)—CH₂—CO—N(CH(CH₃)—CH₂)—CO	28	11	26	(2)
197	O=C=NH—CH(CH₃)—CH₂—CO—N(CH(CH₃)—CH₂)—CO	9	3	5	(1)
171	CH₃—CO—NH—CH(CH₃)—CH₂—CO—NH=CH(CH₃)	–	23	65	(3)
169	NH₂=C(CH₃)—CH₂—CO—N(CH(CH₃)—CH₂)—CO	41	100	70	(4)
128	CH₃—CO—NH—CH(CH₃)—CH₂—C=O	14	48	72	(1)
127	CH₃—CO—N(CH(CH₃)—CH₂)—CO	41	6	10	(2)
112	O=C=N(CH(CH₃)—CH₂)—CO	14	2	2	(1)
86	CH₃—CO—NH=CH(CH₃)	100	7	5	(3)
84	H₃C—CH—CH₂ with ‖ NH—CO ring	65	26	41	(4)

The thermal degradation mechanism of the poly-3-methyl-β-alanine at 340°C was found to be the preferential cleavage of the nitrogen-alkyl bond followed by disproportionation, yielding oligomeric β-alanines with unsaturated and amide end groups (5).

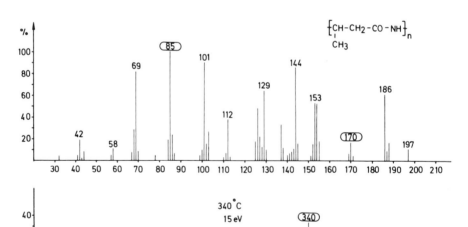

(5)

The mass spectrum shows the "molecule ions" of the oligomeric 3-methyl-β-alanines at m/e (n × 85).

$$CH=CH-CO \left[NH-CH-CH_2-CO \right]_{0-3} NH_2$$
$$\quad\quad |\qquad\qquad\qquad |$$
$$\quad\quad CH_3\qquad\qquad\qquad CH_3$$

m/e   85 , 170 , 255 , 340

FIG. 2. Mass spectrum of poly-3-methyl-β-alanine (I).

This pyrolytic degradation step (5) can be confirmed by the fragmentation reactions of the oligomeric amides:

α_0-fragments (1):

$$CH_3\text{--}CH=CH\left[CO\text{--}NH\text{--}\overset{\overset{\displaystyle CH_3}{|}}{CH}\text{--}CH_2\right]_n C\equiv O^{\oplus}$$

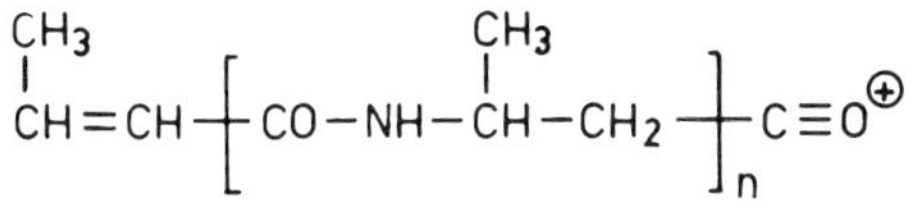

n = 0,1,2   m/e   69, 154, 239

β_0-fragments (2):

$$CH_3\text{--}\overset{\overset{\displaystyle O^{\oplus\cdot}}{\|}}{C}\left[NH\text{--}\overset{\overset{\displaystyle CH_3}{|}}{CH}\text{--}CH_2\text{--}CO\right]_n NH_2$$

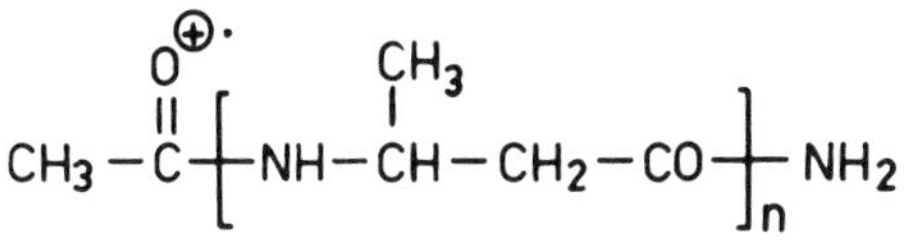

n = 1,2   m/e   144, 229

α_N-fragments (3):

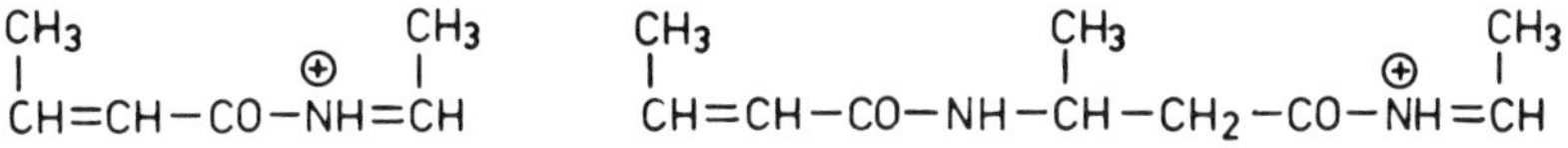

$$\overset{\overset{\displaystyle CH_3}{|}}{CH}=CH\text{--}CO\text{--}NH\overset{\oplus}{=}\overset{\overset{\displaystyle CH_3}{|}}{CH} \qquad \overset{\overset{\displaystyle CH_3}{|}}{CH}=CH\text{--}CO\text{--}NH\text{--}\overset{\overset{\displaystyle CH_3}{|}}{CH}\text{--}CH_2\text{--}CO\text{--}NH\overset{\oplus}{=}\overset{\overset{\displaystyle CH_3}{|}}{CH}$$

m/e 112   m/e 197

"Ketene-cleavage" (4):

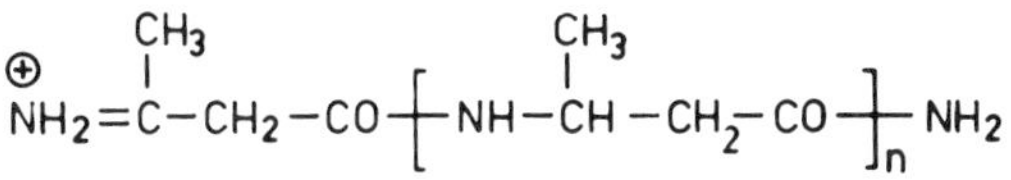

$$\overset{\oplus}{NH_2}=\overset{\overset{\displaystyle CH_3}{|}}{C}\text{--}CH_2\text{--}CO\left[NH\text{--}\overset{\overset{\displaystyle CH_3}{|}}{CH}\text{--}CH_2\text{--}CO\right]_n NH_2$$

n = 0,1,2   m/e   101, 186, 271

The poly-3-ethyl-β-alanine follows the same degradation pathways on pyrolysis [*cf.* (5)], yielding oligomeric β-ethyl-acrylamides (m/e 99, 198, 297 . . . .), and on the fragmentation steps induced by ionization of the hetero atoms.

In addition to the fragmentations described above [(1)–(4)] it is notable that all homologue fragments due to the "ketene-cleavage" (4) are accompanied by peaks with, respectively, two mu higher, which can be accounted for by ammonium ions (6):

(4) 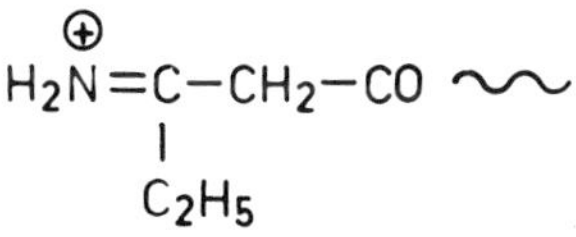 $\overset{\oplus}{H_2N}=\overset{\underset{\displaystyle C_2H_5}{|}}{C}\text{--}CH_2\text{--}CO \sim\!\!\sim$ m/e   115, 214, 313,-----

(6) 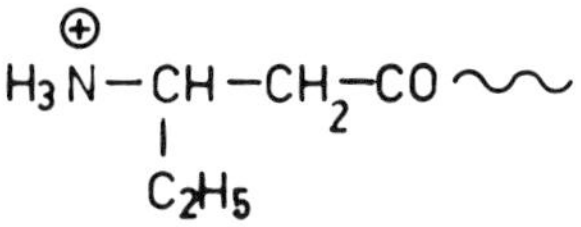 $\overset{\oplus}{H_3N}\text{--}\overset{\underset{\displaystyle C_2H_5}{|}}{CH}\text{--}CH_2\text{--}CO \sim\!\!\sim$ m/e   117, 216, 315-----

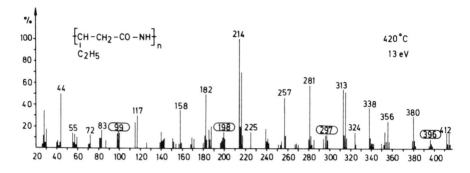

FIG. 3. Mass spectrum of poly-3-ethyl-β-alanine (II).

The mechanism of thermal degradation (5) can be established definitely by comparison of the isomeric poly-β-alanines II, III, and IV:

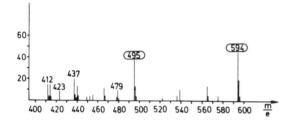

The peaks due to the thermally formed oligomers which are multiples of the monomer unit (99) were found in all mass spectra (Figs. 3–5).

$$\begin{array}{c} CH_3 \\ | \\ C=CH-CO \end{array} \left[\begin{array}{c} CH_3 \\ | \\ NH-C-CH_2-CO \\ | \\ CH_3 \end{array} \right]_n NH_2$$

n	1	2	3	4	5
m/e	198	297	396	495	594

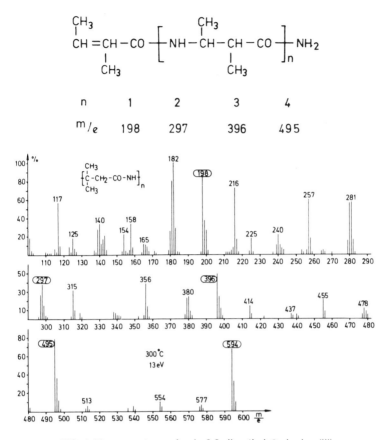

FIG. 4. Mass spectrum of poly-3,3-dimethyl-β-alanine (III).

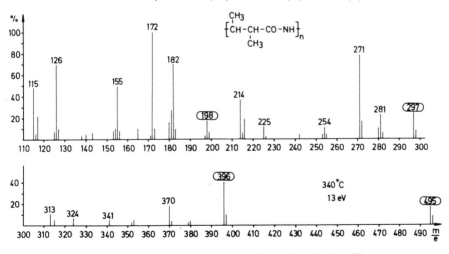

FIG. 5. Mass spectrum of poly-2,3-dimethyl-β-alanine (IV).

However, an exact differentiation of the isomeric polyamides is possible using the fragmentation reactions. The α_N-cleavage (3) allows one to determine selectively the substitution on the β-carbon; thus, in the poly-2,3-dimethyl-β-alanine the corresponding series of fragments appears with 14-μ lower masses.

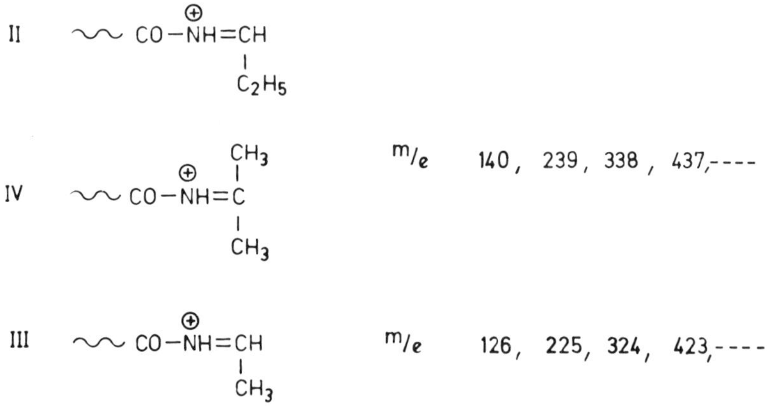

II $\sim\!\!\sim\!$ CO$-$NH$\overset{\oplus}{=}$CH
                    |
                    C_2H_5

IV $\sim\!\!\sim\!$ CO$-$NH$\overset{\oplus}{=}$C$\overset{\displaystyle CH_3}{\underset{\displaystyle CH_3}{|}}$ m/e 140, 239, 338, 437,$----$

III $\sim\!\!\sim\!$ CO$-$NH$\overset{\oplus}{=}$CH$\underset{\displaystyle CH_3}{|}$ m/e 126, 225, 324, 423,$----$

On the other hand, the "ketene-cleavage" (4) allows the differentiation of the polymers II and IV. Figure 6 points out that, because of the lack of any hydrogen on the β-carbon, no rearrangement corresponding to (4) is possible on the fragmentation of the poly-3,3-dimethyl-β-alanine, where

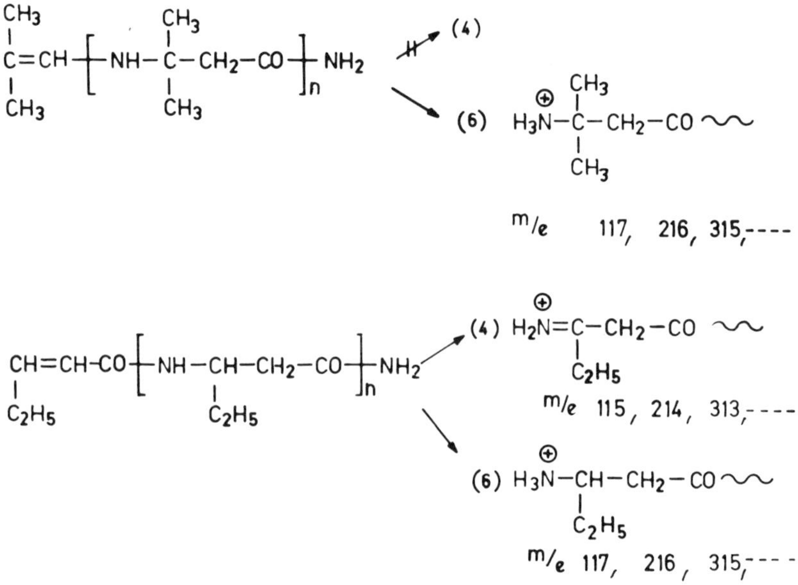

FIG. 6. Differentiation of the isomeric poly-β-alanines (II) and (III) by the "ketene-cleavage" (4).

only the formation of ammonium-ions (6) is found. However, in the mono-substituted isomer II both types of fragmentations (4) and (6) take place. Further investigations of differently substituted and labeled poly-β-alanines (V, VI, VII) confirmed the clear differentiation of this type of polymer in terms of thermal degradation and fragmentation.

The investigation of polymers by direct degradation in the ion source of the mass spectrometer might be of considerable interest. Thus, depending on well-defined experimental conditions, the study of different thermal degradation reactions (statistical degradation, stripping, side chain reactions) should give structure information on polymers, especially if very small amounts of material are required. Both the combination of direct pyrolysis in the mass spectrometer and the pyrolysis-GC–MS coupling might be appropriate for the analysis of even biological polymer systems.

REFERENCES

Beugelmans, R., Williams, D. H., Budzkiewicz, H., and Djerassi, C., (1964): *Journal of the American Chemical Society*, 86:1586.

Kiefer, H. C., Longdon, W. I., Scarpa, I. S., and Klotz, I. M. (1972): *Proceedings of the National Academy of Sciences*, 69:215.

Lüderwald, I., and Ringsdorf, H. (1973): *Makromolekulare Chemie*, in press.

McFadden, W. H., Seifert, R. M., and Wasserman, I. (1965): *Analytical Chemistry*, 37:26.

McLafferty, F. W. (1959): *Analytical Chemistry*, 31:82.

Orth, H. D., and Brümmer, W. (1972): *Angewandte Chemie*, 84:319.

Ringsdorf, H. (1967): *Strahlentherapie*, 132:627.

Ringsdorf, H., Heisler, A. G., Müller, F. W., Graul, E. H., and Rüther, W. (1971): In: *Biological Aspects of Radiation Protection*, edited by T. Sugahara and O. Hug, pp. 138–147. Igaku Shoin Ltd., Tokyo.

Schlipköter, H. W., and Brockhaus, A. (1968): *Deutsche Medizinische Wochenschrift*, 93: 2479.

Spiteller, G. (1968): *Massenspektrometrische Strukturanalyse organischer Verbindungen*, p. 125. Verlag Chemie.

Mass Spectrometry in Biochemistry and Medicine,
edited by A. Frigerio and N. Castagnoli.
Raven Press, New York © 1974

Applications of Mass Spectrometry to the Structure Determination of Pyrrole Pigments

D. E. Games, A. H. Jackson, and D. S. Millington

Department of Chemistry, University College, P.O. Box 78, Cardiff CF1 1XL, U.K.

I. INTRODUCTION

The application of mass spectral methods to structural studies of the major biological classes of tetrapyrroles (porphyrins, porphyrinogens, chlorins, bacteriochlorins, corrins, corrinoids, and bile pigments) has recently been reviewed by Dougherty (1972). Since the quantities of material available for study are usually very small, mass spectral studies are particularly useful in establishing molecular weights, empirical formulae, the nature of peripheral substituents, and in yielding information about the purity of the sample.

In connection with our studies of heme metabolism we have been concerned with developing mass spectral methods for the identification and structural determination of porphyrins which are intermediates in the biosynthetic process. The mass spectra of a large number of these compounds have been discussed (Jackson, Kenner, Smith, Aplin, Budzikiewicz, and Djerassi, 1965; Hoffman, 1965), and the technique has been utilized in the characterization of petroporphyrins (Baker, 1966; Baker, Yen, Dickie, Rhodes, and Clark, 1967; Boylan et al., 1969). More recently the mass spectra of the trimethylsilyl ethers of some hydroxylated di- and tetra-carboxylic porphyrins have been described (Chapman and Elder, 1972).

Our mass spectral investigations of porphyrins have been centered in three areas: 1. improving methods of obtaining spectra, 2. development of methods for the examination of crude mixtures of porphyrins, and 3. development of microscale degradation methods for the structure elucidation of porphyrins, using gas chromatography – mass spectrometry (GC–MS).

II. RESULTS AND DISCUSSION

The porphyrins involved in heme metabolism are relatively involatile and in most cases satisfactory mass spectra may only be obtained by conversion into the more volatile methyl esters (Jackson et al., 1965). In a typical porphyrin mass spectrum the molecular ion is usually the most abundant ion, indicating the stability of the porphyrin nucleus, and the major fragment ions are accounted for by cleavage of peripheral substituents (Jackson et al., 1965). A second series of ions, due to fragmentation of the doubly charged molecules, is also a characteristic feature of porphyrin mass spectra. Our studies in this area have been considerably assisted by the availability of on-line computer facilities and a temperature-programmed direct-insertion probe.

To improve the volatility of porphyrins, with the eventual goal of applying GC–MS, we have investigated the use of *bis*-(trimethylsiloxy) silicon complexes. Alkyl porphyrins derivatized in this way are amenable to GC (Boylan and Calvin, 1967), and the technique was used in the study of the porphyrin content of crude oils (Boylan et al., 1969). Although capillary column GC and MS were separately utilized in these studies, combined GC–MS was not used. We have found that *bis*-(trimethylsiloxy) silicon complexes of alkyl porphyrins are amenable to GC–MS using packed GC columns. Figure 1 shows the total ion current trace of a mixture of alkyl porphyrins derivatized in this way, and Fig. 2 shows the mass spectrum obtained during GC–MS of the *bis*-(trimethylsiloxy) silicon complex of octamethylporphyrin, one of the components of the mixture. Efforts are now being made to extend this technique to porphyrins with polar substituents in their side chains. For example, silicon has been introduced into the nucleus of mesoporphyrin-IX dimethyl ester and the mass spectrum of the derivative is shown in Fig. 3. The volatility of the derivative is considerably enhanced in comparison with mesoporphyrin-IX dimethyl ester,

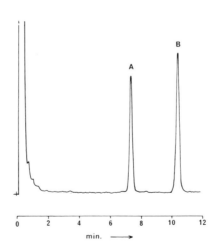

FIG. 1. Total ion current trace of octamethyl (A) and octaethyl *bis*-(trimethylsiloxy) silicon (B) complexes. Conditions: 2 m × 2 mm i.d. glass column, 3% OV 1 on Gas-Chrom Q (100–120 mesh), column temperature 275°C, helium flow rate 30 ml/min.

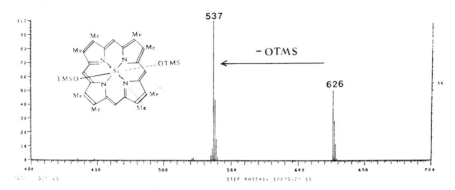

FIG. 2. Mass spectrum of the *bis*-(trimethylsiloxy) silicon complex of octamethylporphyrin.

itself requiring a probe temperature of only 150°C for volatilization whereas the latter compound volatilizes at 250°C.

Isomeric porphyrins cannot generally be distinguished by normal MS (Jackson et al., 1965), but our investigations have shown that in some cases isomers can be distinguished by examination of their metastable ion characteristics. We have studied the metastable decompositions of ions present in the spectra of the methyl esters of the isomeric porphyrins coproporphyrin-III (1) and isocoproporphyrin (2) using the "direct analysis of daughter ions" technique (DADI) (Maurer, Brunnee, Kappus, Habfast, Schroder, and Schulze, 1971). This technique can be used in double-focusing instruments with reversed field arrangements (i.e., with the electrostatic analyzer following the magnetic analyzer). The magnetic field is tuned for ions of certain mass to charge ratio, and, since a proportion of these ions decompose during their flight between the two field sectors, a mass spectrum can be obtained of all the resulting daughter ions by linearly decreasing the electrostatic analyzer voltage with time, while maintaining a constant

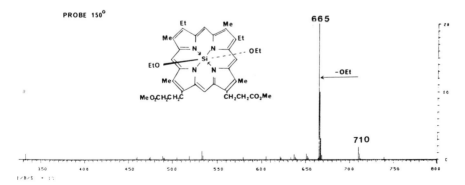

FIG. 3. Mass spectrum of the *bis*-(diethoxy) silicon complex of mesoporphyrin-IX dimethyl ester.

field strength and accelerating voltage. Figure 4 shows the DADI spectra of the fragment ions (m/e 637) produced from both porphyrin methyl esters. As can be seen the ion at m/e 637 from these two isomers shows markedly different metastable spectra, and it can be readily distinguished on this basis. Further investigations of the application of this promising technique with isomeric porphyrins are being undertaken to study its general applicability in this area.

Natural porphyrins usually occur as complex mixtures of a variety of structural types, and their separation into individual compounds can be extremely difficult. We have been developing methods for the examination of mixtures of this type which do not necessitate prior separation of the individual components. Derivatization as *bis*-(trimethylsiloxy) silicon complexes, followed by GC–MS as discussed earlier, provides one possible solution. Examination of crude porphyrin mixtures by low energy electron impact (E.I.) is a further possibility, and this technique has been used to study mixtures of petroporphyrins (Baker, 1966; Baker et al., 1967). We are currently investigating crude mixtures of porphyrins obtained from rat and human porphyric extracts using field ionization (F.I.) and an on-line data system as another alternative. The methylated mixtures are first examined under normal electron impact conditions at 70 eV, and then a

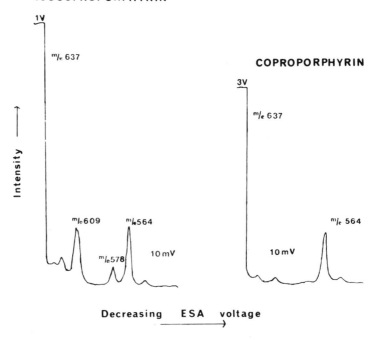

FIG. 4. Direct analysis of daughter ions (DADI) spectra of the fragment ions m/e 637 produced from isocoproporphyrin and coproporphyrin-III tetramethylesters.

field ion spectra of the mixture is obtained. In most cases the field ion spectra of porphyrins show only the molecular ion and the corresponding doubly charged ion, and Fig. 5 shows the F.I. spectrum of a mixture of mesoporphyrin-IX and coproporphyrin-II methyl esters. Hence the field ion spectra of the mixtures will be considerably simplified, and enable all the molecular species present to be readily identified. The relative heights of the ions in the F.I. molecular ion complex are approximately indicative of the quantity of each molecular species present in the mixture, since there is virtually no fragmentation and the porphyrins co-distill from the mixture, thus enabling a quantitative estimation of the porphyrins present to be made. Unfortunately the technique will not enable differentiation of isomeric porphyrins and will be most effective when used in conjunction with chromatographic methods.

The information provided by the mass spectral methods described earlier enables identification of the peripheral substituents on porphyrins, but does not delineate their order on the porphyrin nucleus. Confirmation of the presence of particular substituents and some information about their order is obtainable by oxidative degradation of the porphyrin to maleimides or pyrrole-2,5-dicarboxylic acids, using chromic acid (Morley and Holt, 1961; Grassl, Coy, Seyffert, and Lynen, 1963; Ellsworth and Aronoff, 1968) or permanganate (Nicolaus, Mangoni, and Caglioti, 1956; Nicolaus, Mangoni, and Nicoletti, 1957; Tipton and Gray, 1971), respectively. Reductive degradation of porphyrins to pyrroles (Fischer and Orth, 1937–1941; Chapman, Roomi, Morton, Krajcarski, and MacDonald, 1971) provides similar information. Identification of the degradation products has been effected by a variety of methods, many of which involve prior separation of the compounds, followed by mass spectral identification or GC methods. Our studies have been directed to the improvement of the scope of these techniques by the use of combined GC–MS.

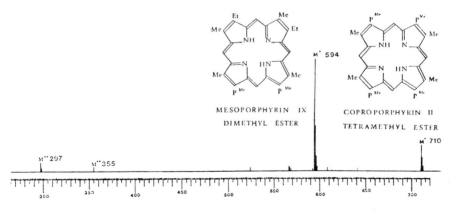

FIG. 5. Field ion spectrum of a mixture of mesoporphyrin-IX and coproporphyrin-II methylesters.

Initially we have studied the reductive degradation of porphyrins to pyrroles using HI and formaldehyde in acetic acid (Chapman et al., 1971). The resulting tetrasubstituted pyrroles are esterified with diazomethane

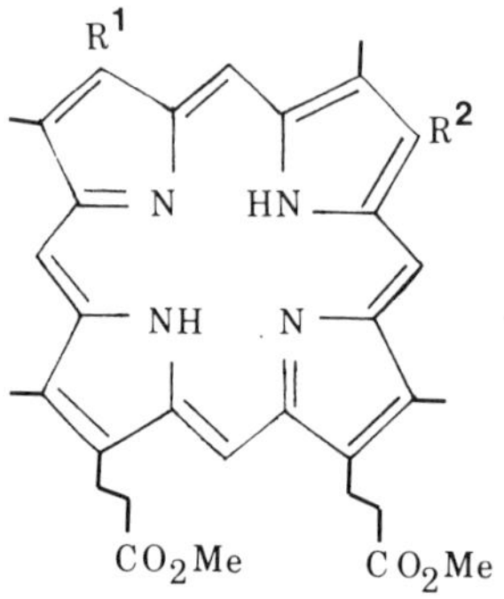

1a $R^1 = R^2 = CH_2CH_2CO_2Me$

1b $R^1 = R^2 = CH_2CH_3$

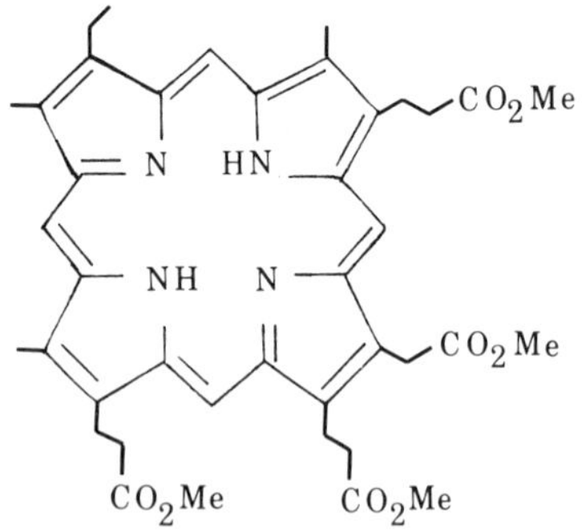

2

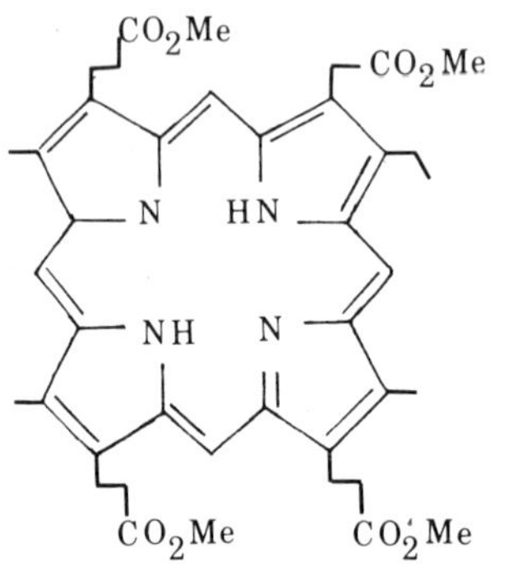

3

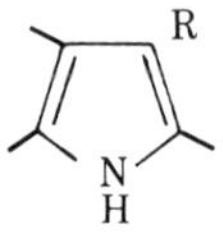

4a $R = CH_2CH_3$

4b $R = CH_2CH_2CO_2Me$

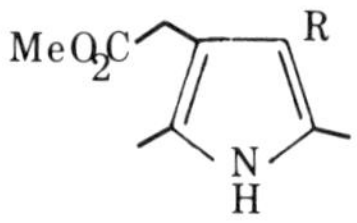

5a $R = CH_2CH_2CO_2Me$

5b $R = CH_2CH_3$

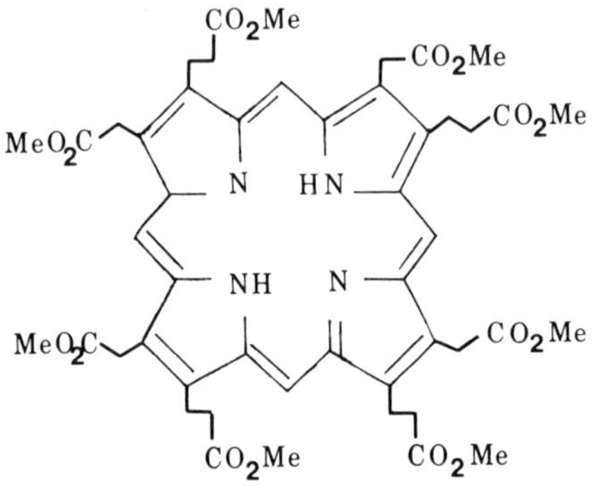

6

and subjected to GC–MS analysis. For example, reduction of the dimethyl ester of mesoporphyrin-IX (1b) under these conditions yielded only the two expected pyrroles (4a) and (4b) in the ratio 1:1. The technique was then applied to the structural investigation of isocoproporphyrin, a porphyrin of abnormal metabolism recently isolated from the feces of a human porphyric (Stoll, Elder, Games, O'Hanlon, Millington, and Jackson, 1973). Proton magnetic resonance studies using the lanthanide "shift reagent" Eu(fod-d_9)$_3$ combined with biosynthetic reasoning established the two possible alternative structures for the tetramethylester of this porphyrin (2) or (3). Reductive alkylation of isocoproporphyrin yielded the pyrroles (4a), (4b), and (5a) in approximate proportions 1:2:1 indicating structure (2) for the compound, since the alternative structure (3) would have yielded the pyrroles (4b) and (5b). The total ion current trace of the pyrroles produced from isocoproporphyrin is shown in Fig. 6 and their mass spectra in Fig. 7. The pyrroles produced were identified both by their GC retention times and their mass spectra, which were identical with those of authentic materials obtained by the reductive alkylation of a mixture of mesoporphyrin-IX (1b) and uroporphyrin (6). Further studies are presently in progress to ascertain the relative merits of the reductive and oxidative methods for ascertaining the peripheral substituents on porphyrins and bile pigments, using combined GC–MS as the method of analysis.

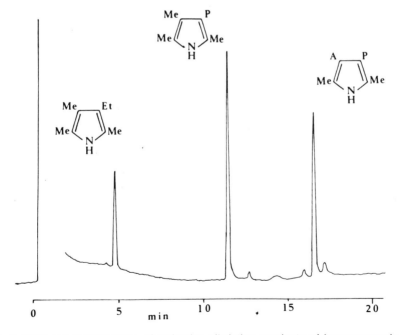

FIG. 6. Total ion current trace of reductive alkylation products of isocoproporphyrin. Conditions: 2 m × 2 mm i.d. glass column, 3%OV 1 on Gas-Chrom Q (100–120 mesh), temperature-programmed at 8°C/min over the range 80 to 250°C, helium flow rate 30 ml/min.

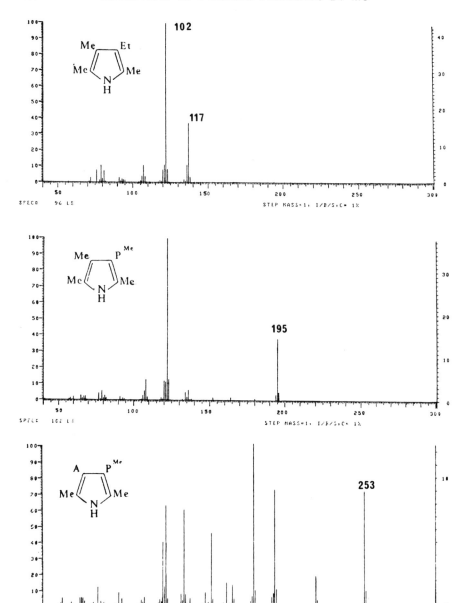

FIG. 7. Mass spectra of the reductive alkylation products isocoproporphyrin.

III. EXPERIMENTAL

Mass spectra (ionizing potential 70 eV, source temperature 220°C) were obtained using a Varian CH5D mass spectrometer on line to a Varian 620i computer. GC–MS results were obtained using a Varian 2740 gas chromatograph coupled to the mass spectrometer by an all glass inlet system incorporating a two stage Watson-Biemann separator. F.I. spectra were obtained using the Varian dual F.I./E.I. source. In the F.I. mode an acetone-conditioned tungsten wire emitter carrying a potential of 9 kV was used and the source temperature was around 200°C.

ACKNOWLEDGMENTS

We thank the Science Research Council for financial assistance toward the purchase of the Varian CH5D mass spectrometer and for financial support for one of us (D.S.M.). Dr. G. H. Elder (Department of Chemical Pathology, Welsh National School of Medicine, Cardiff) kindly provided the crude porphyrin mixtures, and we are grateful to Dr. S. F. MacDonald (NRCC, Ottawa) for helpful discussions about the reductive alkylation of porphyrins.

REFERENCES

Baker, E. W. (1966): Mass spectrometric characterization of petroporphyrins. *Journal of the American Chemical Society*, 88:2311–2315.

Baker, E. W., Yen, T. F., Dickie, J. P., Rhodes, R. E., and Clark, L. F. (1967): Mass spectrometry of porphyrins. II. Characterization of petroporphyrins. *Journal of the American Chemical Society*, 89:3631–3639.

Boylan, D. B., Alturki, Y. I., and Eglinton, G. (1969): Application of gas chromatography and mass spectrometry to porphyrin microanalysis. In: *Advances in Organic Geochemistry*, edited by P. A. Schenck and I. Havenaar. Pergammon Press, New York.

Boylan, D. B., and Calvin, M. (1967): Volatile silicon complexes of etioporphyrin I. *Journal of the American Chemical Society*, 89:5472–5473.

Chapman, J. R., and Elder, G. H. (1972): The mass spectra of the trimethylsilyl ethers of some hydroxylated di- and tetracarboxylic porphyrins. *Organic Mass Spectrometry*, 6:991–1001.

Chapman, R. A., Roomi, M. W., Morton, T. C., Krajcarski, D. T., and MacDonald, S. F. (1971): The analytical reduction of porphyrins to pyrroles. *Canadian Journal of Chemistry*, 49:3544–3564.

Dougherty, R. C. (1972): Tetrapyrroles. In: *Biochemical Applications of Mass Spectrometry*, edited by G. R. Waller. Wiley-Interscience, New York.

Ellsworth, R. K., and Aronoff, S. (1968): Investigations on the biogenesis of chlorophyll a. 1. Purification and mass spectra of maleimides from the oxidation of chlorophyll and related compounds. *Archives of Biochemistry and Biophysics*, 124:358–364.

Fischer, H., and Orth, H. (1937–1941): *Die Chemie des Pyrrols*, Vols. 1 and 2. Akademische Verlag, Leipzig.

Grassl, M., Coy, U., Seyffert, R., and Lynen, F. (1963): Die Chemische Konstitution des Cytohamins. *Biochemische Zeitschrift*, 338:771–795.

Hoffman, D. R. (1965): Mass spectra of porphyrins and chlorins. *Journal of Organic Chemistry*, 30:3512–3516.

Jackson, A. H., Kenner, G. W., Smith, K. M., Aplin, R. T., Budzikiewicz, H., and Djerassi, C.

(1965): Pyrroles and related compounds-VIII. The mass spectra of porphyrins. *Tetrahedron,* 21:2913–2924.

Maurer, K. H., Brunnee, C., Kappus, G., Habfast, K., Schroder, U., and Schulze, P. (1971): Direct analysis of daughter ions arising from metastable decompositions. *Nineteenth Annual Conference on Mass Spectrometry, Atlanta, Georgia.*

Morley, H. V., and Holt, A. S. (1961): Studies on chlorobium chlorophylls. II. The resolution of oxidation products of chlorobium pheophorbide (660) by gas-liquid partition chromatography. *Canadian Journal of Chemistry,* 39:755–760.

Nicolaus, R. A., Mangoni, L., and Caglioti, L. (1956): Acidi pirrolici nella ossidazione delle porfirine. *Annali di Chimica,* 46:793–805.

Nicolaus, R. A., Mangoni, L., and Nicoletti, R. (1957): Altri acidi pirrolici nell'ossidazione delle porfirine. *Annali di Chimica,* 47:178–188.

Stoll, M. S., Elder, G. H., Games, D. E., O'Hanlon, P., Millington, D. S., and Jackson, A. H. (1973): Isocoproporphyrin: Nuclear-magnetic-resonance and mass-spectral methods for the determination of porphyrin structure. *Biochemical Journal,* 131:429–432.

Tipton, G., and Gray, C. H. (1971): Gas chromatographic analysis of pyrrolic acid esters from the potassium permanganate oxidation of bile pigments. *Journal of Chromatography,* 59:29–43.

Mass Spectrometry in Biochemistry and Medicine,
edited by A. Frigerio and N. Castagnoli.
Raven Press, New York © 1974

Mass Spectrometry of Carcinogenic Alkaryltriazenes

G. F. Kolar

Institute for Experimental Toxicology and Chemotherapy, German Cancer Research Center, 6900 Heidelberg, Germany

I. INTRODUCTION

The alkaryltriazenes have aroused renewed scientific interest because of their pronounced and versatile biological activities. Compounds in this class have been shown to be not only potent carcinogens (Druckrey, Ivankovic, and Preussmann, 1967a; Preussmann, Druckrey, Ivankovic, and v. Hodenberg, 1969), teratogens (Druckrey, Ivankovic, Preussmann, and Brunner, 1967b), and mutagens (Vogel, 1971; Fahrig, 1971; Ong and de Serres, 1971; Siebert and Kolar, 1973), but in some instances also promising cytostatic agents (Shealy and O'Dell, 1971).

The general structure of alkaryltriazenes, compounds that can be formally derived from the parent triazene, $HN{=}N{-}NH_2$, is invariably substituted with an aromatic system at N1. This fact already suggests its mode of synthesis. Conventional preparation of these compounds involves the coupling of arenediazonium solutions with secondary or primary aliphatic amines in the presence of excess inorganic base to consume the acid used in the diazotization. According to the class of the aliphatic amine chosen for the passive component in the coupling, the reaction leads to the formation of a dialkaryl ($R_1 = R_2 =$ alkyl) or of a monoalkaryltriazene ($R_1 = H$, $R_2 =$ alkyl), respectively, as shown in Table 1.

A. Preparation

Although the synthesis of triazenes appears to be a straightforward process, the preparation of pure compounds is often very difficult due to formation of by-products and the likelihood of some diazoamino compounds undergoing rearrangement to aminoazo compounds. In general, dialkaryltriazenes can be more readily prepared than the corresponding monoalkyl analogues. A modified synthesis of compounds from aromatic amines, which are notorious for by-product formation, has been developed in our laboratory using solid diazonium fluoroborates or hexafluorophosphates as a convenient source of the required arenediazonium cation

TABLE 1. Fragmentation of unsubstituted 3,3-dialkyl-1-phenyltriazenes

Structure: phenyl-N=N-N(R1)(R2); amine fragment: [HN(R1)(R2)]+

#	R1	R2	M+•	m/e 105	m/e 78	m/e 77	m/e 51	amine fragment
1.	CH_3	CH_3	m/e 149 % RA 11.3	35.8	13.2	100	19.5	m/e 44 7.6
2.	CH_3	OH	m/e 151 % RA 16.4	18.2	14.5	100	14.5	m/e 46 1
3.	3N morpholino		m/e 191 % RA 5.0	37.5	11.2	100	12.5	m/e 86 1
4.	C_3H_7	C_3H_7	m/e 205 % RA 7.7	29.1	11.5	100	8.8	m/e 100 4
5.	CH_3	$CH_2COOC_2H_5$	m/e 221 % RA 6.3	38.1	9.5	100	11.1	m/e 116 1
6.	CH_3	CH_2(phenyl)	m/e 225 % RA 5.0	34.8	11.4	100	12.1	m/e 120 21.2
7.	CH_3	C_2H_4OH	m/e 179 % RA 11.9	28.8	13.6	100	15.7	m/e 74 4

(Kolar, 1972). On the other hand, the coupling of arenediazonium cations with primary aliphatic amines cannot be successfully controlled and mixtures of triazenes, pentazdienes, and unidentified products are obtained.

B. Carcinogenic Activity

The results of screening for carcinogenic activity (Druckrey, Preussmann, Ivankovic, Landschütz, Gimmy, Flohr, and Griessbach, *unpublished results*) revealed that the majority of alkaryltriazenes exhibited predominantly systemic effects and induced tumors in organs distant from the site of application. The only exception was 1-(4-methoxyphenyl)-3,3-dimethyltriazene which released mainly local tumors at the site of subcutaneous injection. Since the triazenes are stabilized diazo compounds, it was conceivable that the observed biological activity of this compound could be associated with the susceptibility of the N2–N3 bond to hydrolytic cleavage with the concomitant liberation of the latent arenediazonium cation. Stability determination of a series of substituted alkaryltriazenes (Kolar and Preussmann, 1971) at pH 7.0 and 37°C showed that the half-lives varied from 1.1×10 min for 1-(4-methoxyphenyl)-3,3-dimethyltriazene to 3.59×10^5 min for 3,3-dimethyl-1-(3-nitrophenyl)-triazene. These results suggested that two different mechanisms were responsible for the carcinogenic activity of the dialkyltriazenes. 1. Enzymic hydroxylation leading to the release of a potential alkyl donor was an essential prerequisite for the biological activation, and hence for the resorptive carcinogenic activity, of the stable compounds (Preussmann, v. Hodenberg, and Hengy, 1969). 2. The local carcinogenic activity could be demonstrated only with the labile 1-(4-methoxyphenyl)-3,3-dimethyltriazene and, therefore, it was in complete agreement with the ready release of the reactive 4-methoxybenzenediazonium cation from the applied carcinogen.

The elucidation of the reaction mechanisms that underlie the carcinogenic activity of alkaryltriazenes demands not only extensive screening of selected compounds in *in vivo* tests but also the study of enzyme systems involved in their activation. A thorough knowledge of the chemical reactivity of these carcinogens is therefore necessary to understand the biotransformations of these compounds relevant to carcinogenesis. Investigations of chemical carcinogens by physical methods can disclose invaluable and complementary information on the nature of chemical processes that are likely to take place in the living cell.

II. MASS SPECTROMETRY

The mass spectra were measured on an A.E.I.–MS 9 mass spectrometer fitted with a direct inlet system, at an ionizing voltage of 70 eV, 8 kV accelerating voltage, and 100 μA ionizing current. The temperature of the ion

source was raised slowly from 20°C until sufficient vapor pressure was attained, and then kept at about 200°C. To decrease the rate of vaporization
of the more volatile compounds, ethereal solutions of the samples were adsorbed on animal charcoal which, after drying, was introduced by the direct
inlet of the instrument. Under these experimental conditions, well-resolved
spectra were obtained from all 18 alkaryltriazenes, and their fragmentations
were confirmed by the presence of metastable transitions. However, no
meaningful spectra were obtained from triazene sulfonic acids which
decomposed before volatilization in the mass spectrometer.

A. Fragmentation of 3,3-Dialkyl-1-aryltriazenes

The mass spectrum of the parent 3,3-dimethyl-1-phenyltriazene, a typical
example of an unsubstituted dialkaryltriazene, is shown in Fig. 1.
The spectrum contains the expected molecular ion at m/e 149 (11.3%)
which, after the loss of 44 mass units, yields the prominent peak at m/e
105 (35.8%) of the benzenediazonium cation. After the loss of nitrogen,
the latter ion affords the intense peak at m/e 77 (100%) of the phenonium
cation which is the base peak of the spectrum. A typical feature is the ion at
m/e 44, corresponding to the dimethylated nitrogen, which is characteristic
for the mass spectra of 1-aryl-3,3-dimethyltriazenes. The basic fragmentation pattern does not substantially change either with the increase of chain

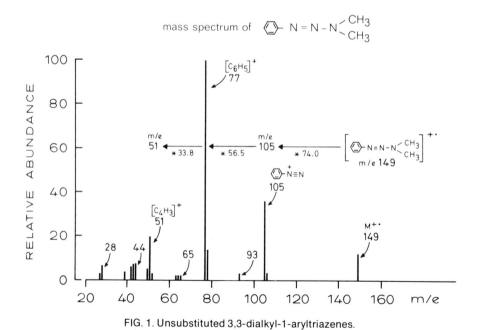

FIG. 1. Unsubstituted 3,3-dialkyl-1-aryltriazenes.

length of the alkyl substituents at N3, or with the replacement of the alkyl groups by other substituents such as an N-hydroxyl group.

The fragmentations of seven 3,3-dialkyl-1-phenyltriazenes are collected in Table 1. The relative abundance of the molecular ion was found to range from 5% in 3-benzyl-3-methyl-1-phenyltriazene and in 3-morpholino-1-phenyltriazene to 16.4% in 3-hydroxy-3-methyl-1-phenyltriazene.

An understanding of the ability of various functional groups to control and direct the mass spectrometric fragmentation of organic molecules is of importance in the use of this technique for structure elucidation. Figure 2 shows the mass spectrum of 1-(4-methoxyphenyl)-3,3-dimethyltriazene as an example of a triazene containing an electron-releasing substituent in the aromatic moiety. Although the basic breakdown of this compound is very similar to that of the unsubstituted dialkaryltriazenes, the relative intensities of some major fragments are distinctly altered. A significant increase in the relative intensity of the molecular ion is a conspicuous feature of the spectra of triazenes substituted with groups bonded through hetero atoms carrying lone pairs of electrons capable of expansion of their valence shell.

The fragmentations of eight substituted 3,3-dimethyl-1-phenyltriazenes are listed in Table 2. A typical example is the fragmentation of 1-(4-dimethylaminophenyl)-3,3-dimethyltriazene and of the already mentioned

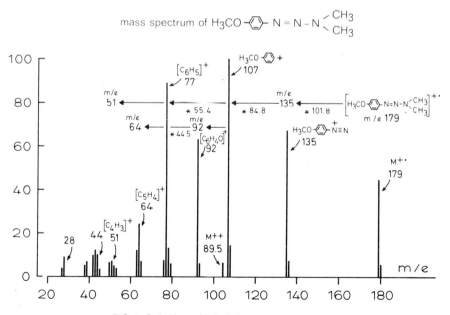

FIG. 2. Substituted 3,3-dialkyl-1-aryltriazenes.

TABLE 2. Fragmentation of substituted 3,3-dimethyl-1-phenyltriazene

Structure: X–C$_6$H$_4$–N=N–N(CH$_3$)$_2$

X		$M^{+\cdot}$	$X\text{–}C_6H_4\text{–}N=N^+$	$[X\text{–}C_6H_4]^+$	m/e 77	m/e 51	m/e 44
1.	p CH$_3$O	m/e 179	135	107	77	51	44
		% RA 44.5	67.0	100	89	7.0	9.5
2.	p HO	m/e 165	121	93	65		44
		% RA 25.0	55.4	94.7	100		24.4
3.	p CH$_3$COO	m/e 207	163	93	65	m/e 43	44
		% RA 12.9	28.1	100	43.8		12.3
			135				
			37.6				
4.	m HO	m/e 165	121	93	65		44
		% RA 19.5	42.8	100	87.9		15.8
5.	p (CH$_3$)$_2$N	m/e 192	148	120	77	51	44
		% RA 57.7	28.8	100	44.4	10.6	15.2
6.	p Cl	m/e 183	139	111	75	50	44
		% RA 13.2	32.9	100	27.6	10.5	11.3
7.	p Br	m/e 227	183	155	76	50	44
		% RA 11.8	33.3	100	48.0	36.3	12.4
8.	p C$_2$H$_5$OOC	m/e 221	177	149	77	51	44
		% RA 23.3	42.7	100	38.3	11.7	22.8

TABLE 3. Fragmentation of unsubstituted 3-alkyl-1-phenyltriazenes

$-N=N-N\overset{H}{\underset{R}{\diagdown}}$

	R	$M^{+\cdot}$	m/e 106	m/e 105	m/e 93	m/e 77	m/e 66	m/e 65	m/e 51
1.	CH_3	m/e 135							
		% RA 10	15.0	18.3	100	53.3	30.0	26.7	15.8
2.	CH_2CH_3	m/e 149							
		% RA 7.6	33.3	24.2	80.3	100	28.8	33.3	24.2
3.	$(CH_2)_3CH_3$	m/e 177							
		% RA 6.2	27.5	26.9	76.2	100	20.0	18.0	15.6

Remarks:

a) Contrary to fragmentation of unsubstituted 3,3-dialkyl-1-phenyltriazenes which show a discernible $N\overset{R^1}{\underset{R^2}{\diagdown}}$ ion, the $N\overset{H}{\underset{R}{\diagdown}}$ ion cannot be detected in the spectra of the corresponding monoalkaryltriazenes.

b) Prominent peaks appear at m/e 106 ($Ph-N=NH^+$) and at m/e 93 ($Ph-NH_2^{+\cdot}$).

1-(4-methoxyphenyl)-3,3-dimethyltriazene which show molecular ions of 57.7% and 44.5% relative abundance, respectively.

B. Fragmentation of 3-Alkyl-1-aryltriazenes

On the other hand, the spectra of 3-alkyl-1-phenyltriazenes revealed some even more typical fragmentations which became of diagnostic importance for this class of compounds. The mass spectrum of the parent compound, 3-methyl-1-phenyltriazene, is shown in Fig. 3.

The spectrum resembles that of the corresponding dimethyl analogue and reveals a molecular ion at m/e 135 (10%), as well as a moderately strong ion at m/e 105 of the benzenediazonium cation. However, an outstanding feature of the spectrum is a very intense peak at m/e 93 (100%) which corresponds to the radical ion of aniline arising by fission of the diazoamino grouping between N1 and N2, followed by hydrogen transfer. Similar fragmentations were established for two higher monoalkaryltriazene homologues which are listed in Table 3.

ACKNOWLEDGMENT

The kind help of Dr. W. Vetter, Research Department, F. Hoffmann-La Roche, Basle, with the measurement of mass spectra is appreciated.

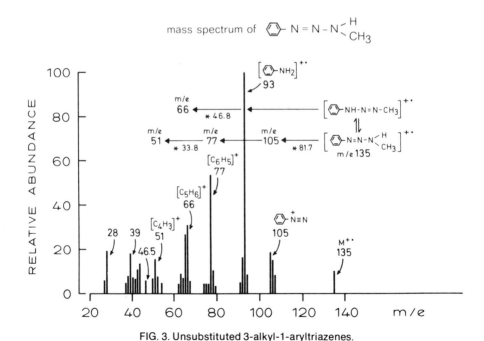

FIG. 3. Unsubstituted 3-alkyl-1-aryltriazenes.

REFERENCES

Druckrey, H., Ivankovic, S., and Preussmann, R. (1967a): Neurotrope carcinogene Wirkung von Phenyl-dimethyl-triazen an Ratten. *Naturwissenschaften*, 54:171.

Druckrey, H., Ivankovic, S., Preussmann, R., and Brunner, U. (1967b): Teratogene Wirkung von 1-Phenyl-3,3-dimethyltriazen, Erzeugung von Gaumenspalten bei BD-Ratten. *Experientia*, 23:1042.

Fahrig, R. (1971): Metabolic activation of aryldialkyltriazenes in the mouse: Induction of mitotic gene conversion in *Saccharomyces cerevisiae* in the host-mediated assay. *Mutation Research*, 13:436.

Kolar, G. F. (1972): Synthesis of biologically active triazenes from isolable diazonium salts. *Zeitschrift für Naturforschung*, 27b:1183.

Kolar, G. F., and Preussmann, R. (1971): Validity of a linear Hammett plot for the stability of some carcinogenic 1-aryl-3,3-dimethyltriazenes in an aqueous system. *Zeitschrift für Naturforschung*, 26b:950.

Ong, T. and de Serres, F. J. (1971): Mutagenicity of 1-phenyl-3,3-dimethyltriazene and 1-phenyl-3-monomethyltriazene in *Neurospora crassa*. *Mutation Research*, 13:276.

Preussmann, R., Druckrey, H., Ivankovic, S., and v. Hodenberg, A. (1969): Chemical structure and carcinogenicity of aliphatic hydrazo, azo, and azoxy compounds and of triazenes, potential *in vivo* alkylating agents. *Annals of the New York Academy of Sciences*, 163:697.

Preussmann, R., v. Hodenberg, A., and Hengy, H. (1969): Mechanism of carcinogenesis with 1-aryl-3,3-dialkyltriazenes. Enzymatic dealkylation by rat liver microsomal fraction *in vitro*. *Biochemical Pharmacology*, 18:1.

Shealy, F. Y., and O'Dell, C. A. (1971): Synthesis, antileukemic activity, and stability of 3-(substituted-triazeno)pyrazole-4-carboxamides. *Journal of Pharmaceutical Sciences*, 60:554.

Siebert, D., and Kolar G. F. (1973): Induction of mitotic gene conversion by 3,3-dimethyl-1-phenyltriazene, 1-(3-hydroxyphenyl)-3,3-dimethyltriazene and by 1-(4-hydroxyphenyl)-3,3-dimethyltriazene in *Saccharomyces cerevisiae*. *Mutation Research*, 18:267.

Vogel, E. (1971): Chemische Konstitution und mutagene Wirkung, VI. Induktion dominanter und rezessiv-geschlechtsgebundener Letalmutationen durch Aryldialkyltriazene bei *Drosophila melanogaster*. *Mutation Research*, 11:397.

Mass Spectrometry in Biochemistry and Medicine,
edited by A. Frigerio and N. Castagnoli.
Raven Press, New York © 1974

A Gas Chromatograph–Mass Spectrometer– Computer System Applied to the Analysis of Sesquiterpene Hydrocarbons

R. Flückiger,[a] Y. Kato, and S. Hishida

Hitachi Ltd., Naka Works, Katsuta, 312 Japan

I. INTRODUCTION

The identification of compounds by using a computer to compare recorded mass spectra with a file of reference spectra has become a widespread technique. For a generally applicable system the reference file has to be large, and the storage space and search time are considerable. However, as the type of a given sample is usually not completely unknown, relatively small reference files containing certain types of compounds could be employed resulting in less expensive systems for individual requirements. For example, if the hexane extracts of wood and fruit oils are investigated and if the gas chromatographically separated compounds are observed in a characteristic retention time region and if the mass fragmentograms run at mass number 204 appear almost identical to the chromatograms obtained with a total ion monitor, one can assume with a high degree of certainty that the examined compounds are sesquiterpene hydrocarbons with molecular composition $C_{15}H_{24}$. In such a case a reference file limited to $C_{15}H_{24}$-compounds may be sufficient. Although a carefully selected abbreviation of the mass spectra does hardly alter the efficiency of a search system even when using large files, it should be possible to compress the mass spectra of a limited file even more. It will be demonstrated that an extreme abbreviation of the reference mass spectra by retaining only the mass numbers of the two most significant peaks can give surprising results.

II. INSTRUMENTATION

A gas chromatograph Hitachi model 063 has been combined with a single focusing mass spectrometer Hitachi model RMU-6L fitted with an improved pumping system by the direct coupling of capillary columns introducing the whole column effluent into the ion source. For on-line data

[a] *Present address: Balzers Aktiengesellschaft für Hochvakuumtechnik und Dünne Schichten FL-9496 Balzers, Fürstentum Liechtenstein*

acquisition and processing and for the file search, a Hitac-10 computer with 8 K core memory and 64 K accessory drum memory has been used. After reducing the raw data to mass number and intensity lists, the unknown spectra could be compared by two independent search programs with a file of 80 sesquiterpene hydrocarbon spectra collected from the literature and our own data. The first program, which will not be discussed in detail, allowed us to abbreviate a recorded mass spectrum in different ways, to calculate a matching index with each reference spectrum, and to print out a list of the reference compounds in the order of decreasing similarity. With the second program, called SIGY, we tried to reduce the search time and the storage space for the reference file to a minimum. For this purpose the reference spectra have been abbreviated by only retaining the mass numbers of the two most significant peaks. By coding these two mass numbers into a 16-bit computer word and by using the storage location as sample code, no more than a single computer word is used for each reference compound. The "most significant peaks" have been determined by a purely intuitive procedure explained in the following section.

III. SIGY ALGORITHM

It is true that the most significant peaks in a mass spectrum are not necessarily those with the largest relative abundances. For example, a large peak at m/e 41 is in no way characteristic for a certain sesquiterpene hydrocarbon, but a relatively large peak at m/e 122 is highly significant, because it will be present for only a few compounds of this type. Accordingly, the significance of a peak has been defined essentially as the normalized and weighted difference between its intensity in ionization percents, stripped additionally from isotopic contributions of the preceding peaks and the average peak height of this mass number in the whole reference file. The isotopic correction avoids giving significance to purely isotopic peaks; the normalization takes into account the individual distribution of the intensities at each mass number; and the weighting discriminates the very small peaks having large statistical and background uncertainties.

With the SIGY algorithm the significance of a certain peak in the spectrum of an unknown compound depends on all the spectra contained in the reference file, and so this method stresses the differences in the spectra rather than the similarities. It is obvious that such an approach is only feasible for relatively small reference files, but in those cases SIGY will show how powerful it can be for distinguishing similar, but not identical spectra. Notice, that SIGY has some resemblance to chemical ionization, except that the surplus of information contained in a mass spectrum is reduced by mathematical rather than physical means.

A point which has to be carefully watched is that a relatively small background peak at an unusual mass number could reach a high significance. It is therefore important to use reference spectra of high quality. It is less important for the unknown spectra, because all pair combinations of the n most significant peaks can be compared with the two most significant peaks of the references, if allowance is given for instrumental and statistical variations.

The search time is short, because after calculating the n most significant peaks of an unknown spectrum, the only remaining task is to find those storage locations corresponding to a combination of the mass numbers of these peaks. Naturally, the two most significant peaks of each reference compound have to be found first for each set of reference spectra using a file set-up program.

IV. RESULTS

The hexane extracts of four citrus fruit oils (orange, lemon, lime, and grapefruit) and two wood oils (Hiba and Chuor Chong) were separated on capillary columns with different stationary phases and the mass spectra processed in the described way. The answers proposed by the computer were verified by visual comparison of the scanned spectra with complete literature and our own reference data and by considering the gas chromatographic retention times.

Both search programs proved to be very efficient and helped to identify quickly more than 90% of the components present. Of course, no distinction could be made between stereoisomers and some other closely related isomers which show practically no differences in their mass spectra. The redundant outputs were then determined as a function of the correct answers retained, by counting the incorrect and correct answers above a certain matching index, or in case of SIGY by considering a distinct number of n. Surprisingly, no significant difference could be observed using SIGY or abbreviation modes such as the two strongest peaks out of every 14 u interval or the eight strongest peaks. For 90% correct answers retained, all these algorithms showed redundant answers on the order of 1 to 2% of the number of spectra contained in the reference file, which seems to be quite reasonable for a reference file of compounds all having the same molecular composition. In contrast, the respective redundances using four and two strongest peaks of a spectrum were 4% and 25%. Whereas in 75% of all cases, the first proposition of the program using the two strongest peaks out of every 14 u interval was unambiguous and correct, this figure was still 69% using SIGY which took 28 times less storage space for the reference file.

V. CONCLUSIONS

It has been observed that the quality of a reference file is of great importance and that sesquiterpene hydrocarbon references obtained by gas chromatography–mass spectrometry (GC–MS) are better suited, because they show generally less impurities. The GC resolution is another critical point, because it is extremely difficult to separate the mass spectra of isomers having their mass spectral peaks at the same mass numbers. The GC–MS–computer system proved to be very effective for correlating the GC peaks of a sample mixture separated on columns with different stationary phases by using the mass spectra from one column as reference file and treating the mass spectra of the other columns as unknowns, thus correlating peaks without even looking at the mass spectra. Finally, it has become evident that a computer-based identification system for limited compound types may be considerably simplified.

Mass Spectrometry in Biochemistry and Medicine,
edited by A. Frigerio and N. Castagnoli.
Raven Press, New York © 1974

Determination of the Structure of Partially Methylated Sugars as O-Trimethylsilyl Derivatives by Gas Chromatography and Mass Spectrometry

Akira Hayashi and Toshiko Matsubara

Department of Chemistry, Faculty of Science and Technology, Kinki University, Kowakae, Higashiosaka, Osaka, Japan

Methylation is a useful technique in the structure analysis of oligo- and polysaccharide and other natural compounds containing carbohydrate monomer units. The value of this method has become greater as a result of improvements in the methylation procedure and developments of new methods for the separation and identification of partially methylated monosaccharides, involving gas chromatography and combined gas chromatography–mass spectrometry (GC–MS).

Methyl ethers, acetates, or trimethylsilyl (TMS) ethers of free glycoses, methyl glycosides, or alditols have been used as convenient derivatives for GC analysis, since free sugars themselves are not sufficiently volatile. TMS derivatives have the advantage that they are readily prepared in nearly quantitative yield under rather mild conditions, thus permitting rapid analysis. Nevertheless, there is no complete GC–MS study of O-TMS partially methylated methyl glycosides.

All types of tetra-, tri-, and di-O-methyl methyl glucosides shown in Table 1 were analyzed as O-TMS derivatives by GC and GC–MS, and their mass spectra were compared with those of untreated partially methylated methyl glucosides.

It was found that mass spectra of TMS derivatives showed more intense and characteristic fragment ions which could be used for the identification of the methylated positions in sugar.

It is evident from these results that the analysis of partially methylated methyl hexosides as O-TMS ether derivatives by GC–MS is a useful aspect of the permethylation method and is more rapid and simple than the case of alditol acetates. Some of the experimental results are shown in Tables 2–4.

The present method was applied to the structure elucidation of the sugar moiety found in a sphingoglycolipid obtained from an oyster mantle.

TABLE 1. *Origin and abbreviation of methyl glucosides studied*

Systematic name	Abbreviation	Origin
Methyl 2,3,4,6-tetra-O-methyl-α-D-glucoside	α-2,3,4,6-Me$_4$	
Methyl 3,4,6-tri-O-methyl-D-glucoside	3,4,6-Me$_3$	Kojibiose
Methyl 2,4,6-tri-O-methyl-D-glucoside	2,4,6-Me$_3$	Laminaribiose
Methyl 2,3,6-tri-O-methyl-D-glucoside	2,3,6-Me$_3$	Lactose
Methyl 2,3,4-tri-O-methyl-D-glucoside	2,3,4-Me$_3$	Melibiose
Methyl 2-O-TMS 3,4,6-tri-O-methyl-D-glucoside	TMS-3,4,6-Me$_3$	
Methyl 3-O-TMS 2,4,6-tri-O-methyl-D-glucoside	TMS-2,4,6-Me$_3$	
Methyl 4-O-TMS 2,3,6-tri-O-methyl-D-glucoside	TMS-2,3,6-Me$_3$	⎫ Trimethylsilylation
Methyl 6-O-TMS 2,3,4-tri-O-methyl-D-glucoside	TMS-2,3,4-Me$_3$	⎬
Methyl 2,3-di-O-methyl-α-D-glucoside	α-2,3,-Me$_2$	⎭
Methyl 2,4-di-O-methyl-D-glucoside	2,4-Me$_2$	Synthesized in our laboratory
Methyl 2,6-di-O-methyl-D-glucoside	2,6-Me$_2$	Dextran
Methyl 3,4-di-O-methyl-D-glucoside	3,4-Me$_2$	Dr. G. Adams
Methyl 3,6-di-O-methyl-D-glucoside	3,6-Me$_2$	Dr. A. Misaki
Methyl 4,6-di-O-methyl-α-D-glucoside	α-4,6-Me$_2$	Dr. G. Adams
Methyl 4,6-di-O-TMS 2,3-di-O-methyl-α-D-glucoside	α-TMS-2,3-Me$_2$	Synthesized in our laboratory
Methyl 3,6-di-O-TMS 2,4-di-O-methyl-D-glucoside	TMS-2,4-Me$_2$	
Methyl 3,4-di-O-TMS 2,6-di-O-methyl-D-glucoside	TMS-2,6-Me$_2$	
Methyl 2,6-di-O-TMS 3,4-di-O-methyl-D-glucoside	TMS-3,4-Me$_2$	⎫ Trimethylsilylation
Methyl 2,4-di-O-TMS 3,6-di-O-methyl-D-glucoside	TMS-3,6-Me$_2$	⎬
Methyl 2,3-di-O-TMS 4,6-di-O-methyl-α-D-glucoside	α-TMS-4,6-Me$_2$	⎭

TABLE 2. *Relative retention times of methylated methyl glucosides and their TMS derivatives*

Methyl glucosides	T^a		T^b	
α-2,3,4,6-Me$_4$	–	1.00		1.00
3,4,6-Me$_3$	1.88	2.22		
2,4,6-Me$_3$	1.95	3.86		
2,3,6-Me$_3$	2.10	2.84		
2,3,4-Me$_3$	1.56	2.19		
TMS-3,4,6-Me$_3$			0.68	0.82
TMS-2,4,6-Me$_3$			0.88	0.98
TMS-2,3,6-Me$_3$			0.70	1.09
TMS-2,3,4-Me$_3$			0.56	0.78
α-2,3-Me$_2$	–	8.14		
2,4-Me$_2$	4.68	6.84		
2,6-Me$_2$	6.01	8.14		
3,4-Me$_2$	4.56	5.40		
3,6-Me$_2$	5.28	6.68		
α-4,6-Me$_2$	–	4.90		
α-TMS-2,3-Me$_2$			–	0.76
TMS-2,4-Me$_2$			0.64	0.76
TMS-2,6-Me$_2$			1.00	1.31
TMS-3,4-Me$_2$			0.53	0.61
TMS-3,6-Me$_2$			0.76	0.91
α-TMS-4,6-Me$_2$			–	1.05

T: Retention Time relative to α-2,3,4,6-Me$_4$
[a] 160°C   [b] 140°C

It has been found that 2,3,4-tri-O-methylfructose, 2,3,4,6-tetra-O-methyl-galactose, 2,3,6-tri-O-methylglucose, 2,4,6-tri-O-methylgalactose, and 4,6-di-O-methylgalactose are present. The presence of 4,6-di-O-methyl-galactose indicates that the oyster sphingoglycolipid has a branched structure in the sequence of its sugar moiety. Yang and Hakomori (1971) have also found the branched structure in the sphingoglycolipid obtained from human adenocarcinomas. However, the positions of branching and the kind of sugar at the branched position are different from those of the oyster lipid. This finding presents a new problem to the biochemical pathway of gly-colipid in marine shellfish.

TABLE 3. *Partial mass spectra of tri-O-methyl methyl glucosides*

Methyl glucosides	45	71	73	74	75	87	88	89	101	102	133	146	159	161	187
α-2,3,4,6-Me$_4$	40	–	28	6	75	2	100	17	66	11	1	–	1	–	4
3,4,6-Me$_3$	54	100	17	63	94	23	59	20	25	11	–	–	4	13	–
2,4,6-Me$_3$	49	100	13	62	27	11	11	8	43	33	–	–	1	–	18
2,3,6-Me$_3$	45	–	–	13	83	15	100	–	16	–	–	–	22	16	15
2,3,4-Me$_3$	12	–	10	6	28	4	100	5	35	–	–	–	–	–	–
TMS-3,4,6-Me$_3$	66	43	66	15	52	4	28	41	19	6	12	100	78	7	–
TMS-2,4,6-Me$_3$	33	–	42	6	25	2	10	20	34	7	56	100	11	–	–
TMS-2,3,6-Me$_3$	35	–	44	7	55	3	100	25	25	–	12	14	89	–	–
TMS-2,3,4-Me$_3$	55	17	69	11	47	–	100	42	39	–	–	–	–	–	–

TABLE 4. *Partial mass spectra of di-O-methyl methyl glucosides*

Methyl glucosides	45	71	73	74	75	87	88	89	101	102	133	146	147	159	204	217
α-2,3-Me$_2$	39	–	39	21	72	36	100	13	32	–	–	–	–	10	–	–
2,4-Me$_2$	51	87	19	100	23	34	42	6	54	25	–	–	–	4	–	–
2,6-Me$_2$	25	–	5	100	8	43	2	9	2	4	–	–	–	–	–	–
3,4-Me$_2$	62	84	38	100	73	45	50	19	17	4	–	–	–	7	–	–
3,6-Me$_2$	66	–	7	100	65	61	–	5	2	–	–	–	–	–	–	–
α-4,6-Me$_2$	53	95	–	100	26	75	6	6	8	–	1	–	–	2	–	–
α-TMS-2,3-Me$_2$	16	–	76	7	70	2	100	27	29	–	23	20	–	72	–	7
TMS-2,4-Me$_2$	27	1	95	11	32	3	13	58	15	3	30	100	22	22	–	3
TMS-2,6-Me$_2$	16	–	51	5	9	1	5	15	3	–	32	100	–	22	–	1
TMS-3,4-Me$_2$	37	27	100	10	60	–	20	77	13	3	10	83	10	70	–	–
TMS-3,6-Me$_2$	20	–	60	6	26	1	1	33	5	–	–	88	–	13	–	100
α-TMS-4,6-Me$_2$	15	15	60	6	11	1	1	15	7	–	50	22	18	26	100	7

ACKNOWLEDGMENT

The authors are grateful to Dr. G. Adams (National Research Council of Canada) for his generous gifts of 2,6- and 3,6-di-O-methylglucose and also to Dr. A. Misaki (Osaka University, Japan) for 3,4-di-O-methylglucose.

REFERENCE

Yang, H., and Hakomori, S. (1971): *Journal of Biological Chemistry*, 246:1192.

Mass Spectrometry in Biochemistry and Medicine,
edited by A. Frigerio and N. Castagnoli.
Raven Press, New York © 1974

Experience with Gas Chromatography–Mass Spectrometry in Clinical Chemistry

Lorentz Eldjarn, Egil Jellum, and Oddvar Stokke

Institute of Clinical Biochemistry, Rikshospitalet, University of Oslo, Oslo, Norway

The gap between what is possible to do for our patients and what in fact is being done is steadily increasing. Limited economical resources are usually claimed to be the reason. Of more significance may be the fact that the ever accelerating speed of frontier research makes it impossible for the clinician to keep up. We should also bear in mind that a large part of the medical profession who today holds the most responsible positions had their basic education before the age of the tricarboxylic acid cycle.

It seems clear that the clinical chemists carry a particular responsibility in helping to bridge this gap between frontier research and practical medicine. It is generally accepted that hospital departments of clinical chemistry are responsible for the production of innumerable tests in a limited spectrum of standard analyses comprising approximately 100 to 200 biochemical parameters. The complexity of this task and the many obstacles with respect to accuracy and precision of data as well as requirements for speed have made departments of clinical chemistry essential parts of all modern hospitals.

However, in our opinion, the clinical chemist carries far greater responsibilities. With his background both in clinical medicine and basic biochemical and biological sciences, his first and foremost duty should be to bridge the gap between bench-side and bed-side medicine.

However, we are all familiar with the difficulties encountered. The hectic and strained routine of modern departments of clinical chemistry leaves little time and peace of mind for such activities. For these reasons the University of Oslo in 1961 founded an Institute of Clinical Biochemistry which geographically is located next to the routine department of clinical chemistry. In addition to research activities, it should organize pre- and postgraduate medical education in clinical chemistry. The close cooperation between the research institute and the routine department has proved extremely efficient in promoting the application of up-to-date biochemical techniques on clinical problems. In this setup GC–MS has proven a most valuable tool. In this chapter we will describe our approach and survey some of the results where GC–MS has been of particular value.

I. THE SEARCH FOR BIOCHEMICAL ABNORMALITIES

During the last 6 to 8 years we have attempted to develop a systematic approach in our search for biochemical abnormalities in clinical conditions. The procedure has at least four essential steps, of which GC–MS is only one.

1. In most cases the work starts with an *alert clinician* who observes particular signs or symptoms and at the same time is informed about our interests and activities. Such an initial screening by clinical judgment seems to be essential in order to obtain a reasonable yield in our work. We appreciate being informed of unusual clinical pictures and pronounced metabolic derangements. Since we have particular interest in inborn errors of metabolism, the clinician should among other things be aware of special or peculiar odors from the patient, lasting acidosis, increased excretion of acid equivalents in the urine, and clear-cut inherited disorders.

2. When we decide to investigate a given condition, we first carry out a *spectrum of ordinary clinical chemical tests* in order to judge the major metabolic pathways. It should be stressed that some of our most exciting observations have been made at this stage—without the use of GC–MS. Thus, on the basis of ordinary tests, four cases of a novel familial magnesium absorption deficiency in newborns were described (Strømme, Nesbakken, Normann, Skjørten, Skyberg, and Johannessen, 1969). The lives of these babies can be saved by increased peroral magnesium administration.

Also at this stage the novel inborn error of metabolism "familial plasma lecitin: cholesterol acyltransferase (LCAT) deficiency" was described (Norum and Gjone, 1967). To the present, seven patients with this disease have been described, all of whom are lacking the LCAT-enzyme. As a consequence most of the serum cholesterol is unesterified and most probably the clinical symptoms are due to this derangement in the cholesterol transport mechanism.

3. With our *GC–MS screening system* we are aiming at profiling as many as possible of the metabolites present in serum or urine samples from the patients. The analytical system has been described in detail elsewhere (Jellum, Stokke, and Eldjarn, 1972). There is nothing particularly fancy about our setup. We are working in the "old-fashioned macro-scale" of 10^{-6} to 10^{-7} g, i.e., we are using flame ionization detectors on our GC systems to visualize peaks of interest. After the addition of internal standards (n-eicosane, trehalose, α-aminooctanoic acid) and acidification, the sample is divided into an ether extract and an aqueous phase. The ether extract is run on Porapak P and OV-17 in order to separate very volatile and somewhat lesser volatile lower alcohols, acetone, acetaldehyde, and lower free fatty acids as well as aromatic hydrocarbons and certain barbiturates (Fig. 1, system A and B).

After methylation with diazomethane, the ether extract is further separated on BDS (System C) and OV-17 (System D) in order to separate

URINE

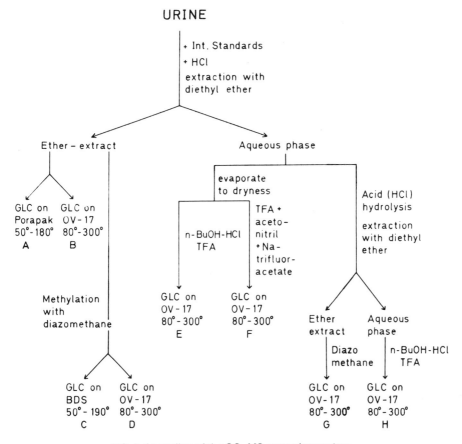

FIG. 1. An outline of the GC–MS screening system.

aliphatic and aromatic acids, certain phenols, and di- and tricarboxylic acids (to the extent they have been extracted).

The aqueous phase is in part converted to n-butanol esters and TFA derivatives of amines and amino acids (System E) and to O-trifluoroacetyl derivatives of carbohydrates (System F).

Some indications as to the presence of high molecular weight compounds and conjugated metabolites are obtained by using the system D and E after acid hydrolysis of a sample of the aqueous phase (Fig. 1, system G and H).

Altogether 500 to 800 components are separated and visualized by these GC-systems. The mass spectra of all major peaks and of all unexpected peaks are recorded on a Varian CH7 mass spectrometer.

Up to now the spectra have been handled manually, but a Varian Spectro-System 100 MS is being installed. It is hoped that this setup with the proper amount of software will increase our efficiency tremendously at this stage.

Manual interpretation of the spectra is difficult and time-consuming. We have therefore programmed an off-line library search using a display terminal connected to a large CDC 6600 computer situated 30 miles away from our hospital. At present 24,000 reference mass spectra characterized by their five most abundant peaks as well as by the molecular ion are on the library file. These spectra stem from available compilations (Stenhagen, Abrahamsson and McLafferty, 1969; Stenhagen, Abrahamsson, and McLafferty, 1970; *Eight Peak Index of Mass Spectra,* 1970), which regrettably contain a lot of compounds of remote biological interest. About 400 spectra are our own recordings.

Matching of an unknown mass spectrum against the reference file can be carried out either by using the five strongest peaks in order of decreasing intensity or at random. Concordance with all peaks, with four, or with three peaks only may be displayed. Also a search for spectra with identical apparent M^+ can be carried out. From our console, new spectra can also be inserted in or deleted from the file.

The search time for the complete file requires only 7 sec. The computer matching has tremendously improved our efficiency. The GC-system permits two to three samples to be run per day. Obviously, the number of unexpected GC-peaks may be quite large. With only manual interpretation of the spectra, we for a long period of time suffered from obstruction at this stage.

4. The last step in our system permits the *further study of the metabolic conditions* discovered in patients. Such facilities are offered by our Institute of Clinical Biochemistry, which through financial support from a number of sources has at its disposal a spectrum of research facilities. This allows a variety of biochemical or physiological studies to be made *in vivo* on man or mammals or *in vitro* on fibroblast cultures, tissue slices, cell subfractions, or purified enzymes.

It thus appears that in the results to be presented that GC–MS has been only one of many techniques. However, in most cases the GC–MS has been the procedure which made the key observation possible.

II. NECESSARY PRETREATMENT OF THE PATIENTS TO BE STUDIED

Although others may be particularly happy when a drug metabolite appears in the chromatogram, such metabolites drive us crazy. In too many instances exciting peaks in the chromatograms have turned out to be what we call "simple" metabolites of a drug. Drug metabolites which have complicated our chromatographic patterns are phthalic acid which could be traced to cellulosephthalate in a drug capsule; phenoxyacetic acid originating from phenoxymethylpenicillin; trichloroacetic acid which stemmed from a rectal suppositorium containing chloral hydrate; and N-acetyltryptophan which could be traced back to an intravenous albumin treatment. For some reason the manufacturer had added this N-acetyltryptophan to the preparation as a stabilizing agent.

Obviously, the minimum requirements for pretreatment of a patient before attempting a detailed study of his metabolic pattern should be: 1. no drugs used for the last week, 2. dietary excesses avoided for the last days, and 3. the urine and serum samples collected in the morning.

We feel sure, however, that as the sensitivity of our system is increased, we will be forced to put the patients on a standardized synthetic diet for the last days before sampling.

On one occasion, unusual metabolites which could be traced to an exogenous origin led us to conclusions of considerable significance. We, as well as other investigators, had previously found small amounts of various furan derivatives (2-furoylglycine, furan-2,5-dicarboxylic acid) in the urine from normal subjects (Pettersen and Jellum, 1972). The compounds were found to be exogenous in origin.

Recently we were confronted with two severely ill babies who excreted large quantities of furan-2,5-dicarboxylic acid and 5-hydroxymethylfuroic acid (Jellum, Børresen, and Eldjarn, 1973). It was shown that these compounds stemmed from a commercial glucose-fructose mixture which had been used for intravenous feeding. The pharmaceutical firms acidify these mixtures before sterilization at 110 to 130°C. At pH lower than 3.5 to 4.0, this treatment inevitably produces large amounts of 5-hydroxymethyl-2-furfural and 2-hydroxyacetylfuran as well as large amounts of 2-keto-3-deoxyglucose. Obviously, the furfural derivative is metabolized to the 2,5-dicarboxylic acid derivatives found in the urine (Fig. 2). It should be

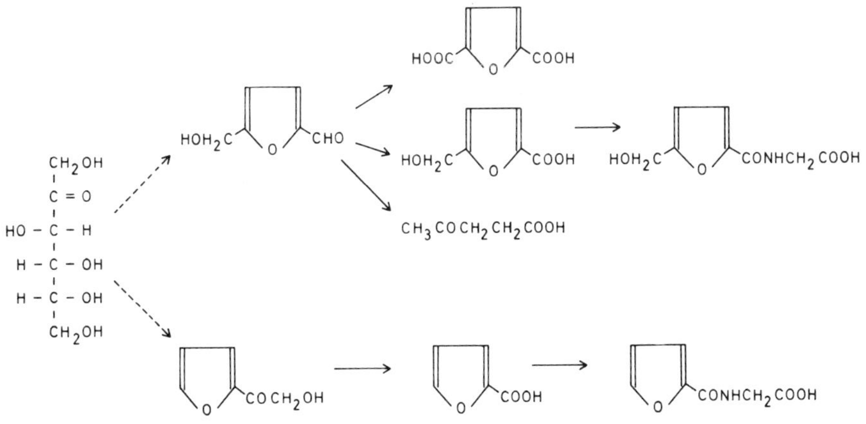

FIG. 2. The formation of furan derivatives from fructose.

stressed that the keto-alcohol 2-hydroxyacetylfuran in the body most probably is oxidized to the corresponding ketoaldehyde which immediately interacts with thiol groups and amines of the cell and thereby possibly causes unwanted toxic effects. Such a view is consistent with the fact that no metabolites of this compound could be found in the urine, indicating that the compound was completely retained in the body.

We concluded that when preparing such fructose-containing mixtures for intravenous nutrition, the pH should be carefully controlled before the heat sterilization. According to our experience, the pH should be kept no lower than pH 4.0 in order to prevent the above unwanted side reactions.

III. STUDIES ON CLINICALLY WELL-DEFINED DERANGEMENT OF METABOLISM

Using our technique, we have decided to reinvestigate carefully a number of well-defined metabolic derangements such as: liver failure, renal failure, various types of acidosis, as well as untreated cases of pronounced endocrine disorders.

As an example we will use our studies on acidosis. In medical terms acidosis or alkalosis is classified as respiratory or metabolic and its degree measured by various parameters such as "pH of blood," "pCO_2," and "standard bicarbonate." There are good reasons, however, to look more carefully into the biochemical basis of various types of acidosis. In such studies we found that, in the urine samples from patients with ketotic acidosis, substantial amounts of the dicarboxylic acids adipic (C_6) and suberic (C_8) acid were found (Pettersen, Jellum, and Eldjarn, 1972). The daily excretions paralleled the degree of ketosis and reached levels of 750 and 150 mg, respectively. In a few cases succinic and glutaric acids were also found in amounts up to 1,200 and 170 mg per day. The dicarboxylic acid excretion was only seen in acidotic patients with ketosis, whereas acidotic patients without ketosis and a number of other patients and controls did not excrete such acids apart from trace amounts of adipic and succinic acid.

A dietary origin of the dicarboxylic acids proper, or of medium-chain fatty acids as possible precursors, was excluded. Two possible pathways for the endogenous formation were considered. 1. Long-chain monocarboxylic acids may be shortened by β-oxidation and then leak out of the mitochondria at a chain length of six to eight carbon atoms, to become ω-oxidized. 2. Long-chain monocarboxylic acids may be ω-oxidized as such and the long-chain dicarboxylic acid shortened in the mitochondria by β-oxidations. For some reason they then leak out at chain lengths of six to eight carbon atoms. In experiments with $1\text{-}^{14}C$ and $16\text{-}^{14}C$ hexadecanoic acid in ketotic rats, strong indications were found for the latter pathway (Pettersen, 1972).

The clinical significance of this deranged fatty acid oxidation in ketotic acidosis is still obscure.

IV. THE DIAGNOSIS OF INBORN ERRORS OF METABOLISM

Since Sir Archibald Garrod in 1908 coined the term "inborn errors of metabolism," the list of such abnormalities has steadily increased and comprises at present approximately 150 diseases. Our understanding of these metabolic defects has been greatly improved by Beadle and Tatum's "one gene, one enzyme theory" and by our present detailed understanding of the genetic transcription mechanism.

Approximately 1,500 enzymes have been described in the mammalian organism. It is generally felt that the number of inborn errors of metabolism should be as large, since no regulator, repressor, or operator gene or cistrone could be imagined to be safe from mutations. The number actually appearing in man may be much lower due to the fact that defects concerned with essential metabolites or structures (ATP-production, DNA-replication, cell division, membrane functions, etc.) are not compatible with a completed pregnancy and thus would be "unborn errors of metabolism."

Our screening system may detect about 40 of the known inborn errors (Table 1). We have detected several of these in the course of the last years, usually as a verification of the clinician's suspicion.

Since inborn errors of metabolism can be expected to occur at any point of the intermediate metabolism, our system should be well suited to search for novel metabolic defects. We have so far investigated approximately 400 urine samples from patients selected by clinicians as suspect of inborn errors. We have also attempted a systematic screening of about 450 mentally retarded children in various institutions. The former approach has proven particularly fruitful and several novel diseases have been described. Also the studies have contributed to our understanding of the biochemical defect in previously described diseased states.

Our GC–MS has played an essential role in studies on *Refsum's disease* (heredopathia atactica polyneuritiformis). This rare but serious disease was delineated as a clinical entity by our professor of neurology in Oslo (Refsum, 1946). Klenk and Kahlke (1963) described the accumulation of phytanic acid, a branched hexadecanoic acid of isoprenoid structure, in a patient suffering from this disease. In long-term studies one of our Oslo patients was kept on a total body-water enriched with D_2O for 6 months (Steinberg, Mize, Avigan, Fales, Eldjarn, Try, Stokkes and Refsum, 1967). Using MS for analytical purposes, we can, in cooperation with Steinberg's group in Bethesda, show that, whereas deuterium was incorporated in cholesterol in the predictable way, trace amounts only accumulated in phytanic acid isolated from serum. This excluded that endogenous synthesis of phytanate took place. Further studies showed the phytol of the

TABLE 1. *Inborn errors of metabolism detectable by the screening system*

Name of disease	Compounds detectable by the GLC-MS methods	Name of disease	Compounds detectable by the GLC-MS methods
Alcaptonuria	homogentisic acid	Hypertryptophanemia	tryptophane
Carnosinemia	carnosine	Hypervalinemia	valine
Congenital lactacidosis	lactic acid	Isovaleric acidemia	isovaleric acid, β-hydroxy-
Cystathioninuria	cystathionine		isovaleric acid, isovaleryl-
Cystinuria	cystine		glycine
Diabetes mellitus	glucose, β-hydroxybutyric acid, acetoacetic acid	Maple syrup urine disease	valine, leucine, isoleucine, α-ketoisovaleric acid,
Essential fructosuria	fructose		α-ketoisocaproic acid,
Essential pentosuria	L-xylulose		α-keto-β-methylvaleric acid
Galactosemia	galactose, amino acids	Methylmalonic acidemia	methylmalonic acid
L-glyceric aciduria	L-glyceric acid, oxalic acid	Non-ketotic hyper-	glycine
Hartnup disease	neutral amino acids	glycinemia	
Histidinemia	histidine, imidazoleacetic acid	Oast-House disease	α-hydroxybutyric acid
		Ornithinemia	ornithine
Homocystinuria	homocystine, methionine	Orotic aciduria	orotic acid
β-Hydroxyisovaleric aciduria & β-methyl-crotonylglycinuria	β-hydroxyisovaleric acid, β-methylcrotonylglycine	Phenylketonuria	phenylalanine, phenyl-pyruvic acid, phenyllactic acid, o-hydroxyphenyl-acetic acid
Hydroxylysinuria	hydroxylysine		
Hydroxyprolinemia	hydroxyproline		
Hyper-β-alaninemia	β-alanine, β-aminoisot-butyric acid, γ-amino-butyric acid	Propionic acidemia	propionic acid
		Pyroglutamic aciduria	pyroglutamic acid (pyrrolidone-2-carboxylic acid)
Hyperlysinemia	lysine	Refsum's disease	phytanic acid
Hypermethioninemia	methionine, α-keto-γ-methiolbutyric acid	Renal glycosuria	glucose
Hyperoxaluria	oxalic acid, glycolic acid, glyoxylic acid	Short chain fatty acidemia	butyric acid, caproic acid
		Tyrosinosis	tyrosine, p-hydroxyphenyl-pyruvic acid, p-hydroxy-phenyllactic acid
Hyperprolinemia	proline		
Hypersarcosinemia	sarcosine		

chlorophyll molecule or preformed phytanic acid from, e.g., butter and cow's milk to be the source (Steinberg, Avigan, Mize, Eldjarn, Try, and Refsum, 1965). By treatment with a diet free from such constituents, the pathological amounts of phytanic acid of our two Oslo patients disappeared within 2 years (Eldjarn, Try, Stokke, Munthe-Kaas, Refsum, Steinberg, Avigan, and Mize, 1966c), and, for more than 7 years, serum and organ samples have shown only the trace amount normally present (Eldjarn, Stokke, and Try, 1974). One of these patients died in 1972 from an inter-current disease (an intrapontine hemorrhage). It was verified in this patient that no phytanic acid above the normal trace levels occurred in either parencymatous organs or lipid stores. Since no clinical relapses of Refsum's disease occurred during the total of 15 patient-years of observation in these two patients, it seems reasonable to conclude that most of the clinical symptoms of this disease are caused by phytanic acid accumulation (Eldjarn et al., 1974).

In the case of phytanic acid the methyl branchings in 3, 7, and 11 position block the ordinary β-oxidation pathway. Following ω-oxidation to a dicar-boxylic acid, β-oxidation can take place, however, from the ω-end. In

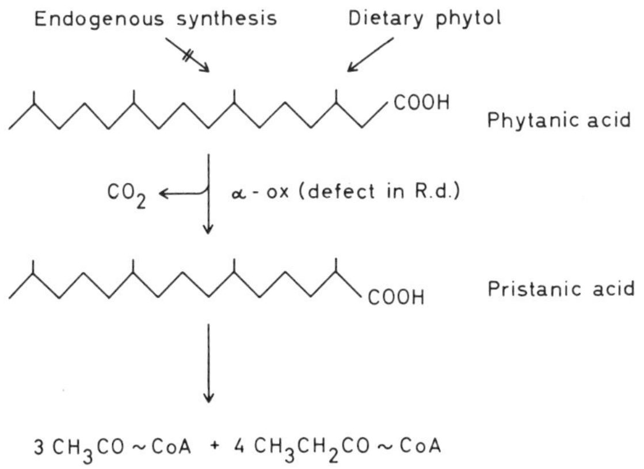

Endogenous synthesis   Dietary phytol

COOH   Phytanic acid

CO_2 ←┤ α - ox (defect in R.d.)

COOH   Pristanic acid

$3\ CH_3CO \sim CoA\ +\ 4\ CH_3CH_2CO \sim CoA$

FIG. 3. The metabolism of phytanic acid in man, and the defect in patients with Refsum's disease.

studies with synthetic ^{14}C-labeled acids of particular chemical structure preventing a β-oxidation from both ends of the molecule (3,6-dimethyl-caprylic acid; 3,14,14-trimethylpentadecanoic acid), we proved that an alternative metabolic pathway must exist in mammals (Eldjarn, Try, and Stokke, 1966b; Stokke, Try, and Eldjarn, 1967b; Try and Stokke, 1969). Later this pathway was shown by us (Eldjarn, Stokke, and Try, 1966a) as well as by Steinberg's group in Bethesda (Avigan, Steinberg, Gutman, Mize, and Milne, 1966) and Shorland and Hansen in New Zealand (Shorland, Hansen, and Prior, 1966) to be an α-oxidation mechanism converting phytanic acid to pristanic acid. Refsum's disease was shown to be an inborn error in this α-oxidation mechanism (Fig. 3).

A. Methylmalonic Acidemia

In 1967 independently of Oberholzer, Levin, Burgess, and Young (1967), we described methylmalonic acidemia in a newborn girl (Stokke, Eldjarn, Norum, Steen-Johnsen, and Halvorsen, 1967). Two siblings of the patient had died in the neonatal period with clinical symptoms strongly resembling those of our patient. Immediately upon the demonstration of large amounts of methylmalonic acid in serum and urine, we instituted cyanocobalamin treatment without apparent effect. A diet low in propionic acid precursors significantly reduced the urinary excretion of methylmalonic acid leading to considerable improvement of the clinical condition. However the patient developed a severe infection and died of septicemia.

Based on the occurrence of large amounts of glycine in the serum of our patient, we pointed out that patients previously diagnosed as having hyper-glycinemia actually may suffer from methylmalonic acidemia (Halvorsen,

Stokke, and Eldjarn, 1968, 1970). This has actually been found to be the case in many instances.

Recently we described a patient who, in addition to methylmalonic acidemia, also showed occurrence of large amounts of β-hydroxy-n-valeric acid and propionic acid in serum (Stokke, Jellum, Eldjarn, and Schnitler, 1973). The patient showed muscular hypotonia, dystrophy, attacks of metabolic acidosis, and respiratory tract infections from the first day of life until death at 10 months. After the death, we were asked to investigate a small sample of serum (1.5 ml) which was the only material left from the patient. In this sample we found in addition to methylmalonic acid large amounts of propionic acid and β-hydroxy-n-valeric acid. The fatty acid pattern of serum showed an increased amount of odd-numbered (pentadecanoic and heptadecanoic) acids. These odd-numbered acids most probably stem from condensation reactions in which propionyl-CoA may compete with acetyl-CoA. Accumulation of propionic and odd-numbered fatty acids have not been described in patients with methylmalonic acidemia in which the enzyme defect is believed to be associated with methylmalonyl-CoA-mutase (Fig. 4). It is possible that our patient has a defect located on some other enzyme in the metabolic sequence that converted propionate to succinate, perhaps at the methylmalonyl-CoA racemase step.

B. β-Methylcrotonyl-CoA-carboxylase Deficiency

In 1970 we were confronted with a 4-month-old girl with symptoms resembling those of Werdnig-Hoffmann's disease. The child's urine had a peculiar smell, resembling that of a cat's urine. She was found to excrete large amounts of β-hydroxyisovaleric acid and β-methylcrotonylglycine. Most likely she was suffering from a deficiency at the β-methylcrotonyl-

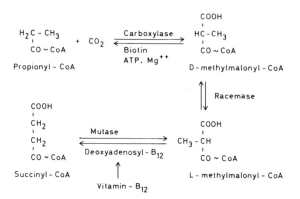

FIG. 4. The main metabolic route for propionyl-CoA in man. Both a defect of the mutase apoenzyme and a defect in the conversion of vitamin B_{12} to deoxyadenosyl-B_{12} have been shown to cause methylmalonic acidemia.

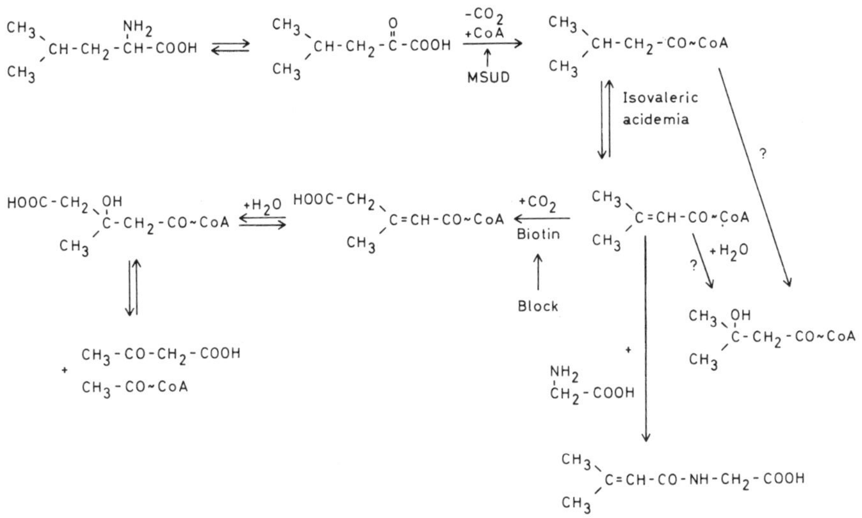

FIG. 5. The leucine degradation pathway. The probable block in the present patient is indicated, as well as the defects in maple syrup urine disease (MSUD) and isovaleric acidemia.

CoA-carboxylase step of leucine degradation (Eldjarn, Jellum, Stokke, Pande, and Waaler, 1970; Stokke, Eldjarn, Jellum, Pande, and Waaler, 1972) (Fig. 5). The institution of a diet containing the minimum requirement of leucine resulted in a rapid drop in the urinary excretion of the abnormal metabolites, particularly of β-hydroxyisovaleric acid. However, the child died from pneumonia at the age of 9 months.

C. Pyroglutamic Acidemia

Recently we found a novel metabolic error in a 20-year-old male suffering from retarded psychomotor development, slight spastic tetraparesis, and metabolic acidosis (Jellum, Kluge, Børresen, Stokke, and Eldjarn, 1970). He was found to excrete daily in the urine 30 to 40 g of L-pyroglutamic acid (L-2-pyrrolidone-5-carboxylic acid).

When a new metabolic error is discovered by the accumulation or excretion of abnormal metabolites in a patient, it is usually possible through our detailed knowledge of the mammalian metabolism to pinpoint the enzyme involved and explain the occurrence of the abnormal metabolites by a failure of this particular enzyme step. This is not true in the case of pyroglutamic acidemia. Although this compound is known to occur in minute amounts as an N-terminal amino acid of the light chain of the γ-type immunoglobulins and of the tripeptide "thyroid-stimulating hormone release factor" as well as of a number of other proteins, we must admit that the

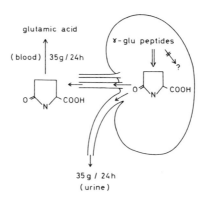

FIG. 6. The dynamics of pyroglutamic acid in our patient with pyroglutamic acidemia.

urinary excretion of the formidable amounts of L-pyroglutamic acid still presents a number of puzzling questions. Where in the mammalian metabolism do these large amounts of L-pyroglutamic acid originate? Are such amounts of the compound normally found as an intermediate in the metabolism of man?

Recently a second patient most probably suffering from the same disease was reported from Sweden (Hagenfeldt, Larsson, and Zetterström, 1974). It should be stressed that pyroglutamic acid is ninhydrine-negative and that these patients will therefore not be detected by the ordinary routine screenings for inborn errors related to amino acid metabolism. Several more cases may therefore have been overlooked.

Our studies (Eldjarn, Jellum, and Stokke, 1972, 1973) have shown that the endogenous production of L-pyroglutamic acid is in fact of the order of 60 to 80 g per 24 hr, i.e., about two times greater than the amount excreted in the urine (Fig. 6). The compound seems to be degraded to glutamic acid only (Stromme and Eldjarn, 1972). It seems unlikely that the diseased state can be explained by different rates of pyroglutamic acid degradation in patients and in normal controls. Our results make it improbable that the pyroglutamic acid formation is related to the urea cycle or to the ammonium ion formation by the kidneys. Of the known routes of formation of pyroglutamic acid only γ-glutamyl substrates seem capable of producing the amount seen in the patient. Accordingly, a metabolic defect could be related to the further metabolism of γ-glutamyl peptides possibly as a derangement of a γ-glutamyl cycle for amino acid absorption in the kidney, as proposed by Orlowski and Meister (1970) (Eldjarn et al., 1974).

V. CONCLUSIONS

GC–MS has proven a most helpful tool in our work to investigate biochemical abnormalities in patients with metabolic derangements. If laboratories of clinical chemistry fulfill the responsibilities in bridging the gap between frontline biological research and bedside medicine, it seems reasonable to foresee that GC–MS equipment will be included in many clinical chemistry laboratories of larger hospitals. The main use of such equipment in clinical chemical laboratories will probably be (a) determination of toxic or noxious compounds in intoxicated emergency cases, (b) the diagnosis of inborn errors of metabolism, (c) routine analyses of compounds occurring in minute amounts (hormones), (d) control of blood levels of drugs, and (e) evaluation and etiological diagnosis of pronounced metabolic derangements.

However, the inclusion of GC–MS in the standard routine setup of the laboratories must await several improvements in the present-day equipment. First, simpler and less expensive machinery with less breakdown time is necessary. Secondly, mass spectrum libraries comprising compounds of biological interest must be collected and be available through data-processed library search. Possibly, specialized libraries should be developed comprising biologically occurring compounds, drugs and drug metabolites, etc.

New techniques which will make it possible to carry out analyses on a picogram scale may furthermore permit clinical chemical investigations on minute tissue biopsies or tissue cultures. Of particular significance to clinical chemistry would be the development of GC–MS systems using nonfragmenting ionization techniques (chemical ionization, field ionization, external ionization), which permit analyses to be performed on mixtures of compounds.

REFERENCES

Avigan, J., Steinberg, D., Gutman, A., Mize, C., and Milne, W. A. (1966): Alpha-decarboxylation, an important pathway for degradation of phytanic acid in animals. *Biochemical and Biophysical Research Communications*, 24:838–844.

Eight Peak Index of Mass Spectra, 1 and 2 (1970): Mass Spectrometry Data Centre, AWRE, Aldermaston, England.

Eldjarn, L., Jellum, E., and Stokke, O. (1972): Pyroglutamic aciduria: Studies on the enzyme block and on the metabolic origin of pyroglutamic acid. *Clinica Chimica Acta*, 40:461–476.

Eldjarn, L., Jellum, E., and Stokke, O. (1973): Pyroglutamic aciduria: Rate of formation and degradation of pyroglutamate. *Clinica Chimica Acta*, 49:311–323.

Eldjarn, L., Jellum, E., Stokke, O., Pande, H., and Waaler, P. E. (1970): β-Hydroxyisovaleric aciduria and β-methylcrotonylglycinuria. A new inborn error of metabolism. *Lancet*, 2:521–522.

Eldjarn, L., Stokke, O., and Try, K. (1966a): Alpha-oxidation of branched chain fatty acids in man and its failure in patients with Refsum's disease showing phytanic acid accumulation. *Scandinavian Journal of Clinical and Laboratory Investigation*, 18:694–695.

Eldjarn, L., Stokke, O., and Try, K. (1974): Biochemical aspects of Refsum's disease and principles for the dietary treatment. In: *Handbook of Clinical Neurology*, edited by P. J. Vinken and G. W. Bruyn. North-Holland Publishing Company, Amsterdam (*in press*).

Eldjarn, L., Try, K., and Stokke, O. (1966b): The existence of an alternative pathway for the degradation of branch-chained fatty acids, and its failure in heredopathia atactica polyneuritiformis (Refsum's disease). *Biochimica Biophysica Acta*, 116:395–397.

Eldjarn, L., Try, K., Stokke, O., Munthe-Kaas, A. W., Refsum, S., Steinberg, D., Avigan, J., and Mize, C. (1966c): Dietary effects on serum phytanic acid levels and on clinical manifestations in heredopathia atactica polyneuritiformis. *Lancet*, 1:691–693.

Hagenfeldt, L., Larsson, A., and Zetterström, R. (1974): Pyroglutamic aciduria. Studies in an infant with chronic metabolic acidosis. *Acta Paediatrica Scandinavia,* 63:1–8.

Halvorsen, S., Stokke, O., and Eldjarn, L. (1968): Methylmalonic acidaemia – hyperglycinaemia. *Lancet,* 1:756.

Halvorsen, S., Stokke, O., and Eldjarn, L. (1970): Abnormal patterns of urine and serum amino acids in methylmalonic acidemia. *Acta Paediatrica Scandinavica,* 59:28–32.

Jellum, E., Børresen, H. C., and Eldjarn, L. (1973): The presence of furan derivatives in patients receiving fructose-containing solutions intravenously. *Clinica Chimica Acta* 47:191–201.

Jellum, E., Kluge, T., Børresen, H. C., Stokke, O., and Eldjarn, L. (1970): Pyroglutamic aciduria – A new inborn error of metabolism. *Scandinavian Journal of Clinical and Laboratory Investigation,* 26:327–335.

Jellum, E., Stokke, O., and Eldjarn, L. (1972): Combined use of gas chromatography, mass spectrometry, and computer in diagnosis and studies of metabolic disorders. *Clinical Chemistry,* 18:800–809.

Klenk, E., and Kahlke, W. (1963): Über das Vorkommen der 3,7,11,15-Tetramethyl-hexadecansäure (Phytansäure) in den Cholerinestern und anderen Lipoidfraktionen der Organe bei einem Krankheitsfall unbekannter Genese (Verdacht auf Heredopathia atactica polyneuritiformis (Refsum-Syndrom). *Zeitschrift für Physiologische Chemie,* 333:133–139.

Norum, K., and Gjone, E. (1967): Familial plasma lecithin: cholesterol acyl transferase deficiency. Biochemical study of a new inborn error of metabolism. *Scandinavian Journal of Clinical and Laboratory Investigation,* 20:231–243.

Oberholzer, V. G., Levin, B., Burgess, A., and Young, W. (1967): Methylmalonic aciduria. An inborn error of metabolism leading to chronic metabolic acidosis. *Archives of Diseases in Childhood,* 42:492–504.

Orlowski, M., and Meister, A. (1970): The γ-glutamyl cycle: A possible transport system for amino acids. *Proceedings of the National Academy of Sciences,* 67:1248–1255.

Pettersen, J. E. (1972): Formation of n-hexanedioic acid from hexadecanoic acid by an initial ω oxidation in ketotic rats. *Clinica Chimica Acta,* 41:231–237.

Pettersen, J. E., and Jellum, E. (1972): The identification and metabolic origin of 2-furoylglycine and 2,5-furandicarboxylic acid in human urine. *Clinica Chimica Acta,* 41:199–207.

Pettersen, J. E., Jellum, E., and Eldjarn, L. (1972): The occurrence of adipic and suberic acid in urine from ketotic patients. *Clinica Chimica Acta,* 38:17–24.

Refsum, S. (1946): Heredopathia atactica polyneuritiformis. A familial syndrome not hitherto described. A contribution to the clinical study of the hereditary diseases of the nervous system. *Acta Psychiatrica Scandinavica,* Suppl. 38.

Shorland, F. B., Hansen, R. P., and Prior, I. A. M. (1966): The effects of phytanic acid composition of the lipids of the rat with further observations on its metabolism. In: *Proceedings of 7th International Congress on Nutrition,* 5:399–407. Vieweg & Sohn, Brauschweig.

Steinberg, D., Avigan, J., Mize, C., Eldjarn, L., Try, K., and Refsum, S. (1965): Conversion of U-C^{14}-phytol to phytanic acid and its oxidation in heredopathia atactica polyneuritiformis. *Biochemical and Biophysical Research Communications,* 19:783–789.

Steinberg, D., Mize, C., Avigan, J., Fales, H. M., Eldjarn, L., Try, K., Stokke, O., and Refsum, S. (1967): Studies on the metabolic error in Refsum's disease. *Journal of Clinical Investigation,* 46:313–322.

Stenhagen, E., Abrahamsson, S., and McLafferty, F. W. (1969): *Atlas of Mass Spectral Data.* Interscience Publishers, New York.

Stenhagen, E., Abrahamsson, S., and McLafferty, F. W. (1970): *Archives of Mass Spectral Data.* Interscience Publishers, New York.

Stokke, O., Eldjarn, L., Jellum, E., Pande, H., and Waaler, P. E. (1972): Beta-methylcrotonyl-CoA carboxylase deficiency: A new metabolic error in leucine degradation. *Pediatrics,* 49:726–735.

Stokke, O., Eldjarn, L., Norum, K., Steen-Johnsen, J., and Halvorsen, S. (1967a): Methylmalonic acidemia. A new inborn error of metabolism which may cause fatal acidosis in the neonatal period. *Scandinavian Journal of Clinical and Laboratory Investigation,* 20:313–328.

Stokke, O., Jellum, E., Eldjarn, L., and Schnitler, R. (1973): The occurrence of β-hydroxy-n-valeric acid in a patient with propionic and methylmalonic acidemia. *Clinica Chimica Acta* 45:391–401.

Stokke, O., Try, K., and Eldjarn, L. (1967*b*): α-Oxidation as an alternative pathway for the degradation of branched-chain fatty acids in man, and its failure in patients with Refsum's disease. *Biochimica Biophysica Acta*, 144:271–284.

Strømme, J. H., and Eldjarn, L. (1972): The metabolism of L-pyroglutamic acid in fibroblasts from a patient with pyroglutamic aciduria: The demonstration of an L-pyroglutamate hydrolase system. *Scandinavian Journal of Clinical and Laboratory Investigation*, 29:335–342.

Strømme, J. H., Nesbakken, R., Normann, T., Skjørten, F., Skyberg, D., and Johannessen, B. (1969): Familial hypomagnesemia. Biochemical, histological and hereditary aspects studied in two brothers. *Acta Paediatrica Scandinavica*, 58:433–444.

Try, K., and Stokke, O. (1969): *Biochemical and Dietary Studies in Refsum's Disease (Heredopathia atactica polyneuritiformis)*. Universitetsforlaget, Oslo.

Mass Spectrometry in Biochemistry and Medicine,
edited by A. Frigerio and N. Castagnoli.
Raven Press, New York © 1974

The Application of Mass Spectrometry to the Study of Thyroxine and Related Compounds in Biological Fluids

R. Hoffenberg,* A. M. Lawson, D. B. Ramsden,* and P. J. Raw*

Divisions of Clinical Investigation and Clinical Chemistry, Clinical Research Centre, Watford Road, Harrow, Middlesex, U.K.

I. INTRODUCTION

The hormone thyroxine (T_4), which is synthesized by the thyroid gland, is the body's major form of circulating covalently bound iodine. It is associated with the control of many different functions, e.g., regulation of basic metabolic rate, and severe deficiency in early life results in cretinism. Thyroxine was discovered by Kendall (1915) and its structure elucidated by Harington and Barger (1927). Recently, however, an increasing amount of evidence has been put forward to suggest that the true hormonal action resides within the deiodination product of thyroxine, 3,3′,5-triiodothyronine (T_3) (see Hoffenberg, 1973).

Quantitation of these hormones has proved difficult, because of their low concentration in serum (in man T_3 2 ng/ml, T_4 80 ng/ml) and the use of such inexact methods as butanol extractable iodide (BEI), protein-bound iodide (PBI), and competitive binding assays. In the past few years radioimmunoassay has been advocated for T_3 (Gharib, Mayberry, and Hockert, 1971) and still more recently for T_4 (Mitsuma, Colucci, Shenkman, and Hollander 1972). Whereas this approach possesses the inherent sensitivity required, a recent survey has shown that large variations in values (up to 400%) have been obtained for identical sera in different laboratories (Gharib, Ryan, and Mayberry, 1972). Further difficulties arise when investigating hepatic metabolism because the liver-produced metabolites, with many of the structural features of the hormones themselves, e.g., 3,3′,5,5′-tetraiodothyroacetic and 3,3′,5-triiodothyroacetic, give considerable cross-reaction with allegedly monospecific antisera. The function of such metabolites is obscure at present, although they possess great hormonal activity in

---

* Present address: Department of Medicine, Queen Elizabeth Hospital, Edgbaston, Birmingham B15 2TJ, U.K.

certain tests, e.g., amphibian metamorphosis test, see Pitt-Rivers and Tata (1959).

The difficulties outlined above have prompted a number of investigators to turn to gas chromatography (GC) as a means of direct quantitation. This has required the development of volatile derivatives as the hormones themselves are amino acids and the metabolites are either amino acids or acids. Although considerable effort has been made in this area, no method is available for the routine measurement of these substances in biological fluids. This is due in great part to the lack of specificity afforded by GC, particularly when considering complex mixtures of components where the gas-liquid chromatographic (GLC) peak to be measured can be obscured or affected by other peaks.

In the present work a preliminary investigation of the combined technique of GC–MS for the estimation of these substances has been made where the resolution capabilities of the gas chromatograph are compounded with the high sensitivity and specificity of the mass spectrometer. The quantitative aspects considered have been principally concerned with the hepatic deaminated analogues of the iodothyronines.

II. MATERIALS AND METHODS

A. Compounds

The compounds used in this study and their trivial notations are shown in Table 1.

TABLE 1. *Compounds used in this study*

	COMPOUND	TRIVIAL NOTATION
Amino Acids	3, 3′, 5, 5′-tetraiodothyronine	T_4
	3, 3′, 5-triodothyronine	T_3
	3, 5-diiodothyronine	T_2
	thyronine	T_0
	3, 5-diiodotyrosine	DIT
	3-monoiodotyrosine	MIT
	tyrosine	TYR
Deamino Acids	3, 3′, 5, 5′-tetraiodothyroacetic acid	T_4A
	3, 3′, 5-triiodothyroacetic acid	T_3A
	3, 5-diiodothyroacetic acid	T_2A
	3-monoiodothyroacetic acid	T_1A
	thyroacetic acid	T_0A
	3, 3′, 5, 5′-tetraiodothyropropionic acid	T_4P
	3, 3′, 5, 5′-tetraiodothyroformic acid	T_4F

B. Derivatives

Trimethylsilyl derivatives were prepared by heating a sample of each compound with N,O-*bis*-(trimethylsilyl)acetamide (BSA) at 60°C for 15 min (Alexander and Scheig, 1968) and their deuterated analogues with d_{18}BSA in pyridine (McCloskey, Stillwell, and Lawson, 1968).

Pivalyl methyl esters, first reported by Stouffer, Jaakonmaki, and Wenger (1966), were synthesized by the method employed by Docter and Hennemann (1971). The methyl ester, obtained by heating the acid with anhydrous HC1 in methanol at 70°C for 10 to 15 min, was treated with excess pivalic acid anhydride in triethylamine (4:1) at 80°C for 1 hr.

Permethyl derivatives were only made of the deamino acids and involved treatment with methyl sulfinyl carbanion in dimethyl sulfoxide under nitrogen and then addition of methyl iodide. After 1 hr, water was added and the derivatives extracted with chloroform.

To get trimethylsilyl (TMS) methyl esters, anhydrous HC1 in methanol was added to the sample which was then heated at 60°C for 15 min in the case of the deamino acids and for 5 hr for the amino acids. The resulting esters were taken to dryness and treated with BSA for 15 min at 60°C. CD_3TMS analogues were obtained by methylating with anhydrous HC1 in CD_3OD and the CD_3 d_9TMS compounds by treating the CD_3 esters with d_{18}BSA in pyridine.

C. Gas-Liquid Chromatography

A Pye 104 instrument was used with a 6-ft 3% OV-1 packed glass column. A flow rate of 60 ml/min and injector and detector temperatures of 310°C were used.

D. Gas Chromatography–Mass Spectrometry

Analysis was carried out on a Varian Aerograph 2700 gas chromatograph coupled to a Varian MAT 731 double-focusing mass spectrometer. The latter was operated at 8 kV and 70 ev. The GLC conditions were similar to those used with the Pye 104 with the carrier gas separator and transfer lines maintained at 290°C. Data were processed by the Varian 100 MS SpectroSystem.

E. Liver and Bile Extracts

The appropriate sample was freeze-dried and extracted with anhydrous HC1 in methanol. This extract was neutralized with pyridine, taken to dryness and re-extracted with a 20:1 mixture of benzene and dry HC1/ methanol. This was again freeze-dried and treated with dry HC1/methanol for 15 min at 60°C and silylated with BSA after removal of the esterifying reagent.

III. RESULTS AND DISCUSSION

In order to assess the extent and nature of the information which is available when applying GC–MS to problems involving these compounds, we have synthesized a variety of their derivatives. Their GC–MS behavior has then been considered in terms of stability, structural identification, detection sensitivity, and possible quantitation.

A. Trimethylsilyl Derivatives

There are marked differences in the spectra of the amino acids and their deamino analogues. Representative spectra of these two classes, namely of T_3-TMS-3 and T_3A-TMS-2, are shown in Fig. 1. The Tris-TMS derivatives were formed by all the amino acids treated. In their spectra the facile cleavage of the β bond of the side chain gives rise to a base peak at m/e 218 due to $TMSN\overset{+}{H}{=}CHCO_2TMS$ with only a small amount of charge retention on the remainder of the molecule. Although the molecular ions are never greater than 1% relative intensity (RI), ions at (M—15) and (M—CO$_2$TMS) serve to confirm the identity of the acid. The deamino analogues, on the other hand, show stable molecular species (50 to 100%

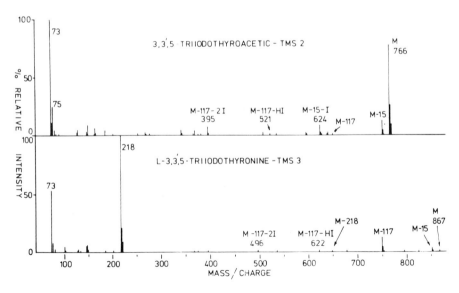

FIG. 1. The mass spectra of the TMS derivatives of triiodothyroacetic acid and triiodothyronine.

RI). In both sets of compounds, series of deiodinated ions of low abundance, formed from primary fragments, are apparent. These are smaller in the case of the amino acids due to competition with the formation of m/e 218. The deiodinated ions serve as useful fingerprint ions but give little structural information. In addition to high-resolution measurement, the d_9TMS analogues assist in establishing structural assignments.

The detection of the amino acids for quantitative purposes would be most sensitively achieved by selectively monitoring m/e 218. As this is common to all such spectra, adequate GLC resolution is necessary to obtain preliminary specificity. This is readily possible on 3% OV-1; however, the degree of specificity depends on establishing the presence of the correct ratios of confirmatory ions. As most systems for monitoring selected ions cannot simultaneously follow ions differing in mass by more than 10 to 30% of the lowest mass, proof of the presence of the acid would depend on rerunning the sample and monitoring other characteristic ions (e.g., M—CO_2TMS).

The M and (M—15) ions of the deamino compounds are suitable ions for their detection.

Unfortunately GC problems are experienced with the TMS derivatives. Funakoshi and Cahnmann (1969) found that flame ionization detector response diminished when dilute solutions of these derivatives (0.5 to 1 μg/ul) were stored for several hours, and Bilous, Koehler, and Windheuser (1971) demonstrated the adverse effects of pyrolyzed material in the GC ignition zone. There is a tendency for the amino TMS to be lost and for partial adsorption of the sample on the GLC column to occur. The former gives rise to a series of ions equivalent to (M—72), (M—72—15), (M—72—CO_2TMS), and (218—72).

Although the unfavorable GLC properties can be minimized by using a carrier and the addition of an excess of an analogous compound to give preferential hydrolysis, the lack of suitable labeled derivatives to act as carriers and as standards makes the TMS derivatives unattractive for quantitative purposes.

B. Pivalyl Methyl Esters

The phenolic and amino groups in the thyronines and the phenolic group in the deamino analogues are converted to the pivalyl esters. Docter and Hennemann (1971) have demonstrated favorable GLC quantitation at the nanogram level of the iodothyronines but found the minimum accurately measurable amount by electron capture detection to be approximately 2 ng. The electron impact-induced fragmentations of these compounds involve, to a great extent, the pivalyl groups themselves giving informative ions in the high mass range. These are more intense in the deamino than in the amino series. Groups of deiodinated ions are also present.

The value of the pivalyl derivatives for quantitative GC–MS applications is still under active consideration.

C. Permethyl Derivatives

The dimethylated deamino acids give good GLC and MS properties with intense M and M—15 ions suitable for sensitive detection. Several series of deiodinated ions are present, originating from the M and (M—CO_2CH_3) ions and some ions, for example (M—CO_2CH_3—2I) and (M—2I) in thyroxine are sufficiently intense and characteristic to be useful for detection.

However, some problems remain in the chemical preparation of these compounds associated with the formation of mixed derivatives due to the methylation proceeding further than desired. This needs to be resolved before the permethyl compounds could be considered for GC–MS measurements.

D. Trimethylsilyl Methyl Esters

These were the derivatives chosen to investigate the quantitative possibilities of detecting the deamino acid analogues. They may also prove suitable for the thyronine series but this has still to be assessed.

The mass spectra of T_4—Me—TMS and T_4A—Me—TMS are shown in Fig. 2 and can be seen to closely resemble the fully silylated compounds. Labeled derivatives were again used to confirm structural assignments.

Cleavage of the β carbon–carbon bond of the side chain in the amino

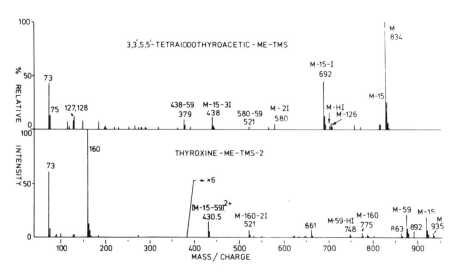

FIG. 2. The mass spectra of the TMS methyl ester derivatives of tetraiodothyroacetic acid and thyroxine.

acids gives a base peak at m/e 160 which could be used, in the same way as the m/e 218 in the TMS derivatives, to monitor these substances but with the added advantage that the deuterated methyl ester TMS isomers would serve as carrier compounds.

The deamino series have intense characteristic ions at M, (M—15), and (M—15—I) with the molecular ions of T_4A, T_4F, and T_4P having 20% Σ values. This makes these ions suitable for selective detection with their high mass values giving additional specificity.

E. Selected Ion Monitoring

The technique of monitoring characteristic ions to determine the presence of a compound in a GC effluent is well established. Increasingly elaborate devices have been devised for the selection, switching, and measurement of the ions of interest (e.g., Sweeley, Elliott, Fries, and Ryhage, 1966; Hammar and Hessling, 1971; Holland, Sweeley, Thrush, Teets, and Bieber, 1973). In the present study the peak-matching unit of the instrument was used for switching between ions. Although it is limited to only two channels, further ions can be monitored within a single GC run by changing the ratio of the decade resistances.

Figure 3 demonstrates this point. A standard mixture of the Me—TMS derivatives of T_4F, T_4A, and T_4P was injected into the GLC. B and C indicate single injections where the low mass channel is focused on the molecular ion of T_4F—Me—TMS (m/e 820) and the second channel switches to m/e 834 (MW T_4A—Me—TMS) in C and m/e 848 (MW T_4P—Me—TMS) in B. Note the response for m/e 820 at the retention time of T_4A and for m/e 834 at T_4P. These arise from the first isotope peak of the M—15 ions of T_4A and T_4P respectively but are unimportant as one is only interested in the molecular ion responses at their individual retention values. In A the peak matcher switches between 820 and 834 and then 820 and 834 for the T_4P peak. This mode of operation is somewhat tedious in that several runs need to be made when checking the specificity of detection by looking at ratios of characteristic ions.

The samples in Fig. 3 were run without using carrier compounds and although this is satisfactory in the 100 to 1,000-ng range the linearity of response below this level is affected by adsorptive losses. Derivatives diluted with dry BSA in the nanogram range were found to be stable for several days.

The carrier compounds selected to minimize adsorption on the GLC column were the deutero methyl ester TMS derivatives (i.e., T_4F—CD_3—TMS, T_4A—CD_3—TMS, and T_4P—CD_3—TMS for T_4F—CH_3—TMS, T_4A—CH_3—TMS, and T_4P—CH_3—TMS respectively).

The molecular ions of these carriers differ by 3 mass units from their protium analogues. Possible interference of an ion from the carrier molecule

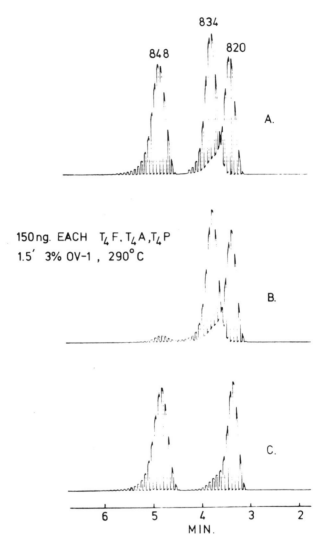

FIG. 3. Selected ion responses from the molecular ions of the TMS methyl ester derivatives of T₄F, T₄A, and T₄P. Notations A, B, and C are explained in the text.

with the molecular ion of the natural one was tested by injecting a 50-ng
sample of the deuterated carrier and switching between M and (M + 3) in
each case. This was found to give rise to a 3% contribution to the molecular
ion of the equivalent of 1 ng of T₄A—Me—TMS and less in the case of
T₄F—Me—TMS and T₄P—Me—TMS.

The amount of carrier used was selected to give a reproducible response
and as a practical level of internal standard to establish the linearity of the
GC–MS response. Figure 4 is a plot of the ratio of the recorded responses

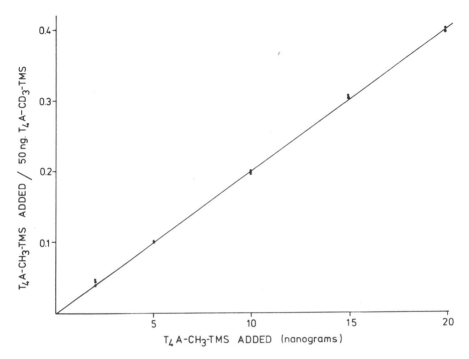

FIG. 4. Plot of the observed ratio of the amount of T_4A—CH_3—TMS added to 50 ng of T_4A—CD_3—TMS against the amount of T_4A—CH_3—TMS added.

of the amount of T_4A—Me—TMS added to 50 ng of carrier-standard versus the amount of T_4A—Me—TMS added. The correlation coefficient of the line is 0.99 and the slope is 0.02.

The response was linear down to 400 pg but with a wider spread in the individual points. Although it is expected that the practical limit of detection can be improved, this is about 100 pg at present (signal to noise 5:1 with a 30% interference from 50 ng of carrier).

F. Biological Samples

The preliminary studies with biological samples, undertaken principally with rat bile and liver, have been encouraging. T_4P and T_4F have been observed in liver homogenates and T_4A and T_4P in the bile. Further work of a confirmatory nature remains to be done in addition to the simplification and the quantitation of the extraction procedure.

IV. CONCLUSIONS

The utility and information available from the mass spectra of the four derivatives prepared, the TMS, the pivalyl methyl esters, the permethylated, and the TMS methyl esters, have been evaluated. They are all suitable for identification purposes in microgram amounts. Detection at lower levels is also possible although the TMS compounds are increasingly susceptible to hydrolysis and adsorption losses. The TMS methyl esters have proved satisfactory for the detection of the deamino acids down to 100 pg with a good linear response in the lower nanogram range when the TMS deutero methyl esters were used as carriers. This now makes possible the study of hepatic deaminated metabolites of thyroxine in biological samples.

ACKNOWLEDGMENT

The technical assistance of Mr. P. N. Fiveash and Mr. L. N. Louis is gratefully acknowledged. P. J. R. is a recipient of an M.R.C. scholarship.

REFERENCES

Alexander, N. M., and Scheig, R. (1968): Gas chromatography of the trimethylsilyl derivatives of iodotyrosines and iodothyronines. *Analytical Biochemistry*, 22:187–194.

Bilous, R., Koehler, E., and Windheuser, J. J. (1971): Interference of glc determination of iodo-amino acids in hydrolysis products of thyroid extracts. *Journal of Pharmaceutical Sciences*, 60:1270–1271.

Docter, R., and Hennemann, G. (1971): Estimation of thyroid hormones by gas-liquid chromatography. *Clinica Chimica Acta*, 34:297–303.

Funakoshi, K. and Cahnmann, H. J. (1969): Gas chromatographic determination of iodinated compounds. *Analytical Biochemistry*, 27:150–161.

Gharib, H., Mayberry, W. E., and Hockert, T. (1971): Radioimmunoassay for triiodothyronine (T_3). *53rd Annual Meeting of the Endocrine Society*, San Francisco.

Gharib, H., Ryan, R. J., and Mayberry, W. E. (1972): Triiodothyronine radioimmunoassay. A critical evaluation. *Mayo Clinic Proceedings*, 47:934–937.

Hammar, C. G., and Hessling, R. (1971): Novel peak matching technique by means of a new and combined multiple ion detector-peak matcher device. *Analytical Chemistry*, 43:298–306.

Harington, C. R., and Barger, G. (1927): Chemistry of thyroxine III. Constitution and synthesis of thyroxine. *Biochemical Journal*, 21:169–181.

Hoffenberg, R. (1973): Triiodothyronine. *Clinical Endocrinology*, 2:75–87.

Holland, J. F., Sweeley, C. C., Thrush, R. E., Teets, R. E., and Bieber, M. A. (1973): On-line computer controlled multiple ion detection in combined gas chromatography–mass spectrometry. *Analytical Chemistry*, 45:308–314.

Kendall, E. C. (1915): The isolation in crystalline form of the compound containing iodin which occurs in the thyroid. *Journal of the American Medical Association*, 64:2042.

McCloskey, J. A., Stillwell, R. N., and Lawson, A. M. (1968): Use of deuterium-labelled trimethylsilyl derivatives in mass spectrometry. *Analytical Chemistry*, 40:233–236.

Mitsuma, T., Colucci, J., Shenkman, L., and Hollander, C. S. (1972): Rapid simultaneous radioimmunoassay for T_3 and T_4 in unextracted serum. *Biochemical and Biophysical Research Communications*, 46:2107–2113.

Pitt-Rivers, R., and Tata, J. (1959): *The Thyroid Hormones*. Pergamon Press, New York.

Stouffer, J. E., Jaakonmaki, P. I., and Wenger, T. J. (1966): Gas-liquid chromatographic separation of thyroid hormones. *Biochimica et Biophysica Acta*, 127:261–263.

Sweeley, C. C., Elliott, W. H., Fries, I., and Ryhage, R. (1966): Mass spectrometric determination of unresolved components in gas chromatographic effluents. *Analytical Chemistry*, 38:1549–1553.

Mass Spectrometry in Biochemistry and Medicine,
edited by A. Frigerio and N. Castagnoli.
Raven Press, New York © 1974

Applications of Quantitative Mass Fragmentography in Pharmacology and Clinical Medicine

Robert E. Finnigan, James B. Knight,
William F. Fies, and Victor L. DaGragnano

Finnigan Corporation, Sunnyvale, California 94086

I. INTRODUCTION

Several recent publications (Strong and Atkinson, 1972; Strong, Parker, and Atkinson, 1973) have described the use of the gas chromatograph–quadrupole mass spectrometer (GC–QMS) system for the quantitative measurement of lidocaine and its pharmacologically active metabolites in the blood plasma of human patients being treated with intravenous lidocaine to correct heart arrhythmias. Atkinson and his colleagues utilized conventional electron impact–quadrupole mass fragmentography (EI–QMF) in their quantitative assays and were able to get precisions of the order of 3 to 7% when the levels of lidocaine and its metabolites were in the range 0.1 to 15 μg/ml of original plasma.

Chemical ionization–quadrupole mass spectrometry (CI–QMS) offers several distinct advantages over EI mass spectrometry which assume particular importance when one attempts the analyses and quantitation of drugs and drug metabolites at very low levels in the body fluids using quadrupole mass fragmentography:

(1) In CI–QMS there is no separator required between the GC and the QMS ion source even when using packed columns with flow rates up to 35 ml/min. This eliminates the most trouble prone element in the GC–MS system. It also eliminates the possibility of adsorption on the separator at very low concentrations of drugs and drug metabolites.

(2) In CI–QMS most often the base peak occurs at the M+1 and/or M-1 position rather than at a lower mass as in EI. With most drugs and drug metabolites this produces an ion which can be monitored using quadrupole mass fragmentography, which is well above the mass range of the interfering biological contaminants.

(3) For many drugs and drug metabolites when using methane or isobutane as reagent gas, the M+1 or M-1 peak often comprises more than 50% of the total ions observed (Milne, Fales, and Axenrod, 1971) (i.e., little or

no fragmentation). This produces a high-level signal in a mass range where there is little or no contribution from biological contaminants.

II. CI–QMS

During the past 3 years, a large number of quadrupole mass spectrometers have been modified for combined EI/CI GC–MS operation. There are several characteristics of the QMS which make it an ideal ion-sorting device for chemical ionization mass spectrometry:

(1) The QMS normally utilizes ions at relatively low energies, i.e., below 10 eV. In CI, ions are formed at near-thermal energies; they need only be accelerated slightly using a repeller voltage for proper introduction into the QMS.

(2) Because the QMS uses low-energy ions, there is no need for high voltages in the ion source of the QMS (as is the case with magnetic instruments). Normally, the highest voltage present is less than 100 V. This becomes of great importance in CI–QMS where it is necessary to operate the ion source at pressures in the range of 1 to 2 torr. There is no danger of electrical arcing in the QMS at these pressures nor is there any more danger to the operator of the system than in EI.

(3) Because of the continuous focusing nature of the quadrupole mass filter, it can be run at higher pressures than could be tolerated in a magnetic sector instrument. Even after ion-molecule collisions in the quadrupole mass filter region, ions are refocused and transmitted or removed at the appropriate masses. This tolerance to higher pressures, results in little or no redesign of the quadrupole filter section for CI operation and permits long-term operation of the QMS system at higher pressures than could be tolerated in a magnetic sector instrument.

(4) Because of the high-pressure tolerance of the ion source and filter, the entire effluent of the GC can be delivered directly to the ion source of the QMS, negating the need for a separator or other interface device. Elimination of the separator is perhaps the single greatest advantage of going to GC–CI–QMS. Separators of all types are trouble prone: they plug, they adsorb the sample at low levels with many compounds, they are subject to breakage, and they are expensive.

The chemical ionization source used with the QMS (Story, 1972) is shown in Fig. 1. It is differentially pumped by a 1,200 l/sec oil diffusion pump which permits flow rates up to 35 ml/min of helium in the GC and flow rates of at least 20 ml/min of methane or isobutane.

Figure 2 shows a schematic of the entire GC–CI–QMS system. As can be seen, in the GC–CI–QMS mode the reactant gas is introduced as the carrier gas at the GC column. The entire GC effluent is then delivered directly to the QMS where the carrier gas becomes the reactant gas. The CI electronics console permits the operator to select any of three carrier/re-

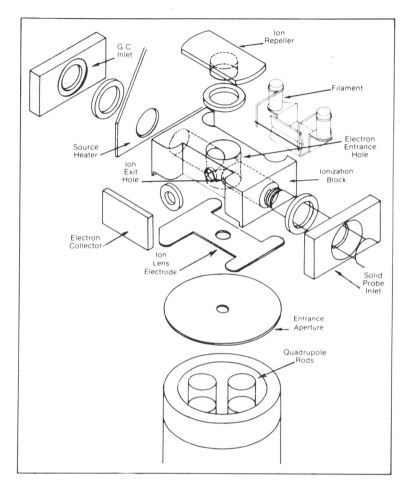

FIG. 1. Exploded view of chemical ionization source for QMS.

actant gases by means of switch-operated solenoids. The reactant gas can alternately be delivered directly to the ion source for CI analysis of solid probe or gas/volatile liquid samples. There is a provision for vacuuming the delivery lines to eliminate cross-contamination of the various reactant gases when switching gases.

Comparison of EI and CI Spectra

Figure 3 shows EI spectrum of the street drug LSD introduced into the QMS from the solid probe. Figure 4 shows the CI spectrum of the same drug using isobutane as the reactant gas. It is noted that major ion in the EI

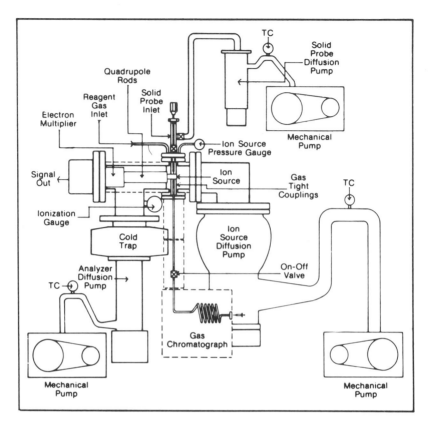

FIG. 2. Vacuum diagram of GC–MS system equipped with chemical ionization source.

spectrum is $M/E = 60$, and that there are few major ions above $M/E = 224$. By contrast, the largest peak in the CI spectrum is $M/E = 324$, the M+1 ion; furthermore, this ion accounts for more than 45% of the total ions observed.

The above spectra were taken on a combination EI–CI source as shown in Fig. 1. In the EI mode the high-pressure isobutane is removed and the repeller voltage is lowered from its setting in the CI mode.

Obviously, it is not that simple to switch from CI to EI when doing GC–MS analyses. One would, at minimum, have to reinstall the separator between the GC and ion source which would take several minutes. Ideally, one would like to exchange the CI source for the EI source as well, in order to have an optimum geometry in each mode of operation.

Rather than carry out this reasonably complex procedure, it has been shown by Foltz (1971) and others (Fentiman, Foltz, and Kinzer, 1973), that good EI spectra can be produced by switching to helium carrier/reactant gas and remaining in the CI mode of operation (CI_{He}). Typical "high pressure" helium spectra and high pressure nitrogen spectra (CI_{N_2}) are shown in

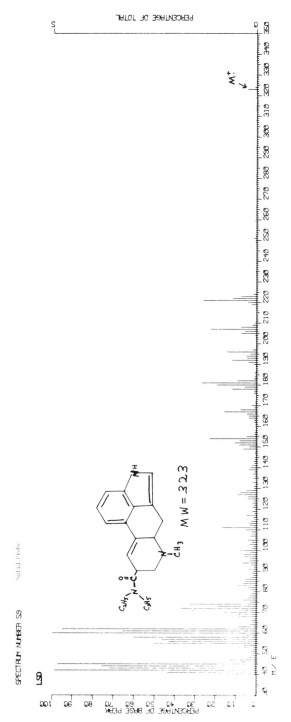

FIG. 3. Electron impact spectrum of LSD street drug.

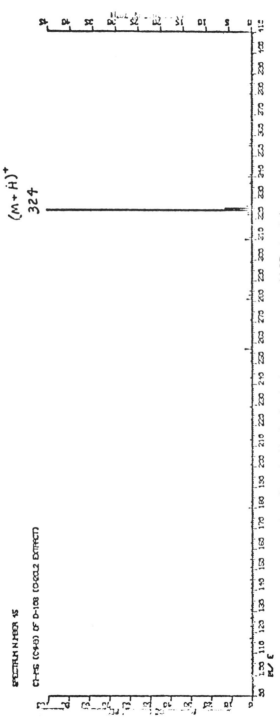

FIG. 4. Chemical ionization isobutane spectrum of LSD street drug.

Fig. 5 for 3-phenyl-1-propanol and are compared to EI spectra taken in the normal way as well as to CI_{Me}. It can be seen that there is generally a close correspondence between the CI_{He} spectra and the EI spectra. In fact, the Battelle-Columbus group has been using this technique for more than 2 years and are doing routine identifications of CI_{He} spectra using the large Battelle library stored on the CDC-6400. This library consists primarily of the Aldermaston library taken from magnetic instruments and utilizes a comparison of the two most intense peaks every 14 amu (Hertz, Hites, and Biemann, 1971) to perform the identification.

III. QMF

In QMF one monitors from one to eight selected ions rather than the entire mass range. This can result in significant enhancement in the ion current measured as shown in Fig. 6 where a comparison is shown between the ion current measured by monitoring the major ion at molecular weight 109 of the parathion spectra and that obtained by measuring the total ion current over the mass range 30 to 300. Enhancements of several hundred times are often easily attained using this technique. In QMF one can monitor *any* eight peaks in any order desired regardless of mass. With magnetic instruments the masses are usually restricted to a mass range of 25% of the largest mass. Chemical ionization lends itself particularly well to QMF since it often produces a single major ion as has been seen earlier.

QMF can be carried out using either the PROMIM (Programmed Multiple-Ion Monitor) or the Model 6000 Interactive Data System (the latter is limited to four ions).

A. Promim

The PROMIM is a hardware programmer which permits the operator to monitor from one to eight ions in any desired order. Controls of the PROMIM permit the operator to set individually on each channel (one channel for each ion) the integration time (1, 10, 100 milliseconds), preamplifier gain setting (10^{-5} to 10^{-9} A/V), filter setting, and bucking voltage (positive or negative voltages available). Any channel may be disabled in order to discontinue monitoring the corresponding ion.

B. Model 6000 Interactive Data System

This dedicated interactive data system (DaGragnano and Hotz, *This Volume*) permits one to do computerized mass fragmentography on any one to four ions selected. The computer optimizes its own integration time based upon signal-to-noise ratio at each selected ion. The mass fragmentograms are plotted on the CRT in real time and stored in disk for postrun analyses.

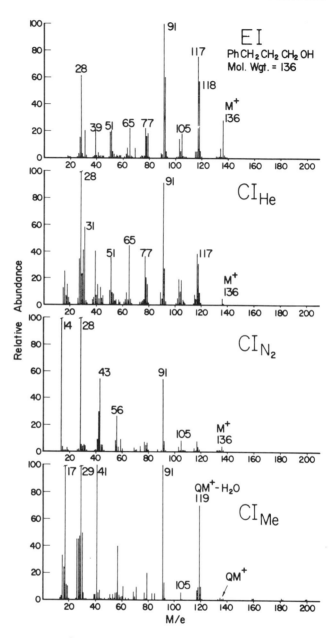

FIG. 5. Mass spectra of 3-phenyl-1-propanol.

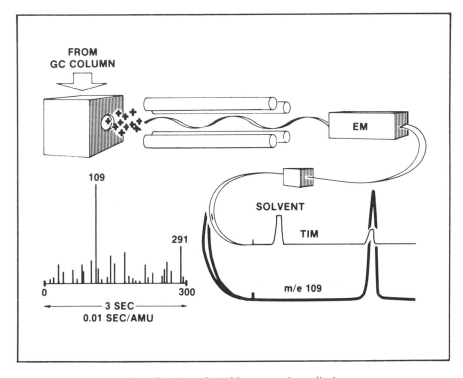

FIG. 6. Single and total ion current monitoring.

Atkinson (1972) and associates (Strong et al., 1973) have carried out quantitative mass fragmentography by use of an internal standard having ions at or near the same mass as the compound being measured. The quantitative value of the unknown is determined by comparing the peak height of its mass fragmentogram to that of the standard and entering the precomputed standard's curves. Bonelli (1972) earlier carried out quantitative mass fragmentography on DDE but used comparison of peak areas rather than peak height of the mass fragmentograms.

Using the interactive data system one can easily and quickly compute and compare either peak areas or peak heights through the use of a CRT display and push-button initiated computations. This accessory greatly enhances the power of the GC–MS instrument.

IV. COMPARATIVE ANALYSIS OF DERIVATIZED PROSTAGLANDIN PGF$_{2\alpha}$ BY EI AND CI

Prostaglandin PGF$_{2\alpha}$ as the methylester, tri-trimethylsilylether (PGF$_{2\alpha}$-ME-TRI-TMS) was analyzed by both EI and CI GC–QMS with an objective of determining the ultimate sensitivity in each mode using QMF. The

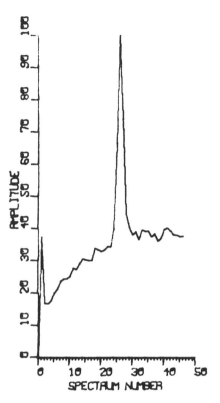

FIG. 7. Reconstructed gas chromatogram for PGF$_{2\alpha}$-Me-TRI-TMS run in Cl$_{He}$ mode.

analysis was done by Dr. Rodger Foltz of Battelle Memorial Institute at Columbus, Ohio, using a Finnigan Model 1015 equipped with a combination differentially pumped EI/CI source and with the System /150 Computer Control System. The samples were provided by Dr. J. Throck Watson of Vanderbilt.

Figure 7 shows the reconstructed gas chromatogram obtained on 1 μg of PGF$_{2\alpha}$-ME-TRI-TMS when taking complete spectra over the range 100 to 600. This EI data was actually obtained while running in the CI mode with helium as the carrier/reactant gas.

Figure 8 shows the complete EI spectrum of PGF$_{2\alpha}$-ME-TRI-TMS. From this spectrum it was determined that M/E = 191 and 423 were the best ions to monitor in the mass fragmentography mode of operation. The base peak at M/E = 73 was also selected for monitoring. Figures 9 and 10 show mass fragmentograms taken using the System/150 Computer System with sample sizes of 40 and 2 ng, respectively. These fragmentograms have been normalized by the computer prior to plotting so they have no quantitative meaning. In the Model 6000 Interactive Data System the mass frag-

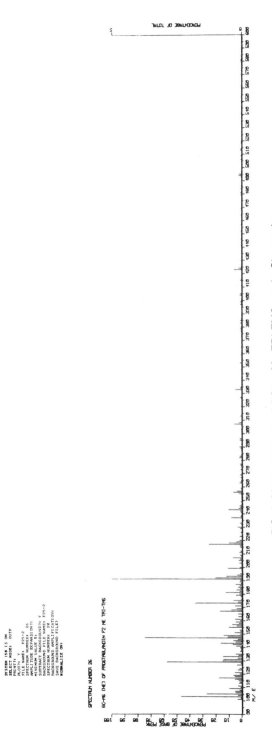

FIG. 8. Mass spectrum of PGF$_{2\alpha}$-Me-TRI-TMS run in CI$_{He}$ mode.

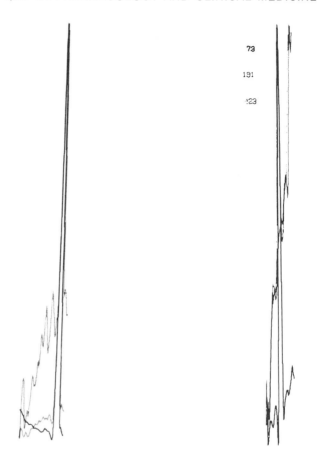

73

191

123

FIG. 9. *Left.* Mass fragmentogram of PGF$_{2\alpha}$-Me-TRI-TMS run in CI$_{He}$ mode. Sample size 40 ng.

FIG. 10. *Right.* Mass fragmentograms of PGF$_{2\alpha}$-TRI-TMS run in CI$_{He}$ mode. Sample size 2 ng.

mentograms do carry quantitative information. It was determined that the sensitivity limit in the CI$_{He}$ mode was approximately 2 ng. Sensitivity in the "pure" EI mode would be somewhat better.

Figure 11 shows the corresponding complete CI$_{Me}$ mass spectrum. From this, it was decided to monitor M/E = 569 and 405 in the mass fragmentography mode. Figures 12 and 13 show the mass fragmentograms at 1 ng and 200 pg respectively. Again, there is no quantitative meaning because of computer normalization. It was determined from this analysis that the sensitivity limit in the CI$_{Me}$ mode was somewhat better than 200 pg without any attempt at optimizing the system.

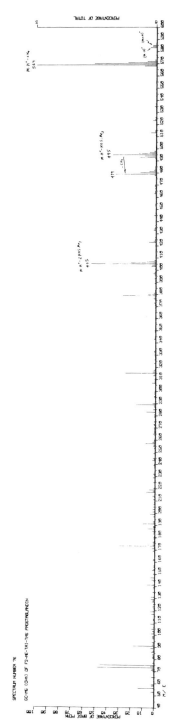

FIG. 11. Mass spectrum of PGF$_{2\alpha}$-Me-TRI-TMS run in Cl$_{Me}$ mode.

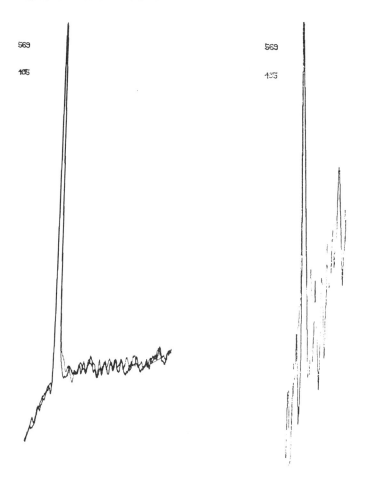

FIG. 12. *Left.* Mass fragmentogram of PGF$_{2\alpha}$-Me-TRI-TMS run in CI$_{Me}$ mode. Sample size 1 ng.

FIG. 13. *Right.* Mass fragmentogram of PGF$_{2\alpha}$-Me-TRI-TMS run in CI$_{Me}$ mode. Sample size 200 pg.

From the above analyses which were carried out in approximately one-half day, we can make several observations:

(1) In the CI$_{Me}$ mode one finds major peaks at higher molecular weights as compared to CI$_{He}$ (or EI). These peaks are of significant amplitude and are highly useful in the QMF mode.

(2) Sensitivity in the CI$_{Me}$ mode was approximately 10 times better than in the CI$_{He}$ mode. This advantage would be somewhat less when comparing CI$_{Me}$ to "pure" EI (i.e., using an optimal EI source with magnetically· constrained electron beam).

If one were to attempt detection and/or quantitation of this compound in the body fluids then these advantages would take on considerably greater importance.

V. QUANTITATIVE ANALYSIS OF PROPOXYPHENE AND ITS METABOLITES IN HUMAN PLASMA BY EI AND CI QMF

A comparative analysis was carried out using EI and CI QMF to quantitatively measure propoxyphene and its metabolite propoxyphene amide at low levels in human blood plasma (R. L. Wolen, *in preparation*). The sample preparation and extraction techniques are described in a paper by Wolen and Gruber (1968).

The experimental apparatus consisted of a Finnigan Model 3100D GC–MS equipped with separate EI and CI sources and the PROMIM with a Rikadenki four-channel recorder to plot the mass fragmentograms.

The molecular structures of propoxyphene, propoxyphene amide, and the internal standard used here, SKF 525A, are shown in Fig. 14. Figure 15 shows the complete EI mass spectrum of propoxyphene which was obtained using the CI_{He} mode. From this mass spectrum it was determined that $M/E = 193$ was the highest ion which could be monitored in the mass fragmentography mode in order to get a reasonably good signal/noise ratio. Figure 16 shows the corresponding CI_{Me} spectrum. As is seen, there are

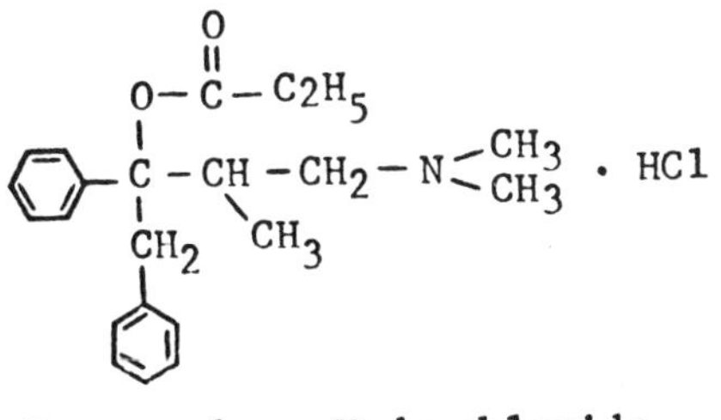

Propoxyphene Hydrochloride
MW 339 + 36.5 = 375.5

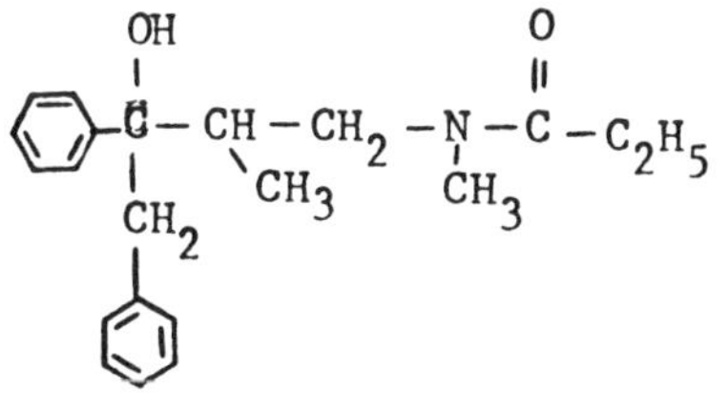

Propoxyphene Amide
MW 325

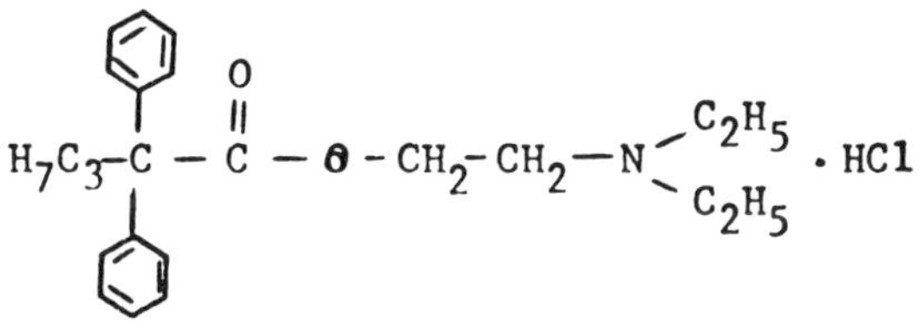

SKF 525A
MW 353 + 36.5 = 389.5

FIG. 14. Molecular structure of propoxyphene, propoxyphene amide, and SKF 525A.

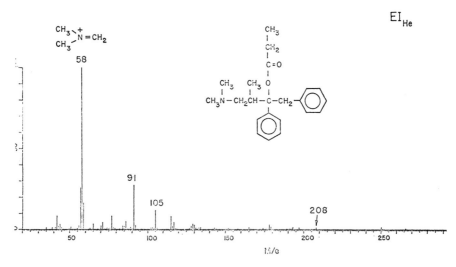

FIG. 15. EI mass spectrum of propoxyphene.

significant ions at both M/E = 338 and M/E = 340, the M−1 and M+1 ions respectively. It was decided to use the M/E = 338 ion for mass fragmentography studies.

Both propoxyphene amide and SKF 525A produced the major ion at the M+1 location in the CI_{Me} mode so this ion was used in the CI mass fragmentography analyses. Table 1 shows a summary of the ions used for each of the compounds in the EI and CI mass fragmentography studies.

Figure 17 shows the mass fragmentograms which were obtained when monitoring the selected ions in the EI mode for various sample sizes of

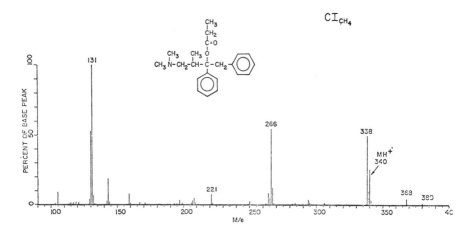

FIG. 16. CI_{Me} mass spectrum of propoxyphene.

TABLE 1. *Summary of ions*

Compounds	Molecular ion	Ions monitored	
		CI	EI
Propoxyphene	339	338	193
Propoxyphene amides	325	326	234
SKF 525A	353	354	165

standards of these compounds injected into the GC. Figure 18 shows the corresponding CI_{Me} mass fragmentograms. The latter exhibit a significantly better signal-to-noise ratio at the lowest levels of standards injected.

From these data a plot of peak height versus concentration was prepared. Figure 19 shows the standard curve which was computed for the CI_{Me} mode. The nonlinearity in the propoxyphene amide 326 and 354 standard curves was a result of adsorbance on the GC column and was eliminated by appropriate adjustment of column conditions.

Using these standard curves, quantitative measurements were made of propoxyphene and propoxyphene amide in human blood plasma specimens. The following results were noted:

(1) When operating in the EI mass fragmentography mode, there was considerable interference from biological contaminants precluding meaningful quantitative assays below the 10 μl level (size of sample injected into GC). This corresponds to .175 μg/ml of original plasma. This is approximately 10 times the minimum sample level analyzed with standards.

(2) In the CI mass fragmentography mode there was little or no discernible difference between the minimum detectable limits when using actual plasma samples as compared to the standards. The biological contaminants did not interfere significantly in the mass range which was being utilized (M/E > 300).

(3) Using CI mass fragmentography, we were able to get at least a factor of 10 higher ultimate sensitivity than in the EI mode when comparing sensitivities with actual blood plasma specimens. This enhancement with CI has been noted with drugs other than propoxyphene in subsequent analyses.

ACKNOWLEDGMENT

We are indebted to Dr. Rodger Foltz and his associates at Battelle Memorial Institute for their pioneering efforts in the application of chemical ionization GC/QMS. From these efforts came much of the differentially pumped three-gas CI system described herein as well as the everyday use of high-pressure helium mass spectrometry to provide EI spectra.

We thank Dr. Robert Wolen of Eli Lilly Company for his help in pre-

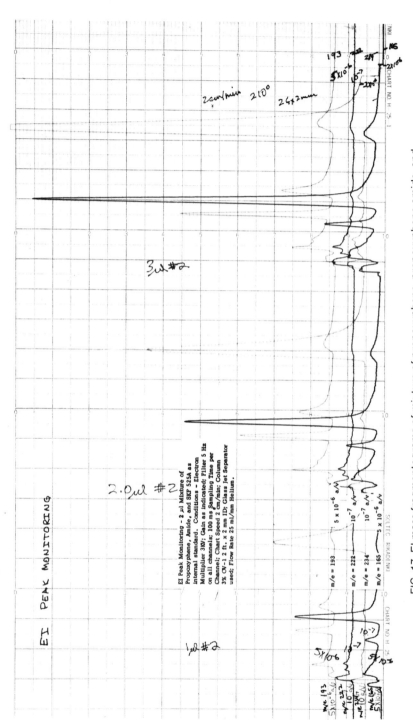

EI PEAK MONITORING

EI Peak Monitoring - 2 μl Mixture of
Propoxyphene, Amide, and SKF 525A as
internal standard. Conditions - Electron
Multiplier 3KV; Gain as indicated; Filter 5 Hz
on all channels; 100 ms Sampling Time per
Channel; Chart Speed 2 cm/min; Column
3% OV-1 2 ft. × 2 mm ID; Glass Jet Separator
used; Flow Rate 25 ml/min Helium.

2.0μl #2

3μl #2

1μl #2

2cm/min 210°

26×2mm

m/e = 193   5 × 10⁻⁶ a/v
m/e = 222   10⁻⁷ a/v
m/e = 234   10⁻⁷ a/v
m/e = 165   5 × 10⁻⁶ a/v

FIG. 17. EI mass fragmentograms of mixture of propoxyphene, propoxyphene amide, and SKF 525A standards.

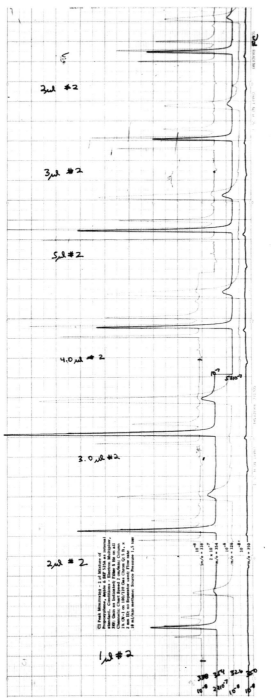

FIG. 18. CI mass fragmentograms of mixture of propoxyphene, propoxyphene amide, and SKF 525A standards.

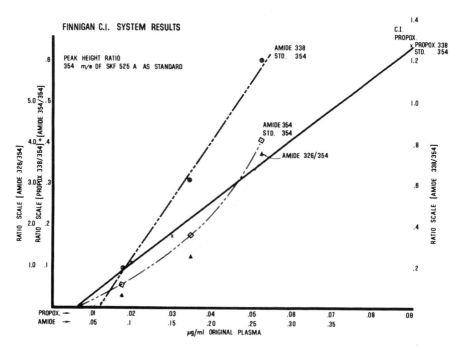

FIG. 19. Standard curves for propoxyphene and propoxyphene amide based upon CI mass fragmentography.

paring and carrying out the experiment above comparing CI and EI analysis of propoxyphene and its metabolite by mass fragmentography. We thank Mr. Ronald Skinner of Finnigan Corporation for much of the experimental work in this study.

REFERENCES

Bonelli, E. J. (1972): Gas chromatograph/mass spectrometer techniques for determination of interferences in pesticide analysis. *Analytical Chemistry*, 44:603–606.

Fentiman, A. L., Jr., Foltz, R. L., and Kinzer, G. W. (1973): Identification of noncannabinoid phenols in marihuana smoke condensate using chemical ionization mass spectrometry. *Analytical Chemistry*, 45:580–583.

Foltz, R. L. (1971): Applications of chemical ionization mass spectrometry. Nineteenth Annual Conference on Mass Spectrometry and Allied Topics, Atlanta, Georgia, May 1971.

Hertz, H. S., Hites, R. A., and Biemann, K. (1971): Identification of mass spectra by computer searching a file of known spectra. *Analytical Chemistry*, 43:681–690.

Milne, G. W. A., Fales, H. M., and Axenrod, T. (1971): Identification of dangerous drugs by isobutane chemical ionization mass spectrometry. *Analytical Chemistry*, 43:1815–1820.

Story, M. S. (1972): Description of a GC/MS quadrupole instrument utilizing a chemical ionization source and no enriching device. Twentieth Annual Conference on Mass Spectrometry and Allied Topics, June 1972.

Strong, J. M., and Atkinson, A. J., Jr. (1972): Simultaneous measurement of plasma concentrations of lidocaine and its desethylated metabolite by mass fragmentography. *Analytical Chemistry*, 44:2287–2290.

Strong, J. M., Parker, M., and Atkinson, A. J., Jr. (1973): Identification of glycinexylidide in patients treated with intravenous lidocaine. *Clinical Pharmacology and Therapeutics,* 14:67–72.

Wolen, R. L., and Gruber, C. M., Jr. (1968): Determination of propoxyphene in human plasma by gas chromatography. *Analytical Chemistry,* 40:1243–1246.

Mass Spectrometry in Biochemistry and Medicine,
edited by A. Frigerio and N. Castagnoli.
Raven Press, New York © 1974

Comparison of Isopentyloxime and Benzyloxime Trimethylsilyl Ethers in the Characterization of Urinary Steroids of Newborn Infants

T. A. Baillie, C. J. W. Brooks, E. M. Chambaz,* R. C. Glass, and C. Madani*

*Chemistry Department, University of Glasgow, Glasgow G12 8QQ, U.K., and * Laboratoire d'Hormonologie, CHR de Grenoble, Grenoble, France*

I. INTRODUCTION

In recent years several laboratories have shown that the urinary steroid excretion of the human newborn differs markedly from the adult pattern with regard to the structure of the individual components. The subject has been well reviewed by Mitchell (1967) and by Mitchell and Shackleton (1969). The major identified metabolites have been reported to possess a Δ^5-3β-ol structure and an oxygen function at C-16; this is in many instances a 16α hydroxy group (Reynolds, 1966; Shackleton, Livingstone, and Mitchell, 1968; Horning, Chambaz, Brooks, Moss, Boucher, Horning, and Hill, 1969). However, additional saturated metabolites, mostly of the pregnane series, have been more recently described in human infant urine by Shackleton, Gustafsson, and Sjövall (1971).

Gas-liquid chromatography (GLC) and gas chromatography–mass spectrometry (GC–MS) have rendered possible the study of individual steroids in metabolic profile separations obtained from biological fluids. Volatile derivatives such as the methyloxime–trimethylsilyl ethers (MO–TMS) introduced by Gardiner and Horning (1966) allowed the separation of the main urinary steroid metabolites in adult and infant urine (Gardiner, Brooks, Horning, and Hill, 1966). In this approach, derivative formation not only has the purpose of stabilizing thermally unstable structures, but can be used for characterizing functional groups of individual steroids by GLC and GC–MS and for improving the chromatographic separations of components of the biological mixture under study. At best, derivatives may yield a true group separation of the metabolites studied, rendering MS studies more informative.

The benzyloxime–TMS (BO–TMS) introduced by Devaux, Horning, and Horning (1971*b*) (*cf.* Devaux, Horning, Hill, and Horning, 1971*a*) afford a complete GC separation of steroids containing reactive ketonic groups from related nonketonic steroids (together with 11-oxosteroids). Their large retention increments preclude the analysis of most diketonic steroids. Furthermore, the late elution of ketosteroid BO derivatives of the pregnane series leads to some difficulties because of the abundance of "background" ionization in MS. The use of O-alkyloximes of intermediate retention time has been proposed by Baillie, Brooks, and Horning (1972) to complement the data obtainable from MO–TMS and BO–TMS. We now report further work on this theme.

II. EXPERIMENTAL

A. Materials

Steroid model compounds were purchased from Schwarz/Mann (Orangeburg, New York). $3\beta,16\beta$-Dihydroxy-5-androsten-17-one, 3β-hydroxy-5,16-pregnadien-20-one, and $3\alpha,6\alpha$-dihydroxy-5β-pregnan-20-one were gifts from Drs. C. H. L. Shackleton and R. W. Kelly; 5-pregnene-$3\beta,17\alpha,20\alpha$-triol was kindly donated by Dr. R. Neher (Ciba).

The following reagents were donated by Dr. D. J. Outred (Beecham Research Laboratories): *sec*-butoxyamine hydrochloride (1-aminoxy-1-methylpropane hydrochloride) and isopentoxyamine hydrochloride (1-aminoxy-3-methylbutane hydrochloride). Other alkoxyamines were commercial samples.

An enzyme preparation containing β-glucuronidase and sulfatase, derived from *Helix pomatia,* was obtained from Industrie Biologique Française, Gennevilliers, France.

1. Isopentoxyamine-1-d₁ Hydrochloride

i. Isoamyl alcohol-1-d₁. To a stirred suspension of lithium aluminium deuteride (1.092 g; 28.8 mmoles) in anhydrous ether (40 ml) under a nitrogen atmosphere was added a solution of isovaleraldehyde (5.006 g; 58.2 mmoles) in anhydrous ether (10 ml). Complete addition took 35 min, and was accompanied by gentle reflux of the reaction mixture. Work-up afforded 5.8 ml (4.736 g; 52.6 mmoles) of isoamyl alcohol-1-d₁. (Yield = 90%.) Infrared absorption (IR) bands at 3,350 cm⁻¹ (s) (broad) $[\nu(\text{O–H})]$; 2,170 cm⁻¹ (s) $[\nu(\text{C–D})]$.

ii. Isoamyl bromide-1-d₁. This was prepared according to the method of Wiley, Hershkowitz, Rein, and Chung (1964). To a stirred solution of dry triphenylphosphine (11.88 g; 45.4 mmoles) in dry dimethylformamide (50 ml) held under a nitrogen atmosphere was added 5.0 ml (4.08 g; 45.4 mmoles) isoamyl alcohol-1-d₁. Bromine (2.4 ml: 1.03 molar equivalent) was then added dropwise, and the temperature of the reaction mixture was maintained below 50°C. Two successive distillations at atmospheric pressure yielded isoamyl bromide-1-d_1, b.p. 117 to 120°C (2.9 ml; 3.28 g; 21.4 mmoles). (Yield = 47.3%.) I.R. showed ν(C–D) at 2,235 cm^{-1} (m).

iii. Isoamyl 1-d_1-benzhydroxamate (Outred, 1972). Benzhydroxamic acid (2.686 g; 19.6 mmoles) was dissolved in AnalaR methanol (25 ml), and the resulting solution vigorously stirred while a solution of sodium hydroxide (0.784 g; 19.6 mmoles) in water (2.6 ml) was added. Isoamyl bromide-1-d_1 (3.0 g; 19.7 mmoles) was then added, and stirring continued for 24 hr. The reaction mixture was kept at room temperature for 3 days, after which it was evaporated to dryness under reduced pressure. The residue was dissolved in ethyl acetate/water (10 ml:10 ml), the organic layer washed with water (3 × 5 ml), and the washings back-extracted with ethyl acetate (5 ml). The combined organic extracts were dried, filtered, and evaporated under reduced pressure to yield the benzhydroxamate (1.557 g; 7.49 mmoles) as a pale yellow oil. (Yield = 38%.) I.R. showed a band at 3,475 cm^{-1} (w) and 3,410 cm^{-1} (w), ν(N–H); 3,025 cm^{-1} (w), ν(aryl-H); 2,155 cm^{-1} (w), ν(C–D); 1,650 to 1,705 cm^{-1} (s), ν(C=O).

iv. Isopentoxyamine-1-d_1 hydrochloride (Outred, 1972). The crude benzhydroxamate (1.50 g; 7.21 mmoles) obtained above was treated with 15 ml of a mixture of 12N-HCl/methanol (1:3), and the resulting clear solution was refluxed for 3 hr. The solvent was evaporated on a rotary evaporator, yielding an oil which slowly solidified. This residue was treated with ether (15 ml), and the solid product collected by filtration, washed well with more ether (15 ml), and dried under vacuum, affording a colorless, amorphous solid (0.779 g; 5.5 mmoles) (Yield = 76.6%), which on recrystallization from ethanol/ether, afforded large translucent plates, double m.p. 115 to 117°C and 143 to 145°.

I.R. showed a band at 2,165 cm^{-1} (w) [ν(C–D)]; nuclear magnetic resonance (NMR) (60 MHz) gave: 9.20τ 1:1 doublet [(J = 5.6 Hz), 6 H (two Me-groups)]; 8.44τ 1:1 doublet [broad, J = 6.4 Hz, 3 H (C$_2$ and C$_3$)]; 5.81τ, 1:3:1 triplet [J = 6.0 Hz, 1 H (C$_1$)]. The signal at 5.81τ corresponded to that at 5.58τ in the unlabeled analogue, which integrated for 2 H.

2. Isopentoxyamine-1-d_2 Hydrochloride

An identical series of reactions to those described above was used, starting from methyl isovalerate in place of isovaleraldehyde. This afforded the dideuterated reagent, the NMR (60 MHz) of which gave: 9.21τ 1:1 doublet [(J = 5.4 Hz), 6 H (two Me-groups)]; 8.44τ singlet [broad, 3 H (C$_2$ and C$_3$)].

3. Pregnenolone-isopentyloxime(-1-d$_1$)-TMS and (1-d$_2$)-TMS

These derivatives were prepared as indicated below using the appropriate, labeled reagents. They gave single GC peaks ($I^{250°}_{OV\text{-}1} = 3130$), and mass spectra consistent with the incorporation of one and two deuterium atoms, respectively (M$^{+\cdot}$: m/e 474, 475).

B. Methods

1. Preparation of Derivatives

Analytical samples of O-alkyloximes were prepared in pyridine solution, using 0.2 to 0.5 mg of reference steroid, according to the procedure of Sakauchi and Horning (1971), and BO derivatives were formed according to Devaux et al. (1971b).

Unhindered hydroxyl groups in reference compounds were converted to TMS derivatives by a noncatalyzed reaction with BSA or BSTFA (Chambaz and Horning, 1969). Hindered hydroxyl groups were silylated with TSIM, either by the noncatalyzed procedure of Devaux et al. (1971b), or by the catalyzed reaction described by Thenot and Horning (1972).

BO–TMS and isopentyloxime–TMS (iPO–TMS) of biological extracts were prepared under the conditions described for MO–TMS by Thenot and Horning (1972).

2. Preparation of Urine Extracts

Urinary samples were collected in plastic bags from normal male and female human newborn infants. An average urinary output of 28.5 ml/24 hr was obtained for 61 babies studied in the first 5 days of life. A fraction of each urine sample was used to constitute a pool.

"Total" steroid extracts were obtained after enzymatic hydrolysis (5 to 50 ml of urine) with a mixture of β-glucuronidase and sulfatase for 48 hr at 37°C in a shaking bath, after addition of 0.1 volume of acetate buffer (0.2 M, pH 5.2). The hydrolyzed urines were extracted with 2 volumes of ether and 2 volumes of ethyl acetate; the pooled organic phases were washed with 0.1 volume of 0.1 N-NaOH and with distilled water, filtered through anhydrous sodium sulfate, and evaporated to dryness in a rotary evaporator: the residues were suitable for derivative formation.

3. GC

A Carlo Erba Model GV instrument was used, with W-shaped 12-ft (3 mm I.D.) glass columns, prepared and packed with 1% stationary phases on silanized 100–120 mesh Gas Chrom Q (Applied Science Laboratories,

State College, Pennsylvania), according to Horning, Horning, Ikekawa, Chambaz, Jaakonmaki, and Brooks (1967). Flame ionization detectors were used, with nitrogen (40 ml/min) as carrier gas. OV-1, OV-101, and SE-30 were used as practically equivalent stationary phases in this work. Separations were carried out with temperature programming from 190 to 300°C (275°C in Glasgow) at 1.2°C/min. n-Alkanes (C_{24} to C_{38}) were used to standardize the retention data for the steroids studied: the values were expressed as Kováts retention indices (Ettre, 1964) [*cf.* methylene units (VandenHeuvel, Gardiner, and Horning, 1964)]. Cholesteryl decylate was added to BO–TMS analyses as an internal standard both for qualitative and quantitative purposes. Cholesteryl butyrate was used for corresponding *i*PO–TMS runs.

4. GC–MS

An LKB 9000 instrument was used, operating at electron energy 70 eV. The column and molecular separator were kept at 250°C, and the flash heater and ion source at 270°C. The helium flow rate was 25 ml/min. The trap current was 60 μA, accelerating voltage 3.5 kV, electron multiplier voltage 2.1 kV; exit and entrance slits were set at 0.1 mm. Spectra were obtained using a recording oscillograph, and with a scan time of approximately 5 sec.

5. High Resolution MS

An AEI MS902 instrument was used, with direct probe sampling, at electron energy 70 eV.

III. RESULTS AND DISCUSSION

GC separations of a reference mixture of typical steroids from newborn infants' urine are illustrated in Fig. 1. In the first week of life the excretion of corticosteroids of dihydroxyacetone and 20,17-ketol types is very limited, with only tetrahydrocortisone being regularly observed. Accordingly, the major urinary steroids may be studied after simple trimethylsilylation (Fig. 1a). Under these conditions the principal components are readily identified by GC–MS, but there is considerable overlap of ketonic and hydroxylic steroid derivatives. The MO–TMS (Fig. 1b) provide a more convenient MS distinction between the two groups but show fortuitous coincidence of retention times, as for example of 5-androstene-3β,16α,17β-triol triTMS with 3α,6α-dihydroxy-5β-pregnan-20-one MO-diTMS. The complete separation of reactive ketones as BO–TMS is illustrated in Fig. 1c. The derivatives of dihydroxy ketones of the C_{21} series require elution temperatures near 280°C, and for more highly substituted steroids the mass spectra recorded

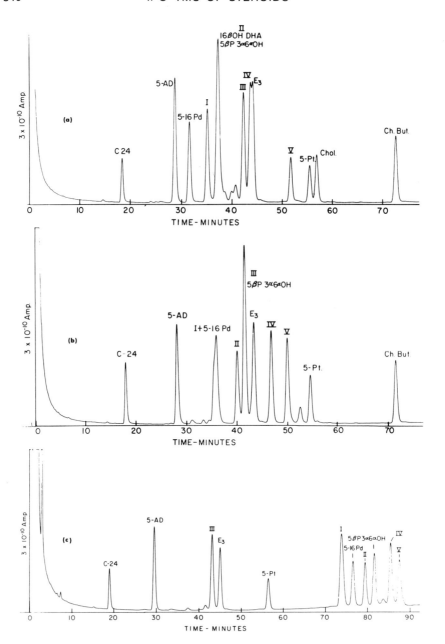

FIG. 1. GC of reference mixtures of typical urinary steroids of newborn infants. 1% SE-30, temperature programmed 190 × 1°C/min. (a) TMS; (b) MO–TMS; (c) BO–TMS. Key: C-24, n-tetracosane; 5-AD, 5-androstene-3β,17β-diol; 5,16-Pd, 3β-hydroxy-5,16-pregnadien-20-one; 5βP3α6αOH, 3α,6α-dihydroxy-5β-pregnan-20-one; E₃, 1,3,5(10)-estratriene-3,16α,17β-triol; 5-Pt, 5-pregnene-3β,17α,20α-triol; Ch.But., cholesteryl butyrate; I, 3β,16α-dihydroxy-5-androsten-17-one; II, 3β,17β-dihydroxy-5-androsten-16-one; III, 5-androstene-3β,16α,17β-triol; IV, 3β,16α-dihydroxy-5-pregnen-20-one; V, 3β,21-dihydroxy-5-pregnen-20-one.

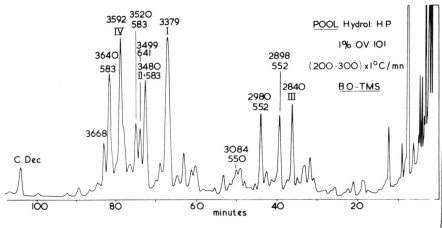

FIG. 2. GC of a "total" urinary steroid extract (as BO–TMS) from a pool obtained from 51 normal newborn infants, aged 1 to 4 days. 1% OV-101, temperature programmed 200 × 1°C/min.

(especially for extracts of urinary metabolites) may be partly obscured by contaminating spectra from the stationary phase and other extraneous sources.

Figure 2 shows the GC separation obtained with a "total" extract (as BO–TMS) derived from pooled urine of 51 newborn infants. Approximate retention parameters could be measured for the major components of the mixture, and combined GC–MS yielded mass spectra for each main metabolite. Several GC peaks appeared as a mixture of metabolites. Identity of retention parameters and mass spectra allowed the identification of some of these steroids, but others could only be partially characterized. The major individual components labeled in Fig. 2 are listed in Table 1. Steroids con-

TABLE 1. *Principal components of the mixture of BO–TMS of urinary steroids derived from pooled urine of newborn infants*

I_{OV-1}	$M^{+\cdot}$	
2,840	522	5-Androstene-3β,16α,17β-triol, triTMS
2,898	552 ⎫	
2,980	552 ⎬	Pregnanetriols, triTMS
3,084	550	5-Pregnene-3β,20α,21-triol, triTMS
3,379	553	3β,16α-dihydroxy-5-androsten-17-one, BO-diTMS
3,414	641	Androstenetriolone, BO-triTMS
	⎧ 553	3β,17β-dihydroxy-5-androsten-16-one, BO-diTMS
3,480	⎨ 583	Pregnanediolone, BO-diTMS
	⎩ 641	Androstenetriolone, BO-triTMS
3,499	641	Androstenetriolone, BO-triTMS
3,520	583	Pregnanediolone, BO-diTMS
3,592	581	3β,16α-dihydroxy-5-pregnen-20-one, BO-diTMS
3,640	583	Pregnanediolone, BO-diTMS

Compare Fig. 2.

taining more than one BO group would have remained undetected because of their high-retention index values (> 3,800).

The possible value of alkyloximes of intermediate retention times in the analysis of hydroxylic, monoketonic, and diketonic steroids has accordingly been further explored. Table 2 includes retention data for a selection of representative steroids as TMS, MO–TMS, iPO–TMS, and BO–TMS. For the identification of steroids in extracts it is of course necessary to compare results from more than one type of stationary phase. For the purpose of this chapter we present only the data of OV-1/OV-101/SE-30 columns. For BO formation the retention increments were 600 to 800 (similar values were observed for OV-17), whereas for iPO the corresponding range was 300 to 450.

In order to apply iPO–TMS (and the analogous secBuO–TMS) to studies of urinary steroids by GC–MS, it was necessary to establish their modes of fragmentation. These were expected to parallel the well-known processes observed for MO–TMS. Comparison of the representative mass spectra in Fig. 3 confirmed this in some respects, e.g., in the characteristic loss of the N-alkoxy radical [(M-73), (M-87)] and the formation of abundant ions retaining the alkyloximino group (e.g., m/e 216, m/e 230). The various oximes

TABLE 2. *Retention index values for representative steroids and derivatives*

Steroids and derivatives	TMS	MO–TMS	iPO–TMS	BO–TMS
1,3,5(10)-Estratriene-3,16α,17β-triol	2,890	–	–	–
5-Androstene-3β,17β-diol	2,615	–	–	–
5-Androstene-3β,16α,17β-triol	2,840	–	–	–
5-Pregnene-3β,17α,20α-triol	2,980	–	–	–
5-Pregnene-3β,20α,21-triol	3,080	–	–	–
5β-Pregnane-3α,20α-diol	2,770	–	–	–
5α-Androstan-17-one[a]	–	2,275	2,600	2,980
3α-Hydroxy-5β-androstan-17-one	2,455	2,520	2,865	3,205
3α-Hydroxy-5β-androstane-11,17-dione	2,540	2,615	2,960	3,300
3β,16α-Dihydroxy-5-androsten-17-one	2,720	2,740	{3,050 3,070}	3,375
3β,16β-Dihydroxy-5-androsten-17-one	2,750	–	3,105	3,445
3β,17β-Dihydroxy-5-androsten-16-one	2,755	2,805	3,135	3,475
3β-Hydroxy-5-pregnen-20-one	2,675	2,785	3,130	3,460
3β-Hydroxy-5,16-pregnadien-20-one	2,655	2,755	3,085	3,430
3β,16α-Dihydroxy-5-pregnen-20-one	2,870	2,930	3,255	3,590
3β,21-Dihydroxy-5-pregnen-20-one	3,010	2,985	3,280	3,630
3α-Hydroxy-5β-pregnan-20-one	2,640	2,710	3,060	3,380
3α,6α-Dihydroxy-5β-pregnan-20-one	2,765	2,850	3,185	3,525
4-Pregnene-3,20-dione	–	2,880	{3,500 3,530}	[b]

[a] Free ketone: 2,195; ethyloxime, 2,335: sec-butyloxime, 2,470.
[b] Not recorded.
Measurements were made by temperature programming (190 × 1°C/min) using 1% OV-1, OV-101, or SE-30 columns. Retention indices were found to be essentially the same on these three phases. Values are rounded off to the nearest five units.

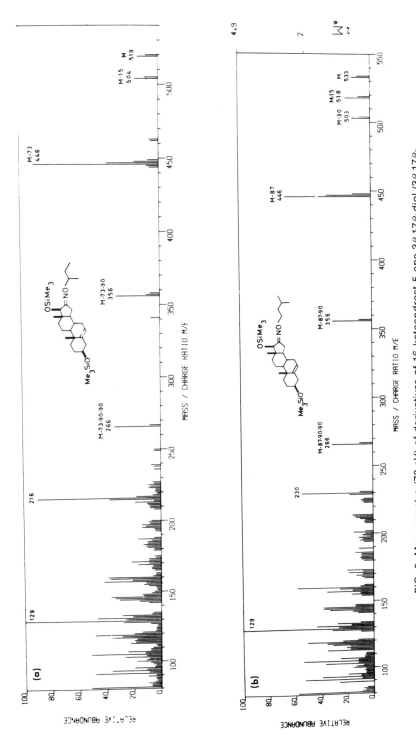

FIG. 3. Mass spectra (70 eV) of derivatives of 16-ketoandrost-5-ene-3β,17β-diol (3β,17β-dihydroxy-5-androsten-16-one): (a) O-sec-BO–diTMS; (b) O-iPO–diTMS.

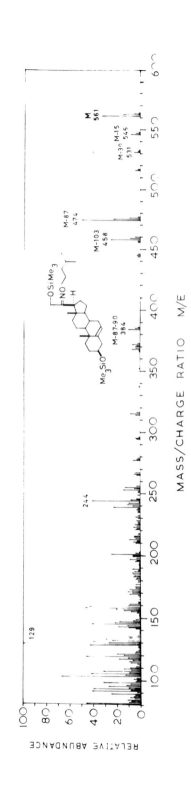

thus afford some common and some distinctive fragment ions, both types being useful in the recognition of ketosteroids in the analysis of mixtures by GC–MS (Horning, Brooks, and VandenHeuvel, 1968; Brooks and Harvey, 1970; Dray and Weliky, 1970).

The mass spectrum of the *i*PO–diTMS of $3\beta,17\beta$-dihydroxy-5-androstene-16-one (Fig. 3b) showed a prominent ion at m/e 503, corresponding to the loss of 30 mass units from the molecular ion. Examination of other data indicated that $[M-30]^{+\cdot}$ ions were absent from the spectra of *sec*BuO–TMS but were formed by many *i*PO–TMS. A further complicating feature was noted in the occurrence of $[M-44]^{+\cdot}$ and $[M-45]^{+\cdot}$ ions, as exemplified in Fig. 4. The incidence and abundance of $[M-30]^{+\cdot}$ and $[M-44]^{+\cdot}$ ions were markedly dependent on the structures of the ketonic substrates, as indicated in Table 3. 17-Oxosteroid *i*PO yielded strikingly prominent ions especially at $[M-30]^{+\cdot}$; 20-oxosteroid *i*PO unsubstituted in ring D gave lesser amounts; and in 16α- or 17α-trimethylsilyloxy-20-oxosteroid *i*PO, no ions of either type were detected.

The molecular formula of the $[M-30]^{+\cdot}$ ion derived from pregnenolone *i*PO was shown by high-resolution MS to be $C_{25}H_{41}NO$ and thus corresponded to the loss of CH_2O from the molecular ion (Fig. 5). It appeared probable that the CH_2O unit originated from the alkoxy group, and this was confirmed by data recorded for derivatives labeled by the route indicated (Fig. 6) with one and two deuterium atoms, respectively. Ions at $[M-31]^{+\cdot}$ and $[M-32]^{+\cdot}$ were observed. The simplest mechanism for the

TABLE 3. *Occurrence of* $[M-30]^{+\cdot}$ *and* $[M-44]^{+\cdot}$ *ions in spectra of iPO derivatives*

Structural type	Number of examples	$[M-30]^{+\cdot}$ %	$[M-44]^{+\cdot}$ %
NO	9	23–70	9–30
CH₂R / NO / ···H	8	1–17	0–3
CH₂R / NO / ···OSiMe₃	7	0	0

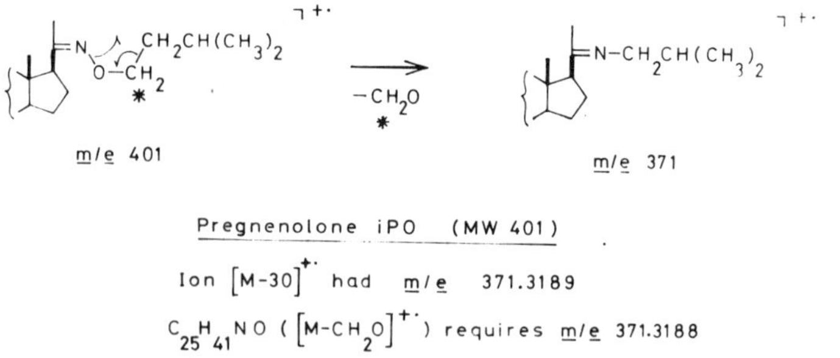
m/e 401
−CH₂O
m/e 371
Pregnenolone iPO (MW 401)
Ion [M-30]+· had m/e 371.3189
C25H41NO ([M-CH2O]+·) requires m/e 371.3188

FIG. 5. Scheme illustrating the origin of the $[M\text{-}30]^{+\cdot}$ ion in the mass spectra of O-isopentyloxime derivatives.* Mono- and dideuteration led to $[M\text{-}31]^{+\cdot}$ and $[M\text{-}32]^{+\cdot}$ respectively.

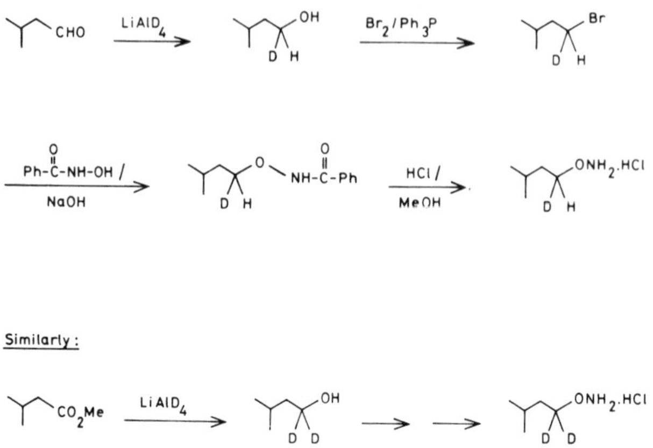
CHO
LiAlD4
OH
Br2/Ph3P
Br
Ph-C-NH-OH /
NaOH
NH-C-Ph
HCl /
MeOH
ONH2.HCl
Similarly:
CO2Me
LiAlD4

FIG. 6. Route employed in the synthesis of mono- and dideuterated isopentoxyamine hydrochloride reagents.

production of these ions would appear to be as in Fig. 5; their absence in the case of 16- and 17-trimethylsilyloxy derivatives is presumably due to steric inhibition of the *cis* conformation of the N-O-CH_2-R grouping. From models, this is clearly very likely with 17α-substituents, but the effect of a 16α-trimethylsilyloxy group would be expected to be less marked.

IV. SUMMARY

GC and MS data have been recorded for *i*PO–TMS of representative steroids of newborn infants' urine. The derivatives are usefully complementary to BO–TMS and are suitable for both mono- and diketonic steroids. An unusual mode of fragmentation observed for many *i*PO–TMS has been demonstrated to occur by loss of the elements of formaldehyde from the alkoxy group.

ACKNOWLEDGMENTS

C.J.W.B. and E.M.C. are greatly indebted to Drs. E. C. and M. G. Horning, whose hospitality, interest, and encouragement led to this collaboration. The work in Glasgow was aided by an SRC studentship (TAB) and by a grant from the Medical Research Council; the LKB 9000 GC–MS was furnished by SRC grants B/SR/2398 and B/SR/8471. We thank Dr. B. A. Knights for additional MS facilities. C.J.W.B. gratefully acknowledges a travel grant from the Wellcome Trust.

The work in Grenoble was possible thanks to the support from the Institut National de la Santé et de la Recherche Médicale, the Délégation Générale à la Recherche Scientifique et Technique, and the Fondation pour la Recherche Médicale Française.

REFERENCES

Baillie, T. A., Brooks, C. J. W., and Horning, E. C. (1972): O-Butyloximes and O-pentyloximes as derivatives for the study of ketosteroids by gas chromatography. *Analytical Letters,* 5:351–361.

Brooks, C. J. W., and Harvey, D. J. (1970): Gas chromatographic and mass spectrometric studies of oximes derived from 20-oxosteroids. *Steroids,* 15:283–301.

Chambaz, E. M., and Horning, E. C. (1969): Conversion of steroids to trimethylsilyl derivatives for gas phase analytical studies. *Analytical Biochemistry,* 30:7–24.

Devaux, P. G., Horning, M. G., Hill, R. M., and Horning, E. C. (1971a): Derivatives for the study of ketosteroids by gas chromatography. Application to urinary steroids of the newborn human. *Analytical Biochemistry,* 41:70–82.

Devaux, P. G., Horning, M. G., and Horning, E. C. (1971b): Benzyloxime derivatives of steroids. A new metabolic profile procedure for human urinary steroids. *Analytical Letters,* 4:151–160.

Dray, F., and Weliky, I. (1970): Identification of O-methyloximes of ketosteroids by gas chromatography, thin-layer chromatography, mass spectra, and kinetic studies. *Analytical Biochemistry,* 34:387–402.

Ettre, L. S. (1964): The Kováts retention index system. *Analytical Chemistry,* 36:31A–37A.

Gardiner, W. L., Brooks, C. J. W., Horning, E. C., and Hill, R. M. (1966): Urinary steroid pattern of the human newborn infant. *Biochimica et Biophysica Acta,* 130:278–281.

Gardiner, W. L., and Horning, E. C. (1966): Gas-liquid chromatographic separation of C_{19} and C_{21} human urinary steroids by a new procedure. *Biochimica et Biophysica Acta,* 115: 524–526.

Horning, E. C., Brooks, C. J. W., and VandenHeuvel, W. J. A. (1968): Gas phase analytical methods for the study of steroids. *Advances in Lipid Research,* 6:273–392.

Horning, M. G., Chambaz, E. M., Brooks, C. J. W., Moss, A. M., Boucher, E. A., Horning, E. C., and Hill, R. M. (1969): Characterization and estimation of urinary steroids of the newborn human by gas-phase analytical methods. *Analytical Biochemistry,* 31:512–531.

Horning, E. C., Horning, M. G., Ikekawa, N., Chambaz, E. M., Jaakonmaki, P. I., and Brooks, C. J. W. (1967): Studies of analytical separations of human steroids and steroid glucuronides. *Journal of Gas Chromatography,* 5:283–289.

Mitchell, F. L. (1967): Steroid metabolism in the fetoplacental unit and in early childhood. *Vitamins and Hormones,* 25:191–269.

Mitchell, F. L., and Shackleton, C. H. L. (1969): The investigation of steroid metabolism in early infancy. *Advances in Clinical Chemistry*, 12:142–215.

Outred, D. J. (1972): *Personal communication.*

Reynolds, J. W. (1966): The identification and quantification of Δ^5-androstenetriol ($3\beta,16\alpha$, 17β-trihydroxyandrost-5-ene) isolated from the urine of premature infants. *Steroids*, 8:719–727.

Sakauchi, N., and Horning, E. C. (1971): Steroid trimethylsilyl ethers. Derivative formation for compounds with highly hindered hydroxyl groups. *Analytical Letters*, 4:41–52.

Shackleton, C. H. L., Gustafsson, J.-Å., and Sjövall, J. (1971): Steroids in newborns and infants: Identification of steroids in urine from newborn infants. *Steroids*, 17:265–280.

Shackleton, C. H. L., Livingstone, J. R. B., and Mitchell, F. L. (1968): The conjugated 17-hydroxy epimers of Δ^5-androstene-$3\beta,17$-diol in infant and adult urine and umbilical cord plasma. *Steroids*, 11:299–311.

Thenot, J.-P., and Horning, E. C. (1972): MO–TMS derivatives of human urinary steroids for GC and GC–MS studies. *Analytical Letters*, 5:21–33.

VandenHeuvel, W. J. A., Gardiner, W. L., and Horning, E. C. (1964): Characterization and separation of amines by gas chromatography. *Analytical Chemistry*, 36:1550–1560.

Wiley, G. A., Hershkowitz, R. L., Rein, B. M., and Chung, B. C. (1964): Studies in organophosphorus chemistry. I. Conversion of alcohols and phenols to halides by tertiary phosphine dihalides. *Journal of the American Chemical Society*, 86:964–965.

Mass Spectrometry in Biochemistry and Medicine,
edited by A. Frigerio and N. Castagnoli.
Raven Press, New York © 1974

Identification and Quantitative Assay of Steroids in Congenital Corticoadrenal Hyperplasia in Newborn Infants by Gas-Liquid Chromatography and Gas Chromatography–Mass Spectrometry

J. Desgrès, R. J. Bègue, G. Curie, J. L. Nivelon,* and P. Padieu

*Laboratoire d'Exploration Fonctionnelle en Chimie Biologique, Hôpital du Bocage, 2 av. de Lattre de Tassigny, et Laboratoire d'Application en Chromatographie Gazeuse et en Spectrométrie de Masse, Faculté de Médecine, 7 bd Jeanne d'Arc 21033 Dijon, France, and *Service de Clinique Médicale Infantile, Hôpital Général, rue de l'Hôpital, 21000 Dijon, France*

I. INTRODUCTION

The steroid metabolism of newborn infants appears to be very distinct from that of an adult. For instance, the high amount of 3β-hydroxy-$\Delta 5$ steroids found in urine is an indication of a reduced, temporary 3β-hydroxy-oxydo-reductase$\Delta 5$-$\Delta 4$ isomerase activity. On the other hand, the liver 16-hydroxylase activity, characteristic of the fetus, is high, and, as a consequence, 16-OH or 16-oxo metabolites are predominant in the urine. Several recent papers have identified many urinary steroids from normal babies using gas chromatography–mass spectrometry (GC–MS) (Horning, Chambaz, Brooks, Moss, Boucher, Horning, and Hill, 1969; Shackleton, Gustafsson, and Sjövall, 1971).

The techniques developed in our laboratory for the routine quantitative assay of urinary steroids from adults were applied to the verification of an enzymatic deficiency in congenital adrenal hyperplasia syndrome. For this purpose, we limited our study to the precise quantification of some C_{19} and C_{21} urinary metabolites, which could be accurately identified by GC–MS.

In this chapter we present evidence concerning the search for 24 steroids and the quantification of those that are present in the urine of four babies suffering congenital adrenal hyperplasia. These results were compared to the normal excretion pattern established from 10 normal babies. The possible application to the detection of such hyperplasia in newborn infants is discussed.

II. MATERIAL AND METHODS

A. Reference Steroids

Reference steroids purchased from Ikapharm (Ramat-Gan, Israel) are listed in Table 1 with trivial names used in the text and abbreviation and numbers used on the chromatograms.

TABLE 1. *Systematic names of studied steroids*

Systematic names	Trivial names	Numbers on the chromatograms
3α-hydroxy-5α-androstan-17-one	Androsterone	1
3α-hydroxy-5β-androstan-17-one	Etiocholanolone	2
3β-hydroxy-5-androsten-17-one	Dehydroepiandrosterone	3
3α-hydroxy-5α-androstan-11,17-dione	11-Oxoandrosterone	4
3α-hydroxy-5β-androstan-11,17-dione	11-Oxoetiocholanolone	5
$3\alpha,11\beta$-dihydroxy-5α-androstan-17-one	11β-Hydroxyandrosterone	6
$3\alpha,11\beta$-dihydroxy-5β-androstan-17-one	11β-Hydroxyetiocholanolone	7
5-androstene-$3\beta,17\beta$-diol	Androstenediol	13
$3\beta,16\alpha$-dihydroxy-5-androsten-17-one	16α-Hydroxydehydroepi-androsterone	16
$3\beta,16\beta$-dihydroxy-5-androsten-17-one	16β-Hydroxydehydroepi-androsterone	16'
$3\beta,17\beta$-dihydroxy-5-androsten-16-one	16-Oxoandrostenediol	17
5-androstene-$3\beta,16\alpha,17\beta$-triol	Androstenetriol-16α	20
5-androstene-$3\beta,16\beta,17\beta$-triol	Androstenetriol-16β	20'
$3\alpha,17\alpha$-dihydroxy-5β-pregnan-20-one	17α-Hydroxypregnanolone	14
5α-pregnane-$3\alpha,20\alpha$-diol	Allopregnanediol	8
5β-pregnane-$3\alpha,20\alpha$-diol	Pregnanediol	9
5β-pregnane-$3\alpha,17\alpha,20\beta$-triol	Pregnanetriol-20β	10
5β-pregnane-$3\alpha,17\alpha,20\alpha$-triol	Pregnanetriol-20α	11
$3\alpha,17\alpha,20\alpha$-trihydroxy-$5\beta$-pregnan-11-one	Pregnanetriolone	22
3β-hydroxy-5-pregnen-20-one	Pregnenolone	15
$3\beta,11\alpha$-dihydroxy-5-pregnen-20-one	11α-Hydroxypregnenolone	23
$3\beta,16\alpha$-dihydroxy-5-pregnen-20-one	16α-Hydroxypregnenolone	19
5-pregnene-$3\beta,17\alpha,20\alpha$-triol	Pregnenetriol-17α	21
5-pregnene-$3\beta,16\alpha,20\alpha$-triol	Pregnenetriol-16α	21'
5β-cholestan-3α-ol	Epicoprostanol	STD

B. Case Histories

Normal newborn infants were 4 to 15 days old.

The case histories of four babies suffering congenital adrenal hyperplasia are summarized in Table 2.

TABLE 2. *Clinical history of babies with congenital adrenal hyperplasia syndrome*

Patient	Sex	Birth date	Age at clinical diagnosis (in days)	Clinical observation	Salt-losing syndrome	Family history
BUT. V N. 28.72	M	29.12.71	15	Macrogeni tosomia, important hyper-pigmentation	+++	1st child
BOU. C N. 143.72	M	03.02.72	12	Slight hyperpig-mentation	++	1 healthy brother
NAR. O. N. 773.72	M	25.08.72	9	idem	++	1 healthy brother
DEM. S. N. 787.72	F	12.09.72	5	Ambiguous genitalia, Female pseudo-hermaphroditism, Type 3 of Prader	+	1st child

C. Analytical Methods

The levels of 17-oxosteroids from Zimmerman reaction and 17-hydroxysteroids from Porter and Silber reaction were routinely assayed on 24-hr urine samples.

Plasma cortisol was quantitated by competitive protein binding (Pham-Huu-Trung, 1970).

The GC procedure was carried out according to the technique previously described (Desgrès, Bègue, and Padieu, 1973) and summarized in Table 3 with the following modifications: in normal infants $1/3$ of the 24-hr urinary samples was assayed whereas $1/10$ was used with the deficient infants and the steroids were analyzed as trimethylsilyl derivatives (TMS) using pyridine as solvent instead of chloroform in order to silylate completely $3\alpha,17\alpha$-dihydroxy-5β-pregnan-20-one.

III. RESULTS

A. Qualitative Study

Steroids were considered as identified when they had the same methylene unit (MU) (Horning, Ikekawa, Chambaz, Jaakonmaki, and Brooks 1967) and mass spectra as the reference standard on OV-I and OV-17 stationary phases. The characterized metabolites in both urines are listed in Table 4. In normal infants only a few steroids were found in reasonable amount, whereas 16 of the 24 tested steroids could be quantitated in congenital adrenal hyperplasia. The main steroids of biological interest in newborn infants could be separated on an OV-I column (Fig. 1).

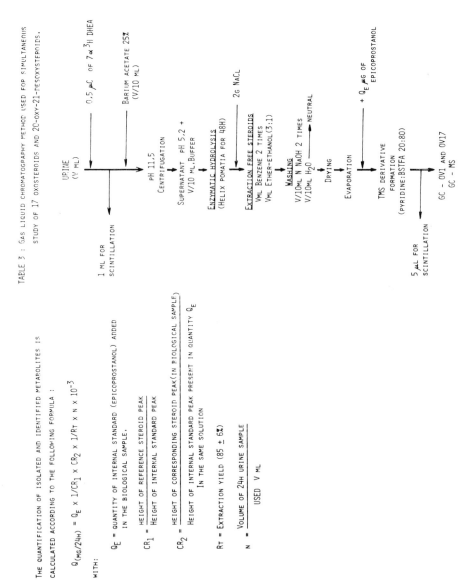

TABLE 3 : GAS LIQUID CHROMATOGRAPHY METHOD USED FOR SIMULTANEOUS STUDY OF 17 OXOSTEROIDS AND 20-OXY-21-DEOXYSTEROIDS.

THE QUANTIFICATION OF ISOLATED AND IDENTIFIED METABOLITES IS CALCULATED ACCORDING TO THE FOLLOWING FORMULA :

$$Q_{(MG/24H)} = Q_E \times 1/CR_1 \times CR_2 \times 1/R_T \times N \times 10^{-3}$$

WITH:

Q_E = QUANTITY OF INTERNAL STANDARD (EPICOPROSTANOL) ADDED IN THE BIOLOGICAL SAMPLE.

CR_1 = $\dfrac{\text{HEIGHT OF REFERENCE STEROID PEAK}}{\text{HEIGHT OF INTERNAL STANDARD PEAK}}$

CR_2 = $\dfrac{\text{HEIGHT OF CORRESPONDING STEROID PEAK(IN BIOLOGICAL SAMPLE)}}{\text{HEIGHT OF INTERNAL STANDARD PEAK PRESENT IN QUANTITY } Q_E}$
IN THE SAME SOLUTION

R_T = EXTRACTION YIELD (85 ± 6%)

N = $\dfrac{\text{VOLUME OF 24H URINE SAMPLE}}{\text{USED V ML}}$

UPINE (V ML)

0.5 μC OF 7α ³H DHEA

BARIUM ACETATE 25% (V/10 ML)

1 ML FOR SCINTILLATION

PH 11.5
CENTRIFUGATION

SUPERNATANT PH 5.2 + V/10 ML BUFFER

ENZYMATIC HYDROLYSIS
(HELIX POMATIA FOR 48H)

2G NACL

EXTRACTION FREE STEROIDS
VML BENZENE 2 TIMES
VML ETHER-ETHANOL (3:1)

WASHING
V/10ML N NAOH 2 TIMES
V/10ML H₂O ⟶ NEUTRAL

DRYING

EVAPORATION

+ Q_E μG OF EPICOPROSTANOL

TMS DERIVATIVE FORMATION
(PYRIDINE:BSTFA 20:80)

5 μL FOR SCINTILLATION

GC - OV1 AND OV17
GC - MS

TABLE 3: Gas liquid chromatography method used for simultaneous study of 17 oxo-steroids and 20-oxy-21-deoxysteroids

TABLE 4. *MU values and identification number of steroids from urine of 10 normal babies and four adrenogenital syndromes*

No.	Steroids	MU (OV 1)	MU (OV 17)	Identification number from 10 normal babies	Identification number from four adreno-genital syndromes
1	Androsterone	24.35	26.42	0	4
2	Etiocholanolone	24.52	26.66	10	4
4	11-Oxoandrosterone	25.10	28.10	0	1
5	11-Oxoetiocholanolone	25.21	28.22	3	4
13	Androstenediol	25.86	26.45	10	4
6	11β-Hydroxyandrosterone	25.88	29.17	0	4
7	11β-Hydroxyetiocholanolone	26.02	29.42	1	2
16	16α-Hydroxydehydroepi-androsterone	27.00	28.85	10	4
16'	16β-Hydroxydehydroepi-androsterone	27.69	28.90	4	4
17	16-Oxoandrostenediol	27.36	29.15	10	4
20	Androstene triol-16α	28.32	28.48	10	4
14	17α-Hydroxypregnanolone	26.29	27.66	2	4
8	Allopregnanediol	27.32	27.83	0	3
19	16α-Hydroxypregnenolone	28.53	30.27	10	4
11	Pregnanetriol-20α	28.95	29.90	0	4
21	Pregnenetriol-17α	29.58	30.88	0	3
21'	Pregnenetriol-16α	29.73	29.85	10	4
22	Pregnanetriolone	30.00	31.62	0	4

Most of the steroids listed in Table 4 give well-known mass spectra. The description of mass spectra will be limited to those of more important biological interest.

1. Normal Babies

i. 3β,16α-Dihydroxy-5-pregnen-20-one. The silyl ether of the compound isolated from urine gives a mass spectrum distinguished by a base peak at m/e 129, molecular ion at m/e 476, and prominent peaks at m/e 172, 157, 461 (M-15), 386 (M-90), 371 (M-90-15), 296 (M-2 × 90), and 281 (M-2 × 90-15).

ii. 3β,16α-Dihydroxy-5-androsten-17-one. The mass spectrum of the silyl ether of this compound gives a base peak at m/e 214 (M-144-90), molecular ion at m/e 448, and prominent peaks at m/e 129 and m/e 304 (M-144).

iii. 3β,16β-Dihydroxy-5-androsten-17-one. The trimethyl silyl ether of this compound gives a mass spectrum closely similar to the previous one. The identification was ascertained as 3β,16β-dihydroxy-5-androsten-17-one

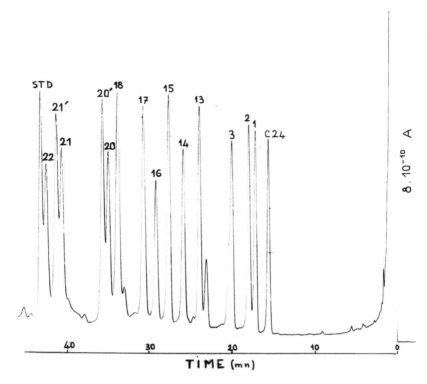

FIG. 1. Separation of reference steroids as TMS derivatives on 10-ft 1% OV-I column with temperature programming at 1.2°C/min from 180°C. The compounds are identified in Table 4.

by the MU determination and comparison to the data of Shackleton, Kelly, Adhikary, Brooks, Harkness, Sykes, and Mitchell (1968).

2. *Characteristics of Deficient Babies*

i. *3α,17α-Dihydroxy-5β-pregnan-20-one.* The mass spectrum shows a molecular ion at m/e 478 and a base peak at m/e 255 (M-2 × 90-43). Peaks were present at m/e 388 (M-90), m/e 373 (M-90-15), and m/e 298 (M-2 × 90).

ii. *3α,17α,20α-Trihydroxy-5β-pregnan-11-one.* The spectrum is characterized by the base peak at m/e 117 and the peaks m/e 494 (M⁺), m/e 287 (M-90-117), and m/e 404 (M-90).

B. Quantitative Study

Typical urinary profiles from normal infants and 21-hydroxylase deficient infants are presented in Figs. 2 and 3. The amounts of the main steroid metabolites from normal and deficient infants are compared in Table 5.

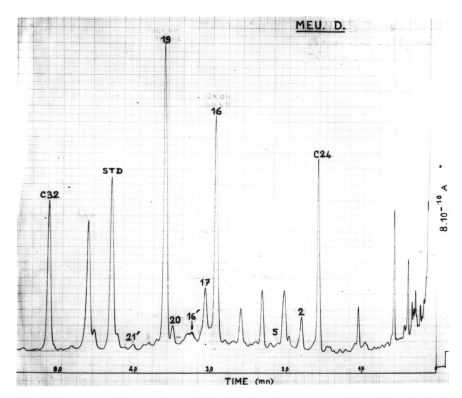

FIG. 2. Urinary steroid separation as TMS derivatives for normal newborn infant for day 6 of life. GC conditions are as for Fig. 1. The compounds are identified in Table 4.

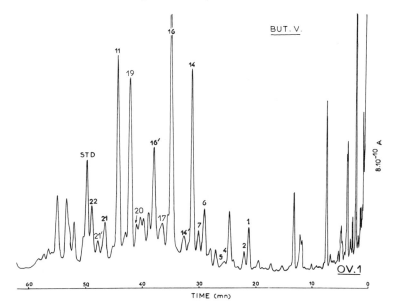

FIG. 3. Urinary steroid separation as TMS derivatives for baby with congenital adrenal hyperplasia syndrome. GC conditions as for Figs. 1 and 2. The compounds are identified in Table 4. The peak 14′ is not an identified isomer of $3\alpha,17\alpha$-dihydroxy-5β-pregnan-20-one (peak 14).

TABLE 5. *Comparison of steroid amounts between normal and four 21-hydroxylase deficient babies*

Steroids amounts (μg/24 hr)	Newborn infants (age)				
	Normal babies[a] (4–15 days)	BUT. V. (15 days)	BOU. C. (12 days)	NAR. O. (9 days)	DEM. S. (5 days)
17-Oxosteroids (Zimmermann)	430 (190–600)	5,700	1,800	5,750	2,000
17-Hydroxysteroids (Porter and Silber)	290 (150–420)	280	840	<100	<100
Plasma cortisol (μg/100 ml)	3.5	4.5	6	4	6
Androsterone	—	450	70	175	60
Etiocholanolone	35 (10–70)	80	20	60	30
11-Oxoandrosterone	—	150	—	260	65
11-Oxoetiocholanolone	15 (10–25)	60	340	175	30
Androstenediol	30 (10–70)	[b]	[b]	[b]	[b]
11β-Hydroxyandrosterone	—	950	420	475	110
11β-Hydroxyetiocholanolone	50	650	—	—	50
16α-Hydroxydehydro-epiandrosterone	670 (120–1950)	4,200	2,390	2,410	515
16β-Hydroxydehydro-epiandrosterone	165 (40–400)	1,810	520	1,600	290
16-Oxoandrostenediol	205 (60–415)	585	1,020	1,520	330
Androstenetriol-16α	130 (50–350)	740	580	675	210
17 α-Hydroxypregnanolone	25	2,700	840	600	280
Allopregnanediol	—	—	550	960	120
16α-Hydroxypregnenolone	500 (120–1080)	2,630	3,885	3,290	1,850
Pregnanetriol-20α	—	2,950	1,110	710	160
Pregnenetriol-17α	—	550	330	400	—
Pregnenetriol-16α	35 (10–85)	400	75	160	90
Pregnanetriolone	—	1,150	860	750	350

[a] The first value indicates the average of the amounts of the identified steroids. Values in parentheses indicate the extreme amounts.
[b] Not determined.

In Table 6 we show steroid levels obtained during therapy. In one case the treatment was stopped for 48 hr, and consequently an enzymatic deficient pattern was observed, proving the biological efficiency of the therapy.

IV. DISCUSSION

The main attempt of this chapter is to show the possibility of detection of congenital adrenal hyperplasia in newborn infants through the quantification of some urinary steroids.

Normal newborn infants excrete variable amounts of urinary steroids. These levels and particularly those of 16α-hydroxydehydroepiandrosterone, 16β-hydroxydehydroepiandrosterone, 16-oxoandrostenediol, and 16α-hydroxypregnenolone are much higher in congenital adrenal hyperplasia.

In these enzymatic deficient babies, an excretion of 17-oxosteroids

TABLE 6. *Effect of therapeutic treatment on the main urinary steroids on the newborn infant BUT. V*

Steroid amounts[a]	Before treatment (15 days)	During treatment (34 days)	During interruption of treatment for 48 hr (36 days)
Androsterone	450	20	40
Etiocholanolone	80	10	20
11β-Hydroxyandrosterone	950	100	280
17α-Hydroxypregnanolone	2,700	160	2,300
16α-Hydroxydehydroepiandrosterone	4,200	30	585
16β-Hydroxydehydroepiandrosterone	1,810	55	1,380
16α-Hydroxypregnenolone	2,630	30	1,750
Pregnanetriol-20α	2,950	170	3,000
Pregnanetriolone	1,150	50	850
16-Oxoandrostenediol	585	—	300
Pregnenetriol-17α	550	traces	240
Pregnenetriol-16α	400	traces	190

[a] Given in micrograms per 24 hr.
Age of the infant is indicated in parentheses.

characteristic of the adult occurs. The predominance of 5α isomers and the ratio of 11-hydroxy to 11-deoxy-17-oxosteroids, which appears to be higher than 1, are abnormal, but confirm the previous results obtained with older deficient children (Makin, 1970; Clayton, Edwards, and Makin, 1971; Bègue, Desgres, Gustafsson, and Padieu, *in preparation*). This shows that the deficient enzyme is the 21-hydroxylase.

As a consequence of 21-hydroxylase deficiency, plasma cortisol is lowered and the corticoadrenal activity stimulated. Thus urinary 17α-hydroxyprogesterone catabolites (particularly 17α-hydroxypregnanolone, pregnanetriol-20α, and pregnanetriolone) are excreted at high levels. These catabolites are not detected, or are at trace levels in urine from normal infants. From these results, we may suggest that the 21-hydroxylase deficiency may stimulate the 17α-hydroxylase activity leading to the biosynthesis of androgens and saturated 17α-hydroxylated pregnanesteroids.

The levels of steroids studied in deficient babies present important variations, and it appears difficult with this small sample to correlate clinical signs with the biological observations. However, the deficiency appears much more pronounced for the babies suffering dehydration and important losses of salt, than for those showing simple virilism. This could suggest a variable inhibition of 21-hydroxylase activity (Migeon and Beaulieu, 1965) or support the hypothesis of two distinct 21-hydroxylases (Gustafsson, Gustafsson, and Olin, 1972), one involved in the biosynthesis of aldosterone and the other one in cortisol biosynthesis.

The levels found in treated babies show a regulation of steroids metabolism.

Our data on plasma cortisol are not significant enough to be interpreted but are in agreement with previous studies (Migeon and Beaulieu, 1965), which show the difficulty of correlating blood aldosterone and cortisol levels with the clinical syndrome. An assay of plasma 17α-hydroxyprogesterone may solve this problem.

However, our results prove that our simple and quick procedure can be applied to urinary samples from 5- to 15-day-old babies to confirm clinical diagnosis. Because fetal and maternal imprinting may influence the urinary steroid profile in the first days of life (Horning, Hung, Hill, and Horning, 1971), we are now investigating the urinary profile of babies just after birth in order to define the minimal age from which it is possible to detect this enzymatic deficiency. Early detection could be very useful when one child in a family is already affected by this congenital disease.

ACKNOWLEDGMENT

Financial help is gratefully acknowledged through research grants from: Délégation Générale de la Recherche Scientifique et Technique, Action Complémentaire Coordonnée Développement périnatal; Centre National de la Recherche Scientifique, ERA 267; Enseignement Supérieur Vème et VIème Plan; Fondation pour la Recherche Médicale Française; Centre Hospitalier Régional de Dijon.

REFERENCES

Bègue, R. J., Desgrès, J., Gustafsson, J. A., and Padieu, P. *In preparation:* Application de la chromatographie gaz liquide et de la chromatographie gaz liquide–spectrométrie de masse à l'analyse qualitative et quantitative des stéroides neutres urinaires dans les hyperplasies surrénales congénitales par déficit en 21-hydroxylase.

Clayton, B. E., Edwards, R. W. H., and Makin, H. L. J. (1971): Congenital adrenal hyperplasia and other conditions associated with a raised urinary steroid 11-oxygenation index. *Journal of Endocrinology,* 50:251–265.

Desgrès, J., Bègue, R. J., and Padieu, P. (1973): Analyse qualitative et quantitative des 17-oxostéroides et des 20-hydroxy-21-deoxysteroides urinaires par chromatographie gaz liquide et par spectrométrie de masse. *Clinica Chimica Acta (in press).*

Gustafsson, J. A., Gustafsson, S., and Olin, P. (1972): Steroid excretion patterns in urine from two boys in the neonatal period with congenital adrenal hyperplasia due to 21-hydroxylase deficiency. *Acta Endocrinologica,* 71:353–364.

Horning, E. C., Horning, M. G., Ikekawa, N. I., Chambaz, E. M., Jaakonmaki, P. J., and Brooks, C. J. W. (1967): Studies of analytical separations of human steroids and steroid glucuronides. *Journal of Gas Chromatography,* 5:283–292.

Horning, M. G., Chambaz, E. C., Brooks, C. J. W., Moss, A. M., Boucher, E. A., Horning, E. C., and Hill, R. M. (1969): Characterization and estimation of urinary steroids of newborn human by gas-phase analytical methods. *Analytical Biochemistry,* 31:512–531.

Horning, M. G., Hung, A., Hill, R. M., and Horning, E. C. (1971): Variations in urinary steroid profiles after birth. *Clinica Chimica Acta,* 34:261–268.

Makin, H. L. J. (1970): The gas-liquid chromatography of steroid formates: An application in congenital adrenal hyperplasia. *Journal of Endocrinology,* 47:55–64.

Migeon, C. J., and Beaulieu, E. E. (1965): Hyperplasie congénitale des surrénales. Etude biologique. *Rapport de 7ème Réunion des Endocrinologistes de Langue Française*, pp. 153–174. Masson Ed., Paris.

Pham-Huu-Trung, M. T. (1970): Dosage du cortisol plasmatique par liaison compétitive à la transcortine. *Annales de Biologie Clinique*, 28:145–152.

Shackleton, C. H. L., Gustafsson, J. A., and Sjövall, J. (1971): Steroids in new-borns and infants. Identification of steroids in urine from new-born infants. *Steroids*, 17:265–280.

Shackleton, C. H. L., Kelly, R. W., Adhikary, P. M., Brooks, C. J. W., Harkness, R. A., Sykes, P. J., and Mitchell, F. L. (1968): The identification and measurement of a new steroid 16β-hydroxy-dehydroepiandrosterone in infant urine. *Steroids*, 12:705–716.

Mass Spectrometry in Biochemistry and Medicine,
edited by A. Frigerio and N. Castagnoli.
Raven Press, New York © 1974

Determination of Tetrahydrocannabinol in Blood Plasma of Cannabis Smokers

S. Agurell, B. Gustafsson, B. Holmstedt, K. Leander,
J.-E. Lindgren, I. Nilsson, and M. Åsberg

Central Military Pharmacy, Karolinska Hospital, S-104 01 Stockholm 60, Sweden

Since a full account of this method, including synthetic procedures, will appear elsewhere (Agurell, Gustafsson, Holmstedt, Leander, Lindgren, Nilsson, Åsberg, 1973), the following account summarizes only the essential details. A technique for the identification and quantitative determination of nonlabeled Δ^1-tetrahydrocannabinol (Δ^1-THC) and/or its metabolites in blood, urine, and other body fluids of persons taking *Cannabis* has long been desired both for pharmacokinetic and forensic purposes. Studies with [^{14}C]-labeled Δ^1-THC in humans indicate that Δ^1-THC levels of 25 to 50 ng/ml plasma may be reached during *Cannabis* smoking (Galanter, Wyatt, Lemberger, Weingartner, Waughan, and Roth, 1972). However, so far no gas-liquid chromatography (GLC) or other specific procedures have appeared in the literature. Mass fragmentography (MF) has found an increasing use in the determination of small amounts of drugs and endogenous compounds (Borgå, Palmér, Sjöqvist, and Holmstedt, 1972).

I. EXPERIMENTAL

To a sample of human plasma (5.0 ml) deuterated Δ^1-THC (1,000 ng) (Δ^1-THC-d_2) was added as internal standard. The plasma was extracted four times with an equal amount of light petroleum containing 1.5% isoamyl alcohol in a glass-stoppered tube.

After centrifugation, the light petroleum layers were drawn off and evaporated to dryness under a stream of nitrogen. The residue was chromatographed on a Sephadex LH-20 column (1 × 40 cm — void volume 15 ml) using light petroleum:chloroform:ethanol (10:10:1) as eluent (0.15 to 0.18 ml/min). The fraction corresponding to the elution volume of Δ^1-THC (28 to 34 ml) was collected and evaporated to dryness under nitrogen. The residue was transferred to a conic vial (0.3 ml) and dissolved in absolute ethanol (25 to 50 μl).

This solution was subjected to MF (LKB 9000, 3% OV-17/Gas Chrom Q 100-120 mesh, 230°C) using a flexible gas chromatography–mass spectrometry–laboratory computer system for on-line data collection and processing. A program to sample simultaneously four different masses and record the resulting data is part of this system (Elkin, Pierrou, Ahlborg, Holmstedt, and Lindgren, 1973).

A standard curve was prepared by adding known amounts (0, 10, 20, etc., ng/ml) of Δ^1-THC to blank plasma samples and carrying out the described procedure. The standard curve was made by plotting peak height Δ^1-THC (m/e 299)/peak height Δ^1-THC-d_2 (m/e 301) against known amounts of added Δ^1-THC in ng/ml of plasma.

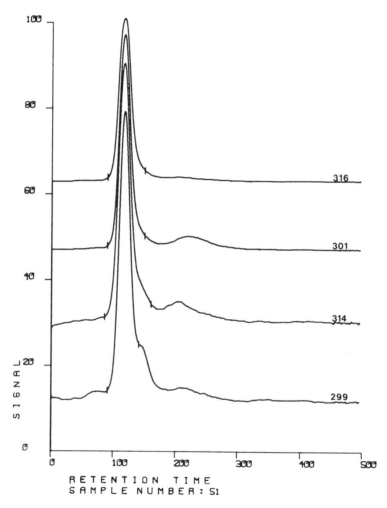

FIG. 1. Mass fragmentogram of internal standard (m/e 301, 316) and Δ^1-THC (m/e 299, 314) in purified extract of a plasma sample taken 10 min after smoking 10 mg Δ^1-THC. Plasma level of Δ^1-THC 26 ng/ml; retention time of Δ^1-THC 2.9 min.

II. RESULTS AND DISCUSSION

After addition of deuterated Δ^1-THC as internal standard to the plasma sample, the present procedure involves an initial extraction of Δ^1-THC followed by separation on Sephadex LH-20 and MF. The mixture of Δ^1-THC and Δ^1-THC-d_2 is analyzed by MF using the ratio: peak height m/e 299 (from Δ^1-THC) versus m/e 301 (from Δ^1-THC-d_2). The molecular peaks of nonlabeled (m/e 314) and labeled (m/e 316) Δ^1-THC are also recorded (Fig. 1).

The specificity in determining Δ^1-THC is also ensured by the GLC retention time and may, if necessary, be improved by the registration of further fragments in the mass spectrum of Δ^1-THC. We have so far in a limited number of control samples encountered no background interference at the retention time of Δ^1-THC when recording the selected four mass numbers. The method can be used at least down to 1 ng Δ^1-THC per ml plasma.

Δ^1-THC is readily identified in blood plasma after a cigarette containing 10 mg Δ^1-THC is smoked. Maximum levels of 20 to 25 ng/ml are reached within 10 min but decline rapidly to less than 5 ng/ml within 2 hr. The present method, which apparently is the first method to identify and accurately determine Δ^1-THC after smoking, will be used for pharmacokinetic studies in man but may also serve forensic purposes.

REFERENCES

Agurell, S., Gustafsson, B., Holmstedt, B., Leander, K., Lindgren, J.-E., Nilsson, I., Sandberg, F., and Asberg, M. (1973): Quantitation of Δ^1-tetrahydrocannabinol in plasma from cannabis smokers. *Journal of Pharmacy and Pharmacology*, 25:554–558.

Borgå, O., Palmér, L., Sjöqvist, F., and Holmstedt, B. (1972): Mass fragmentography as a means of identification and quantification of drugs in biological fluids. In: *Symposium on the Basis of Drug Therapy in Man*, Fifth International Pharmacology Congress, San Francisco. Vol. 3, C/9. Karger, S., Basel.

Elkin, K., Pierrou, L., Ahlborg, U. G., Holmstedt, B., and Lindgren, J.-E. (1973): Computer-controlled mass fragmentography with digital signal processing. *Journal of Chromatography*, 81:47–55.

Galanter, M., Wyatt, R. J., Lemberger, L., Weingartner, H., Waughan, T. B., and Roth, W. T. (1972): Effects on humans of Δ^9-tetrahydrocannabinol administered by smoking. *Science*, 176:934–935.

Mass Spectrometry in Biochemistry and Medicine,
edited by A. Frigerio and N. Castagnoli.
Raven Press, New York © 1974

Gas Chromatography–Mass Spectrometry Applied to the Study of Urinary Acids in a Patient with Periodic Catatonia

C. R. Lee and R. J. Pollitt

M.R.C. Unit for Metabolic Studies in Psychiatry, Middlewood Hospital, P.O. Box 134, Sheffield, S6 1TP, U.K.

Evidence of disordered metabolism in patients with various forms of mental illness has been sought for many years. This search has given rise to modest advances, but many of the major diseases are still poorly understood. Each improvement in methodology has prompted fresh studies, and our laboratory, among others, is engaged in the application of gas chromatography–mass spectrometry (GC–MS) in this field.

I. PROFILE TECHNIQUES AND THEIR APPLICATION

Using GC methods, it is relatively easy to obtain complex "metabolic profiles" of steroids, organic acids, volatile compounds, etc., from various biological materials (Horning and Horning, 1971; Teranishi, Mon, Robinson, Cary, and Pauling, 1972). Each profile consists of a GC trace containing possibly more than 100 recognizable peaks, many more components being present in minor quantities or poorly resolved from the major constituents. With the aid of MS and other methods, it is usually possible to identify the compound responsible for any particular peak so that abnormalities in many areas of metabolism can be detected and characterized. These techniques have already proved successful in identifying new inborn errors of metabolism (for example, see Jellum, Stokke, and Eldjarn, 1971), where the changes are large, but it may be difficult to detect variations from normal in the minor components of many of these profiles. Indeed, the effects of individual constitution, bacterial flora, and diet make it difficult to define normality except in very gross terms. For these reasons we are concentrating on situations where single individuals can be studied longitudinally with careful attention to dietary control, medication, etc. One of the conditions in which this is possible is periodic catatonia.

Regularly occurring catatonic stupor or catatonic excitement is nowadays a rare condition, although it was apparently more common before the advent of modern psychopharmacology and in the more restricted environment of the mental institutions of previous times. It has been studied by a number of workers and in particular detail by R. Gjessing (1960, and previous papers) and L. R. Gjessing (Takahashi and Gjessing, 1972, and previous papers). In some patients there are regular changes in water, electrolyte, and nitrogen balance in the course of the illness, but the underlying mechanism is obscure. We have been studying a 76-year-old woman with a 40-year history of periodic catatonia in collaboration with the Medical Research Council Brain Metabolism Unit in Edinburgh. This patient has an interval period of about 40 days followed, after a period of increased irritability, by a stuporose period of about 30 days. During the study, a repeating daily diet was given and any refusals carefully noted. Medication was restricted to 200-mg secobarbital daily except for two short periods of penicillin treatment. All urine was deep frozen soon after voiding, and collections were continued over three complete cycles, a period of nearly 8 months.

For GC, acids were extracted using a column of Dowex 1 × 8 resin in the formate form eluting with 12 M formic acid, and converted to their trimethylsilyl derivatives or trimethylsilylmethoxime derivatives. The extracts were examined on 9-ft columns of OV-101, OV-17, OV-25, and OV-225, in each case programming from 70 to 270°C at 2°C/min. Evaluation of the profiles proved quite difficult as a number of the peaks varied apparently at random. Furthermore, although we were expecting to see consistent differences between the stuporose phases and the intervals, we did not at this stage find any. It did, however, become evident that some of the constituents were changing sharply at the beginning of each stuporose period. Four potentially interesting peaks were identified as due to the trimethylsilyl derivatives of resorcinol (*m*-dihydroxybenzene), malic acid, tartaric acid, and 3,4-dihydroxybutyric acid. This last compound has not previously been reported in urine, and was identified by comparison with the published mass spectrum (Petersson, 1970). A sample of the authentic acid cochromatographed with the urinary compound.

Because of the complexity of the chromatographic trace and the presence of overlapping constituents, particularly in the case of malate, it was not possible to measure these compounds accurately directly from the trace. Therefore, as a preliminary exercise, we measured the excretions of malate, tartrate, and dihydroxybutyrate by the internal isotopic standard method.

II. INSTRUMENTATION

A Perkin-Elmer 270 GC–MS was used, initially in the repetitive scan mode without modification (Lee and Pollitt, 1972). In this mode the magnet current is returned to zero at the end of the scanning period before being reset to the original start scan value. On repetitive scan over short mass range the resetting time exceeds the scanning time, and with the unmodified instrument the scanning speed is variable only in coarse steps. Using a scan speed of 10 sec/decade, a repeat scan time of 2.7 sec was obtained for 20 a.m.u., giving 15 to 20 scans over the major portion of a GC peak. In the absence of noise, 15-point readings over a Gaussian peak give a theoretical error of less than 0.5% on integration (Dymond and Kilburn, 1967). In practice, a standard deviation of 3% was obtained for tartaric acid standards, injecting between 70 to 700 pmoles (as the trimethylsilyl derivatives) for each determination.

Part way through this study we improved the system by means of simple modifications. The signal from a ramp generator was connected, via a blocking diode, across a low resistance in series with the earthy end of the "scan start" potentiometer. This potentiometer applies a variable stabilized voltage to the DC feedback amplifier which drives the magnet. A DC offset control on the ramp generator is used to forward bias the diode, so that the generator signal (of about 1V. peak-peak) modulates the standing magnet current determined by the "scan start" setting. The internal scan generator is disconnected by a switch. During setting up, the magnet current was monitored on an oscilloscope; the response was satisfactory up to about 20 Hz, apart from ringing during flyback. The oscilloscope was then connected to the electron multiplier amplifier, a signal from the ramp generator being used to drive the X-plates. With this system, it is possible to reduce the range scanned to a few a.m.u., but the first third of the scan is slightly distorted due to hysteresis effects in the magnet. The range covered by the scans was selected using the normal scan start position control, which can easily be reset to cover a range of compounds for each GC run. The scan start setting may drift somewhat but this can be adjusted during the early part of a peak, using the oscilloscope to monitor the signal which is simultaneously recorded. This system, with a repetition time of 2 sec, enables a passband of less than 10 Hz (as opposed to 100 Hz in the unmodified system) to be used without significant attenuation, giving approximately 10-fold increased sensitivity.

III. QUANTITATION OF METABOLITES

For these determinations, deuterated malate (d_3) and tartrate (d_2) were prepared by reduction of oxaloacetate and dihydroxyfumarate respectively with sodium amalgam in D_2O and 3,4-dihydroxybutyrate (2-d_2) by hydrolyzing 3,4-dihydroxybutyronitrile in $NaOD/D_2O$. Small amounts of these compounds were added from a standard solution to each urine sample. The acids were extracted and derivatized as before and examined by the repetitive scan method. In each case the M-15 peaks were used to compare the ratio of deuterated standard to natural compound. Most of these runs were performed on OV-101 columns but a number of them were repeated on

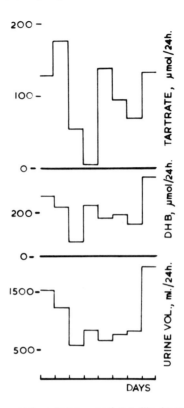

FIG. 1. Daily urine volume and D,L-tartrate and 3,4-dihydroxybutyrate excretion over a switch period.

OV-17 with very similar results, indicating that interference from other urinary components was insignificant.

The results from a short portion of one of the cycles are shown in Fig. 1. The drop in tartrate excretion from an average value of 100 μmoles/24 hr to 4 μmoles for a single 24-hr period is particularly striking, and similar falls of 1 to 3 days duration occurred at the beginning of each stuporose period, and once also shortly after the stupor had ended. The fall in 3,4-dihydroxybutyrate excretion was smaller and did not occur with exactly the same timing as the fall in tartrate excretion. The malate excretion changed irregularly and increasingly out of step with the mood changes as the study progressed. Detailed discussion of these results is not appropriate here; it is sufficient to say that, striking as these changes are, their relationship with, even relevance to, the mental changes is still unclear.

IV. GENERAL DISCUSSION

A. Clinical Aspects

The technique of studying single patients in a longitudinal manner is potentially a very powerful one in this field, particularly as many of the previous attempts at comparing a mentally ill population with a normal control population have resulted in spectacular failure. The apparent simplicity of the longitudinal study is deceptive, however. The most obvious example is the matter of dietary control. Assume that complete dietary standardization is possible (which in many psychiatric patients it is not) and the patient is kept on a constant intake throughout the whole of the experimental period. R. Gjessing has shown that the metabolic rate increases in the irritable phase just prior to stupor, and raised basal metabolism may be present well into the stuporose phase. If the patient is in long-term calorie balance, as should be the case if the diet has been correctly chosen, there will probably be a switch to the consumption of depot fat in the period of raised metabolism, and a compensatory increase in fat synthesis in the interval period, when the metabolic rate falls. R. Gjessing showed changes in respiratory quotient associated with these phenomena. The change from lipogenesis to lipolysis is accompanied by quite marked changes in the intracellular concentrations of several metabolites, such as oxaloacetate, and it is reasonable to expect some of these changes to be reflected in urinary excretion. Their relationship to the basic disease process is, however, very indirect. Other factors, such as the influence of mental state and environmental temperature and humidity on sweating, and hence on urine volume, and the difficulty of maintaining patients, particularly older ones, for longer periods without changes in medication, all serve to add uncertainty to this type of study. In addition, patients with regularly recurring psychosis are relatively rare. Nevertheless, such conditions do offer the best prospects for detailed metabolic study at this stage.

B. The Use of GC–MS

The necessity to use GC–MS techniques for quantitation in this type of study might be questioned. The isotope method is more time-consuming than straight GC, as it requires constant operator attention during the scans, but the rather tedious sample preparation is the same in each case. In addition, with the degree of GC resolution we were using (7,000 theoretical plates/ column), an accurate estimation of peak area, particularly for the lower levels of excretion, was just not possible. Enzymatic methods are available for malate in urine, but for a mixture such as urine, complete specificity may be doubtful at low malate levels (Guilbault, Sadar, and McQueen, 1969) and, once a GC–MS run is contemplated, the addition of extra compounds involves less work than setting up a separate enzyme assay. In this type of extensive study, where the area of metabolism of interest is not predictable, the GC–MS provides not only a method of identification but also a whole range of *ad hoc* quantitative methods without which the approach would be much less feasible. Often the preparation of a deuterated standard is easy enough to justify carrying out prospective studies. For example, when we suspected that glutarate excretion was increasing in patients taking lithium salts, we also added standards of methylmalonate (d_3), succinate (d_4), adipate (d_4), pimelate (d_4), and suberate (d_4) in addition to the glutarate (d_4), and were able to estimate all these acids and fumarate in a single GC–MS run requiring 40 min (Lee and Pollitt, 1973). The repetitive scan method can be applied to a range of compounds of very different molecular weights in a single run. We are currently extending our range of standards to include the major Krebs cycle metabolites. This method can be used on a completely unmodified instrument, but even the more complicated set-up involves only a fraction of the cost of an accelerating voltage alternator or similar peak-switching device, although it may be less sensitive. Very recently, Baczynski, Duchamp, Zieserl, and Axen (1973) described a repetitive scan method using computer acquisition, which is sensitive down to the sub-nanogram range.

The general approach of using GC–MS in this type of study has the great advantage that it is not dependent on any preconceived theory. As in the study just presented, it can reveal changes in areas of metabolism not normally considered in connection with psychiatric disorders.

ACKNOWLEDGMENTS

We are grateful to Dr. L. G. Murray and the staff of the M.R.C. Brain Metabolism Unit, Edinburgh, for organizing the study of the patient with periodic catatonia, which will eventually form part of a joint publication, and to Professor G. Ashcroft (Edinburgh) and Professor F. A. Jenner (Sheffield) for initiating the project. We thank Dr. B. Alfredsson (Göteborg) for the gift of an authentic sample of 3,4-dihydroxybutyric acid.

REFERENCES

Baczynski, L., Duchamp, P. J., Zieserl, J. F., and Axen, U. (1973): Computerized quantitation of drugs by gas chromatography–mass spectrometry. *Analytical Chemistry,* 45:479–482.

Dymond, H. F., and Kilburn, K. D. (1967): Characterization of tobacco smoke by gas chromatography and a digital computer. *Gas Chromatography,* 6:353–375.

Gjessing, R. (1960): Beiträge zur Somatologie der periodischen Katatonie. X. Mitteilung. Pathogenetische Erwägungen. *Archiv für Psychiatrie und Zeitschrift Neurologie,* 200:366–389.

Guilbault, G. G., Sadar, S. H., and McQueen, R. (1969): A fluorimetric enzymic method for the assay of mixtures of organic acids. *Analytica Chimica Acta,* 45:1–12.

Horning, E. C., and Horning, M. G. (1971): Human metabolic profiles obtained by GC and GC/MS. *Journal of Chromatographic Science*, 9:129–140.

Jellum, E., Stokke, O., and Eldjarn, L. (1971): Screening for metabolic disorders using gas-liquid chromatography, mass spectrometry and computer techniques. *Scandinavian Journal of Clinical and Laboratory Investigation*, 27:273–285.

Lee, C. R., and Pollitt, R. J. (1972): The estimation of glutarate in urine using gas chromatography–mass spectrometry with an internal isotopic standard. *Biochemical Medicine*, 6:536–542.

Lee, C. R., and Pollitt, R. J. (1973): The effect of lithium salts on the urinary excretion of some dicarboxylic acids. *Biochemical Society Transactions*, 1:108–109.

Petersson, G. (1970): Mass spectrometry of aldonic and deoxyaldonic acids as trimethylsilyl derivatives. *Tetrahedron*, 26:3413–3428.

Takahashi, S., and Gjessing, L. R. (1972): Studies of periodic catatonia. VI. Longitudinal study of catecholamine metabolism, with and without drugs. *Journal of Psychiatric Research*, 9:293–314.

Teranishi, R., Mon, T. R., Robinson, A. B., Cary, P., and Pauling, L. (1972): Gas chromatography of volatiles from breath and urine. *Analytical Chemistry*, 44:18–20.

Mass Spectrometry in Biochemistry and Medicine,
edited by A. Frigerio and N. Castagnoli.
Raven Press, New York © 1974

An Interactive Display-Oriented Data System for Gas Chromatography–Mass Spectrometry

Victor L. DaGragnano and H. P. Hotz

Finnigan Corporation, Sunnyvale, California 94086

The use of small computers to control the data acquisition with gas chromatograph-mass spectrometers (GC–MS) has become common. However, these systems have been rather inflexible in their handling of data. Data have had to be plotted or listed in order to be interpreted, a process that has required considerable time. We have designed a fully integrated data system which provides live display of the spectra on an oscilloscope. Since the data displayed are resident in the computer's memory, it is straight forward to provide data manipulation directly. To avoid tedious diaglogues via teletype, our computer functions are controlled principally by push buttons. The push buttons are read by the computer, which then executes the desired function. Outputs appear as alphanumerics on the oscilloscope or, optionally, typed on the teletype.

The organization of the data system and its relation to the GC–MS is shown in Fig. 1. The data system controls a setting of the quadrupole mass spectrometer by means of mass-set voltage produced by a digital to analogue

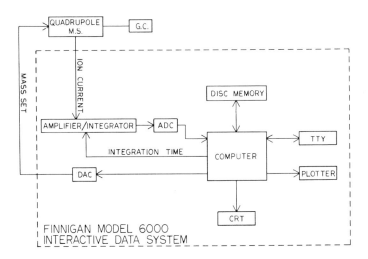

FIG. 1. Block diagram of the data system.

convertor (DAC). Data are acquired by taking the output of the electron multiplier through a preamp and amplifier/integrator to an analogue to digital convertor (ADC). These digitized amplitudes are stored in the computer's memory in the form of counts. Thus a mass spectrum consists of a series of counts for each channel, with each channel representing one atomic mass unit. The integration time is determined by the computer and is timed by the computer's real-time clock. Live display of spectra is maintained on the cathode ray tube (CRT) during acquisition. Past spectra and additional programs are stored in the disc memory. Hard copy data output may either be obtained on the teletype or on the digital plotter.

Core memory provides space for 2,048 channels of data. During acquisition, this is divided into two halves, one for the gas chromatogram and one for the mass spectrum. Halves of these regions may be displayed at will and small segments may be chosen to be expanded across the entire CRT screen. Any two sections may be displayed simultaneously. After acquisition, one region of memory may be transferred to or stripped from any other region of memory. Thus, data can be manipulated quite freely and flexibly between the core memory divisions. This flexible handling of data in core memory provides the basis for all data handling in the system.

For data acquisition, one enters the parameters for the sample run into the system. Since a number of parameters must be entered, and alphanumeric information must be supplied, this is done by a means of a dialogue. Promptings are supplied by alphanumerics on the CRT and responses are provided by the operator on the teletype. An example of the information for a run is shown in Fig. 2. In general, the parts of the printing prior to the colon are the parts supplied by the computer and the parts following the colon have to be supplied by the operator. Once entered, the parameters remain in the machine and the operator only has to change the parameters he wishes. Figure 2 is an actual printout by the computer of a set of current parameters for a run. Once the parameters have been entered, the computer will take control of the mass spectrometer and execute sequential mass scans recording the data as obtained. Upon the completion of each mass scan, the data are compacted and stored on the disc. The mass scan is integrated and one channel is added to the gas chromatogram in the appropriate region of core

```
FILE :UNC
TITLE:UNIV. OF NORTH CAROLINA SAMPLE BY GC-EI
MASS RANGE :30-199;200-400
INTEG. TIME:4,8
SECONDS PER SCAN :4
THRESHOLD:1
INST. RANGE SETTING:H
MAX. RUN TIME 66
```

FIG. 2. Example of run parameters listing.

memory. One may select the live display region during acquisition just as at other times. Thus one can watch for a desired feature in a mass spectrum or the development of the gas chromatogram during acquisition. Two buttons are provided to control data acquisition. One is for acquire only and the data are not stored on the disc. One may then change to an acquire and save mode by pushing a second button. Mass spectra will then be saved beginning with the one currently being acquired. The computer will not allow acquisition of data with a file name identical to that of data previously acquired. Thus there is no danger of confusing data in the disc files.

A second data acquisition mode is provided, which we call the mass fragmentography mode. In this mode, complete mass spectra are not acquired, but only mass fragmentograms are obtained. Up to four mass fragmentograms may be obtained simultaneously. Upon conclusion of data acquisition, these data are automatically written on disc for future reference. In this mode, the computer chooses the integration time so that maximum precision of the data is ensured.

After data have been acquired, data may be recalled from disc memory to core memory by entering the file name into the acquisition parameters. When data from a past file are recalled, all of the parameters of that run are restored to core so that they may be read or a similar run may be taken. Gas chromatograms, except mass fragmentograms, are not stored on disc. Recalling a gas chromatogram results in a reconstructed chromatogram. This may be either the full chromatogram as acquired or a limited mass gas chromatogram. The mass limitations are determined by intensifying a range as set by the lever wheel switches. Any mass spectrum of the run may be recalled by pushing the button with the number of that spectrum intensified. If more than one spectrum is intensified, the average of those spectra is obtained.

The ability to intensify is controlled by a push button and two lever wheel switches. One switch controls the start channel and the other the number of channels of span. The span may be in either direction from the start channel. The use of intensified channels provides a means of focusing the computer's attention to particular channels in the data, and this procedure is used very much as a light pen is used in larger interactive data systems. One simple use of these switches is to determine the size of a fragment lost in the mass spectrum. One might start with a certain peak, set the direction of intensification downward, and increment the span until it just reaches the next lower peak. The number of mass units lost from the first peak to get the second would then be one less than the reading on the span lever wheel switch.

We provide three push buttons allowing one to determine the amplitude of either mass spectra mass peaks or gas chromatograph peaks. The first of these is total intensity, which is particularly useful for mass spectra. When this is pushed, the total of the channels intensified is output. The output is either in the form of the absolute counts stored in the data system or it may be converted to read in volts output of the preamplifier, corrected for whatever integration time was used in its acquisition.

The area push button allows a background to be subtracted from the total. Three button pushings are required to obtain this output. On the first and second pushings, background is determined by fitting (least squares fit) to the channels intensified. The peak is then intensified and the button pushed for the third time. The result is then output and consists of the net total above the least squares straight line fit to the background, the calculated number in the peak channels below the least squares fit line to the background, and a size. The size is the net area multiplied by a factor which is the ratio of a size chosen for a standard peak area to the total counts previously obtained in that standard peak. Thus a normalized, or calibrated answer may be obtained for quantification of the analysis.

The ratio of the totals of any two peaks may be obtained by two pushings of the ratio push button. The first peak is intensified and the button pushed. Then the second peak is intensified and the button pushed again. When the second pushing occurs, the counts in the intensified channels of the second peak are divided by the counts in the intensified channels of the first peak and the ratio output. These may of course be either single channels or multiple channels in either peak, and it may thus be used either for mass spectra or gas chromatograms.

Data output is provided either on the teletype or on the digital plotter. A labeling provision is provided so that one may input one line of text to be applied to the data output. This line of text, which will appear on the CRT, is put at the top of the listing or plot and is followed by a line of the title of the run. When one of the output buttons is pushed, the data output are those shown on the CRT display. Normalization is provided by normalizing either to the highest channel within the display region, or to the first channel intensified if one is intensified. The teletype output may also be in absolute counts. As an example of the flexibility of output one might go through the following sequence of steps. One might recall a mass spectrum from near the top of a gas chromatogram peak and save it by transferring it to the GC memory region. One could then recall a mass spectrum from a background region adjacent to the gas chromatogram peak, and subtract this background spectrum from the peak spectrum to obtain a net spectrum. The interesting portion of this mass spectrum could be expanded to fill the full screen and the resulting spectrum plotted. Of course, if the plot does not show what one desired, one would not have to plot it, since one can preview it on CRT. Selected portions of the plot may be multiplied by any desired factor so that small features may be made more visible. The plot will automatically indicate these scale factors when this is done. If one does not wish to wait for the output at the time one sets it up, one may add it to a lineup for delayed output. This is done by pushing the delayed output push button and then indicating the kind of output desired. Later, such as when one is ready to leave for the day, one may instruct the data system to perform all of the delayed outputs. Gas chromatograms may also be plotted. When mass fragmentography is used, the resulting plots are somewhat different. The four mass fragmentograms are plotted above each other.

Automatic comparison of spectra has come to be a useful function of computers. We provide a library search routine in our data system. The search routine provides for storing multiple libraries of spectrum codes on the disc. The operator may add to or delete from these libraries at will. He may also output or input a library from paper tape so that additional libraries may be exchanged with other users. The search scheme is one in which the mass of the most intense peak in each 14 atomic mass unit range is listed. Fifty-two ranges beginning with mass 34 are used. A spectrum obtained on the data system may be automatically coded, the code edited, and the code searched against the library codes. The 10 best fits are then output, together with the number of mismatches for each. Provision is made for altering parameters in the search so that missing peaks in the library are, or are not, significant and unknown missing peaks in the spectrum are, or are not, significant.

In conclusion, let us emphasize that the data system is display oriented. During acquisition, that which is displayed are the input data. On output, what you see is what will be reduced to hard copy. Data manipulation is of the displayed data and other data are called into display for use. The display orientation of the system thus allows significantly more rapid data manipulation and ensures that only significant data will be outputed.

Mass Spectrometry in Biochemistry and Medicine,
edited by A. Frigerio and N. Castagnoli.
Raven Press, New York © 1974

Determination of Configurations by Gas Chromatography

C. J. W. Brooks, J. D. Gilbert, and Mary T. Gilbert

Chemistry Department, University of Glasgow, Glasgow, G12 8QQ, U.K.

I. NEW CHIRAL REAGENTS FOR ALCOHOLS AND AMINES

The application of gas chromatography (GC) to the analytical separation of diastereomeric compounds is well known, and often permits the assignment of configurations on the basis of retention correlations. For the resolution of enantiomers, it is necessary either to employ chiral stationary phases (Gil-Av and Feibush, 1967) or to prepare diastereomeric derivatives that are separable using conventional phases. Enantiomeric hydroxy and amino compounds may be converted by acylation with suitable chiral reagents to diastereomeric esters and amides, respectively (Halpern and Westley, 1965). Many successful separations have been achieved (with a variety of reagents), and the configurations of the original enantiomers can usually be determined through established correlations of retention data based on studies of reference compounds (Gil-Av, Charles-Sigler, Fischer, and Nurok, 1966; Pollock and Oyama, 1966).

We recently found that several commonly used chiral-acylating agents were ineffective for distinguishing between the enantiomeric methyl 13-hydroxy stearates. Similar difficulties were noted by Annett and Stumpf (1972), and, more recently, by Hammarström and Hamberg (1973), who studied hydroxy esters and alkanols with the hydroxyl group at various positions. It seemed possible that the differences between the chromatographic behavior of diastereomeric esters might be enhanced by using reagents possessing a rigid molecular skeleton with a directly attached carboxyl group. One reagent of this type, 3β-acetoxy-5-etienic acid, had al-

I.

II.

379

ready been applied (Anders and Cooper, 1971) to the GC resolution of
enantiomeric alcohols with up to 13 carbon atoms, but its high molecular
weight necessitates the use of rather high column temperatures for the derived esters. Trials of two lower terpenoids, *(R)-(+)-trans*-chrysanthemic
acid (I) and drimanoic acid (II) (Appel, Brooks, and Overton, 1959), showed
them to be effective for distinguishing many types of enantiomeric alcohols
and amines (Brooks, Gilbert, and Gilbert, 1973). In the partly complementary work of Murano (1972*a,b*) chiral alcohols such as *(R)*-menthol and
(+)- or (−)-octan-2-ol have been used for the GC analysis of enantiomeric
chrysanthemic acids on QF-1 and other stationary phases.

In the present chapter, we illustrate the application of *(R)-(+)-trans*-
chrysanthemoyl derivatives in analytical separations. Experimental details
are given by Brooks et al. (1973). All the chromatograms illustrated were
recorded using a 5-m column packed with 1% SE-30 on Gas Chrom Q.

The separation of (+)- and (−)-2-octyl chrysanthemates is shown in Fig.
1, whereas corresponding results for the esters of (+)- and (−)-pantolactone
(4,5-dihydro-3-hydroxy-4,4-dimethyl-2(3H)-furanone) are depicted in Fig.
2. It is noteworthy that the *(R)-(R)*-ester in the latter example is eluted
before the *(S)-(R)*-ester, and constitutes one of the exceptions to the useful
empirical rule that *(R)-(S)* or *(S)-(R)* forms frequently have shorter retention
times than their *(R)-(R)* or *(S)-(S)* diastereomers (*cf.* Westley and Halpern, 1969).

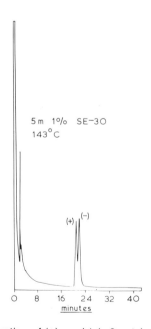

5 m 1% SE-30
143°C

FIG. 1. GC separation of (+)- and (−)- 2-octyl chrysanthemates.

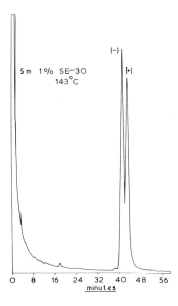

FIG. 2. GC separation of (±)-pantolactone chrysanthemates.

Application of the reagent to amines is illustrated in Fig. 3 for (+)- and (−)-phenylalanine methyl ester. In this case also the *(R)-(R)* isomer is eluted first. The satisfactory separation suggests that chrysanthemamides may also be suitable for the study of substituted phenylalanines.

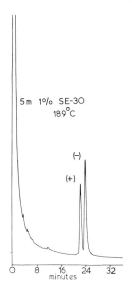

FIG. 3. GC separation of the *N*-chrysanthemoyl derivatives of (+)- and (−)-phenylalanine methyl ester.

Mass spectrometry has been used to verify the structures of the esters and amides studied. The latter compounds afford moderately abundant molecular ions, and it would be feasible to carry out the analysis of diastereomers by "single-ion detection" with at least the same sensitivity as may be achieved with flame ionization detectors.

II. A GC MODIFICATION OF HOREAU'S METHOD OF PARTIAL KINETIC RESOLUTION

The well-known method, developed by Horeau (1961), for determining absolute configurations of secondary alcohols *via* reaction with racemic α-phenylbutyric anhydride is widely applicable and has a remarkably high degree of reliability. We have recently extended its scope through the use of GC to estimate the excess (+)- or (−)-anhydride in the form of diastereomeric amides (Brooks and Gilbert, 1973; Gilbert and Brooks, 1973). With this procedure, it is possible to obtain valid results from 1- to 10-μmole samples of alcohols. Horeau's method is applicable to enantiomeric or epimeric alcohols, provided that there is a distinct steric difference between the two alkyl or other substituents of the carbinol group.

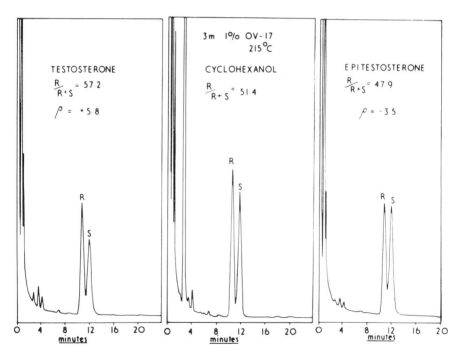

FIG. 4. Gas-liquid chromatograms of the *(R)*-α-phenylethyl-amides of excess (−)-*(R)*- and (+)-*(S)*-α-phenylbutyric acid after acylation of testosterone, cyclohexanol, and epitestosterone with (±)-α-phenylbutyric anhydride.

Application of the method at the 10-μmole level is illustrated in Fig. 4. After 1.5 hr reaction at 40°C in dry pyridine (7 μl) with 20 μmole of ($\pm$)-α-phenylbutyric anhydride, (R)-$(+)$-α-phenylethylamine (6 μl:40 μmole) was added. After 15 min, the solution was diluted with dry ethyl acetate (600 μl) and samples (1 μl) were analyzed by GLC. Relative areas of diastereomer peaks were conveniently evaluated by the approximation of the product of peak height and retention time. A correction was necessary in this rapid method of estimation, and was effected by subtraction of the apparent percentage area observed for the reaction of the achiral alcohol cyclohexanol. The resulting values, represented as ρ in Fig. 4, were a measure of the excess or deficiency of (R)-$(-)$-acid. Alcohols designated as Type I by Horeau (often equivalent to R-forms) react preferentially with (R)-$(-)$-acid. The results for testosterone (Type II) and epitestosterone (Type I) are in accordance with their known 17-configurations.

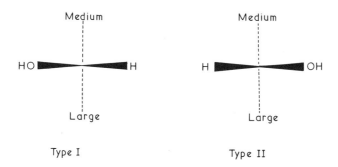

Type I   Type II

The procedure exemplified here is likely to be useful in determining the configurations of secondary alcohols arising in small quantities in biochemical and drug metabolism work, for example, by reduction of carbonyl groups, by hydroxylation of methylenic centers and by hydration of epoxides. An important feature of Horeau's method is its applicability to alcohols that are unsuitable for gas-phase analysis because of high molecular weight, as well as to alcohols containing other hydroxylic or amino functions, provided that the additional groups either react very rapidly with, or are inert toward, α-phenylbutyric anhydride.

ACKNOWLEDGMENTS

We thank the Medical Research Council for a research grant (to C. J. W. B. and W. A. Harland). We are indebted to Dr. S. W. Head and to Roussel-UCLAF for gifts of (+)-*trans*-chrysanthemic acid, and to Prof. Dr. H. H. Appel for supplies of drimenol. M. T. G. acknowledges an SRC studentship.

REFERENCES

Anders, M. W., and Cooper, M. J. (1971): Gas chromatographic resolution of optically active alcohols as 3β-acetoxy-Δ^5-etienates. *Analytical Chemistry*, 43:1093–1094.

Annett, R. G., and Stumpf, P. K. (1972): L-(−)-Menthyloxycarbonyl derivatization of hydroxy acid methyl esters. *Analytical Biochemistry*, 47:638–640.

Appel, H. H., Brooks, C. J. W., and Overton, K. H. (1959): The constitution and stereochemistry of drimenol, a novel bicyclic sesquiterpenoid. *Journal of the Chemical Society*, 3322–3332.

Brooks, C. J. W., and Gilbert, J. D. (1973): Absolute configurations of secondary alcohols. A gas chromatographic modification of Horeau's method. *Chemical Communications*, 194–195.

Brooks, C. J. W., Gilbert, M. T., and Gilbert, J. D. (1973): New derivatives for gas-phase analytical resolution of enantiomeric alcohols and amines. *Analytical Chemistry*, 45:896–902.

Gil-Av, E., Charles-Sigler, R., Fischer, G., and Nurok, D. (1966): Resolution of optical isomers by gas liquid partition chromatography. *Journal of Gas Chromatography*, 4:51–58.

Gil-Av, E., and Feibush, B. (1967): Resolution of enantiomers by gas liquid chromatography with optically active stationary phases. *Tetrahedron Letters*, 3345–3347.

Gilbert, J. D., and Brooks, C. J. W. (1973): Absolute configuration of secondary alcohols: Refinement and extension of a gas-chromatographic modification of Horeau's method. *Analytical Letters*, 6:639–648.

Halpern, B., and Westley, J. W. (1965): High sensitivity optical resolution of D,L-amino acids by gas chromatography. *Biochemical and Biophysical Research Communications*, 19:361–363.

Hammarström, S., and Hamberg, M. (1973): Steric analysis of 3, ω4, ω3 and ω2 hydroxyacids and various alkanols by gas-liquid chromatography. *Analytical Biochemistry*, 52:169–179.

Horeau, A. (1961): Principe et applications d'une nouvelle methode de determination des configurations dite "par dedoublement partiel." *Tetrahedron Letters*, 506–512.

Murano, A. (1972a): Gas chromatographic separation and determination of optical isomers of chrysanthemic acid. *Agricultural and Biological Chemistry*, 36:917–923.

Murano, A. (1972b): Determination of the optical isomers of insecticidal pyrethroids by gas-liquid chromatography. *Agricultural and Biological Chemistry*, 36:2203–2211.

Pollock, G. E., and Oyama, V. I. (1966): Resolution and separation of racemic amino acids by gas chromatography and the application to protein analysis. *Journal of Gas Chromatography*, 4:126–131.

Westley, J. W., and Halpern, B. (1969): Determination of the configuration of asymmetric compounds by gas chromatography of diastereoisomers. In: *Gas Chromatography 1968*, edited by C. L. A. Harbourn, pp. 119–128. Institute of Petroleum, London.

INDEX

A

3-β-Acetoxy-5-etienic acid, 379

Acidosis, 292

26-Acetylamino-1, 4-furostadien-3-one, 95

26-Acetylamino-5-α-furostan-3-β-ol, 93, 95

Acrolein, 20

Actin, amino acids in, 119-129

Acto-myosin, N-methylamino acids in, 119-129

Alclofenac, 29

Aldophosphamide, 19

Aldosterone, 146, 147

Alkaryltriazenes, 267-274

Allopregnandiol, 349-358

4-Allyloxy-3-chlorophenylacetic acid. *See* Alclofenac

Amino acids, assay, 119-129

26-Amino-5-α-cholestane-3-β-16-β-22-triol, 93, 96

Amobarbital
   oxidation in liver, 10

Androstenediol, 342, 349-358

Androstenetriol, 341, 342, 349-358

Androstenetriolone, 341

Androsterone, 159, 349-358

Ant
   Pharaoh's, pheromones from, 197-217

Apomorphine, 102

Azaphenoxazine series, 229-234

B

Barbiturate metabolism, 9

Benzethonium chloride, 183-195

3-Benzyl-3-methyl-1-phenyltriazene, 271

Benzyloxime-TMS steroid deriv., 335-347

Bulbocapnine, 102

2-Butyl-5-pentylpyrrolidine, 211

C

Campesterol, 147

Cannabidiol, pyrolysis, 219-226

Cannabielsoic acid, 222

Cannabielsoin, 224

Cannabis smokers, plasma analysis, 361-363

Carbamazepine, MS-GC studies, 66-74

Carbamazepine-10, 11-epoxide, 67, 69

(N-Carboxymethyl)-4-allyloxy-3-chlorophenylacetamide, 29, 48

Carboxyphosphamide, 19

Carcinogenic alkaryltriazenes, 267-274

Catatonia, urinary acids, 365-371

Catechol amines
   dopamine metabolism, 100-103
   quant. analysis, 111

Cetylpyridinium chloride, 183-195

Cetyltrimethylammonium bromide, 183-195

Chiral reagents, 379-383

5-β-Cholestan-3-α-ol, 349-358

5-β-Cholestan-3-β-ol, 147

Cholesterol, 132, 137, 142, 143, 147

Chrysanthemic acid, resolution, 380